Probability and Its Applications

Published in association with the Applied Probability Trust

Editors: S. Asmussen, J. Gani, P. Jagers, T.G. Kurtz

Probability and Its Applications

The *Probability and Its Applications* series publishes research monographs, with the expository quality to make them useful and accessible to advanced students, in probability and stochastic processes, with a particular focus on:

- Foundations of probability including stochastic analysis and Markov and other stochastic processes
- Applications of probability in analysis
- Point processes, random sets, and other spatial models
- Branching processes and other models of population growth
- Genetics and other stochastic models in biology
- Information theory and signal processing
- Communication networks
- Stochastic models in operations research

For further volumes:
www.springer.com/series/1560

Oswaldo L.V. Costa • Marcelo D. Fragoso •
Marcos G. Todorov

Continuous-Time Markov Jump Linear Systems

Springer

Oswaldo L.V. Costa
Polytechnic School
Department of Telecommunications and Control Engineering
University of São Paulo
São Paulo, Brazil

Marcos G. Todorov
Department of Systems and Control
National Laboratory for Scientific Computing
Petrópolis, RJ, Brazil

Marcelo D. Fragoso
Department of Systems and Control
National Laboratory for Scientific Computing
Petrópolis, RJ, Brazil

ISSN 1431-7028 Probability and Its Applications
ISBN 978-3-642-34099-4 ISBN 978-3-642-34100-7 (eBook)
DOI 10.1007/978-3-642-34100-7
Springer Heidelberg New York Dordrecht London

Library of Congress Control Number: 2012955154

Mathematics Subject Classification: 93E20, 60J27, 93E11, 93E15, 93E03, 60G35

Printed on acid-free paper

Springer is part of Springer Science+Business Media (www.springer.com)

Preface

The intent of this book is to present, in a unified and rigorous way, recent results in the control theory of continuous-time Markov Jump Linear Systems (MJLS). This is neither a textbook nor a complete account of the state-of-the-art on the subject. Instead, we attempt to provide a systematic framework for an understanding of the main concepts and tools underpinning the control theory of MJLS, together with some basic relevant results. We follow an approach that combines probability and operator theory, which we have named here as *the analytical point of view*.

It is worth mentioning that the topics treated here represent just a fraction of the several important results nowadays available for MJLS. The limited size of a book, in addition to the continued rapid advance in the theory of MJLS, makes it unfeasible to cover all the aspects in this field. Therefore, some degree of specialization is no doubt inevitable and, perhaps, desirable. For instance, although fitting under the MJLS field, some important topics and different approaches, such as MJLS with delay, adaptive control of MJLS, and hidden Markov chain filtering, are not considered in this book. We apologize for those who find that some of their favorite topics are missing. The analytical point of view, together with our personal affinities, has determined the choice of topics that, to some extent, are related to our previous contributions with coauthors on the subject.

Another important point of view regarding MJLS, which has achieved a considerable degree of success in applications, is known in the specialized literature as *multiple model*. Departing from the multiple model approach and tracing as much as possible a parallel with the control theory for the linear case, the analytical point of view adopted in this book allows us to obtain explicit solutions to some important stochastic control problems. A basic step in this approach, which indeed permeates most of the book, is the fact that the mean square stability results are written in terms of the spectrum of an augmented matrix. On the other hand, it is worth pointing out that the MJLS carry a great deal of subtleties that distinguish them from the linear case. In fact, MJLS have a certain degree of complexity which does not allow one to recast several results from the linear theory and, therefore, cannot be seen as a trivial extension of the linear case.

The motivation for writing this book was at least twofold. Most of the material presented here is scattered throughout a variety of sources, which includes journal articles and conference proceedings papers. Considering that this material constitutes, by now, a coherent body of results, we have felt compelled to write an introductory text putting together systematically these results. In addition, this seemed an opportunity to introduce, as far as possible in a friendly way, a bent of the MJLS theory that we have named here the analytical point of view, contributing to encourage and open up the way for further research on MJLS.

Although the book is mainly intended to be a theoretically oriented text, it contains several illustrative examples that show the level of maturity reached on this field. For those wishing to delve into topics which were not contemplated in this book, a substantial number of references (not meant to be exhaustive) are included. Most of the chapters end with historical remarks, which, we hope, will be both informative and interesting sources for further readings. The notation is mostly standard, although, in some cases, it is tailored to meet specific needs. A glossary of symbols and conventions can be found at the end of the book.

The book is targeted primarily for advanced students and practitioners of control theory. It may be also a valuable reference for those in fields such as communication engineering and economics. In particular, we hope that the book will be a valuable source for experts in linear systems with Markovian jump parameters. For specialists in stochastic control, the book provides one of those few cases of stochastic control problems in which an explicit solution is possible, providing interesting material for a course while introducing the students to an interesting and active research area. Moreover, we believe that the book should be suitable for certain advanced courses or seminars. As background, one needs an acquaintance with the linear control theory and some knowledge of stochastic processes and modern analysis.

A brief description of the book content goes as follows. Chapter 1 motivates the class of MJLS through some application-oriented examples and presents an outline of the problems. Chapter 2 provides some background material needed in the following chapters. Chapter 3 deals with the mean-square stability for MJLS, while Chap. 4 deals with the quadratic optimal control problem with complete observations. In Chap. 5 the infinite horizon quadratic optimal control problem is revisited but now from another point of view, usually known in the literature of linear systems as H_2 control. Chapter 6 deals with the finite-horizon quadratic optimal control problem and the H_2 control problem for continuous-time MJLS under partial observations. In this case the state variable $x(t)$ is not directly accessible to the controller, but, instead, it is assumed that only an output $y(t)$ and the jump process $\theta(t)$ are available. Chapter 7 considers the best linear mean-square estimator for continuous-time MJLS assuming that only an output $y(t)$ is available (no knowledge of $\theta(t)$ is assumed). Chapter 8 is devoted to the H_∞ control of continuous-time MJLS in the infinite horizon setting. Design techniques, expressed as linear matrix inequalities optimization problems, for continuous-time MJLS are presented in Chap. 9. Chapter 10 presents some numerical applications from the theoretical results introduced earlier. Finally, the associated coupled algebraic Riccati equations and some auxiliary results are considered in Appendices A and B.

The authors are indebted to many people and institutions which have contributed in many ways to the writing of this book. We gratefully acknowledge the support of the Laboratory of Automation and Control, LAC/USP, at the University of São Paulo, and from the National Laboratory for Scientific Computing, LNCC/MCT. This book owes much to our research partners, to whom we are immensely grateful. Many thanks go also to our former PhD students. We acknowledge with great pleasure the efficiency and support of Marina Reizakis, Annika Elting, and Sarah Goob, our contacts at Springer. We are most pleased to acknowledge the financial support of the Brazilian National Research Council, CNPq, under grants 301067/2009-0, 302501/2010-0, and 302676/2012-0; FAPERJ under the grant No. E-26/170-008/2008, and USP project MaCLinC.

Last, but not least, we are very grateful to our families for their continuing and unwavering support. To them we dedicate this book.

São Paulo, Brazil O.L.V. Costa
Petrópolis, Brazil M.D. Fragoso
Petrópolis, Brazil M.G. Todorov

Contents

Chapter 1
Introduction

Recent advances in technology have led to dynamical systems with increasing complexity, which in turn demand for more and more efficient and reliable control systems. It has been widely recognized that the requirements of specific behaviors and stringent performances call for the inclusion of possible failure prevention in a modern control design. Therefore, either due to security reasons or efficiency necessity, a system failure is a critical issue to be considered in the design of a controller in modern technology. In view of this, dynamical systems that are subject to abrupt changes have been a theme of increasing investigation in recent years and a variety of different approaches to analyze this class of systems has emerged over the last decades. A particularly interesting class of models within this framework is the so-called *Markov jump linear systems* (MJLS), which is the subject matter of this book. The goal of this first chapter is to highlight, in a rather informal way, some of the main characteristics of MJLS, through some illustrative examples of possible applications of this class of systems.

1.1 Markov Jump Linear Systems

One of the main challenges when modeling a dynamical system is to find the best trade-off between the mathematical complexity of the resulting equations and the capability of obtaining a tractable problem. Thus, it is of overriding importance to reach a proper balance in the settling of this issue while keeping in mind the performance requirements. For instance, in many situations the use of robust control techniques and some classical sensitivity analysis for time-invariant linear models can handle in a simple and straightforward way the control problem of dynamical systems subject to changes in their dynamics. However, if these changes significantly alter the dynamical behavior of the system, these approaches may no longer be adequate to meet the performance requirements. Within this scenario, the introduction of some degree of specialization on the modeling of the dynamical system in order to accommodate these changes is inevitable and perhaps even desirable.

O.L.V. Costa et al., *Continuous-Time Markov Jump Linear Systems*,
Probability and Its Applications, DOI 10.1007/978-3-642-34100-7_1,

The modeling of dynamic systems subject to abrupt changes in their dynamics has been receiving lately a great deal of attention. These changes can be due, for instance, to abrupt environmental disturbances, to actuator or sensor failure or repairs, to the switching between economic scenarios, to abrupt changes in the operation point for a non-linear plant, etc. Therefore, it is important to introduce mathematical models that take into account these kind of events and to develop control systems capable of maintaining an acceptable behavior and meeting some performance requirements even in the presence of abrupt changes in the system dynamics.

In the case in which the dynamics of the system is subject to abrupt changes, one can consider, for instance, that these changes are due to switching (*jump*) among well-defined models. To illustrate this situation, consider a continuous-time dynamical system that is, at a certain moment, well described by a model $\mathcal{G}_1$. Suppose that, after a certain amount of time, this system is subject to abrupt changes that cause it to be described by a different model, say $\mathcal{G}_2$. More generally, one can imagine that the system is subject to a series of possible qualitative changes that make it to switch, over time, among a countable set of models, for example, $\{\mathcal{G}_1, \mathcal{G}_2, \ldots, \mathcal{G}_N\}$. One can associate each of these models to an *operation mode* of the system, or just *mode*, and say that the system *jumps* from one mode to the other or that there are *transitions* between them. A central issue on this approach is how to append the jumps into the model. A first step is to consider, for instance, that the jumps occur in a random way, i.e., the mechanism that rules the switching between the aforementioned models is random. In addition, one can assume that this process (hereinafter $\theta(t)$) just indicates which model, among the $\{\mathcal{G}_i, i = 1, 2, \ldots, N\}$, is running the system, i.e., $\theta(t) = i$ means that $\mathcal{G}_i$ is running the system. Furthermore, it would be desirable to have some a priori information on the way in which the system jumps from one operation mode to another (the transition mechanism). A random process which bears these features and has been used with a great deal of success for modeling these situations is the Markov chain.

Within this framework, a particularly interesting class of models is the so-called Markov jump linear systems (from now on MJLS). Since its introduction, this class of models has an intimate connection with systems which are vulnerable to abrupt changes in their structure, and the associated literature surrounding this subject is now fairly extensive. Due, in part, to a large coherent body of theoretical results on these systems, MJLS has been used recently in a number of applications on a variety of fields, including robotics, air vehicles, economics, and some issues in wireless communication, among others. For instance, it was mentioned in [266] that the results achieved by MJLS, when applied to the synthesis problem of wing deployment of an uncrewed air vehicle, were quite encouraging. As mentioned before, the basic idea is to consider a family of continuous-time linear systems, which will represent the possible modes of operation of the real system. The modal transition is given by a Markov chain represented, as before, by $\theta(t)$, which is also known in the literature as the operation mode (or the Markov state). This class of systems will be the focus of investigation of the present book. We will restrict ourselves in this book to the case in which all operation modes are continuous-time linear models and the jumps from one mode of operation to another follow a continuous-time

Markov chain taking values in a *finite set* $\{1, \ldots, N\}$. In this scenario, it is possible to develop a unified and coherent body of concepts and results for stability, filtering, and control as well as to present controller and filter design procedures.

In its most simple form, continuous-time MJLS are described as

$$\dot{x}(t) = A_{\theta(t)} x(t). \tag{1.1}$$

For a more general situation, known as the case with partial observations, continuous-time MJLS are represented by

$$\mathcal{G} = \begin{cases} dx(t) = A_{\theta(t)}(t)x(t)\,dt + B_{\theta(t)}(t)u(t)\,dt + J_{\theta(t)}(t)\,dw(t), \\ dy(t) = H_{\theta(t)}(t)x(t)\,dt + G_{\theta(t)}(t)\,dw(t), \\ z(t) = C_{\theta(t)}(t)x(t) + D_{\theta(t)}(t)u(t), \\ x(0) = x_0, \qquad \theta(0) = \theta_0, \end{cases} \tag{1.2}$$

with $x(t)$ standing for the state variable of the system, $u(t)$ the control variable, $y(t)$ the measured variable available to the controller, $z(t)$ the output of the system, and $w(t)$ is a Wiener process.

Although MJLS seem, at first sight, just an extension of linear systems, they differ from the latter in many instances. This is due, in particular, to some peculiar properties of these systems that can be included in the class of complex systems (roughly speaking, a system composed of interconnected parts that as a whole exhibit one or more properties not obvious from the properties of the individual parts). In order to give a small glimpse of this, let us consider a typical situation in which one of these peculiar properties is unveiled. Consider a homogeneous Markov chain θ with state space $\mathcal{S} = \{1, 2\}$ and transition rate matrix

$$\Pi = \begin{bmatrix} -\beta & \beta \\ \beta & -\beta \end{bmatrix}, \quad \beta > 0.$$

Some sample paths of θ, which in this case corresponds to a telegraph process with exponentially distributed waiting times, are shown in Fig. 1.1. Clearly, an increase in the transition rate gives rise to more frequent jumps (fast switching), with the converse applying in the slow switching case.

In this setting, consider system (1.1) evolving in $\mathbb{R}^2$, with initial condition $x(0) = [1\ -1]'$ and state matrices

$$A_1 = \begin{bmatrix} \frac{1}{2} & -1 \\ 0 & -2 \end{bmatrix}, \qquad A_2 = \begin{bmatrix} -2 & -1 \\ 0 & \frac{1}{2} \end{bmatrix}. \tag{1.3}$$

The state trajectory of the system, corresponding to the realizations of θ depicted in Fig. 1.1, is shown in Fig. 1.2. From these figures it is evident that the overall system behavior may vary considerably, depending on how fast the switching occurs. Even more remarkable is the fact that the state converges to the origin in the case $\beta = 3/2$,

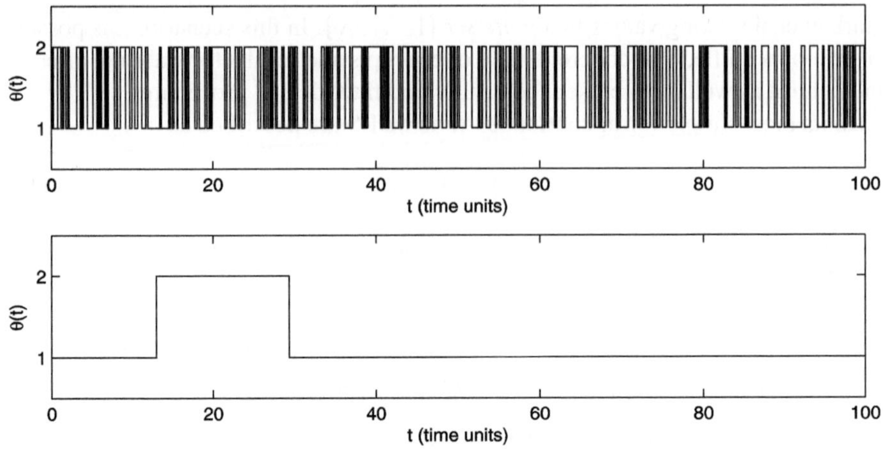

Fig. 1.1 Sample paths of θ for $\beta = 3/2$ (*top*) and $\beta = 3/100$ (*bottom*)

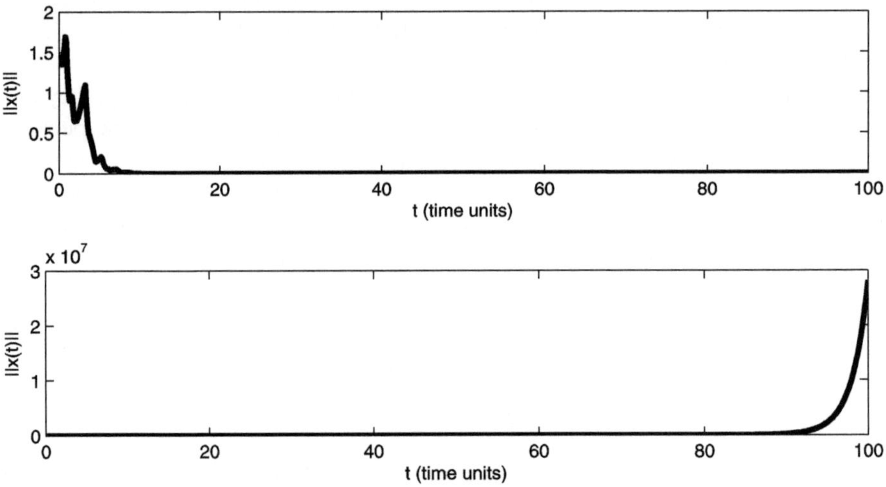

Fig. 1.2 Sample paths of $\|x(t)\|$ corresponding to $\beta = 3/2$ (*top*) and $\beta = 3/100$ (*bottom*) in (1.3)

in spite of neither A_1 nor A_2 being stable. Suppose now that the system matrices in (1.1) are replaced by

$$A_1 = \begin{bmatrix} -1 & 10 \\ 0 & -1 \end{bmatrix}, \qquad A_2 = \begin{bmatrix} -1 & 0 \\ 10 & -1 \end{bmatrix}, \tag{1.4}$$

which are both stable, i.e., have all the eigenvalues with negative real parts. In this case the sample paths of $\|x(t)\|$ corresponding to the same trajectories of θ are depicted in Fig. 1.3. As these two situations suggest, Markovian switching between stable (unstable) systems may produce unstable (stable) dynamics. In fact, as it will

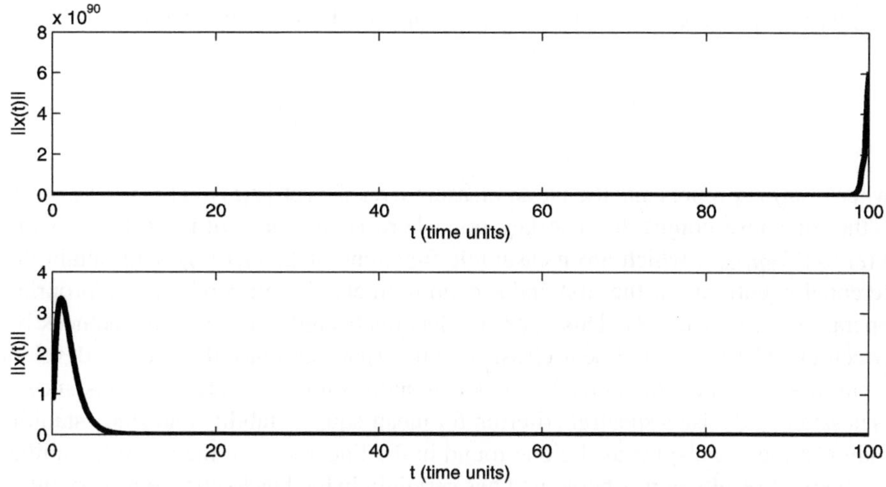

Fig. 1.3 Sample paths of $\|x(t)\|$ corresponding to $\beta = 3/2$ (*top*) and $\beta = 3/100$ (*bottom*) in (1.4)

be explicitly shown in Chap. 3, in the first case stability is equivalent to $\beta > 4/3$, whereas in the second situation the overall system is stable if and only if $\beta < 1/24$.

Some features which distinguish MJLS from the classical linear systems are:

- The stochastic process $\{x(t)\}$ alone is no longer a Markov process.
- As seen in the previous examples, the stability (instability) for each mode of operation does not guarantee the stability (instability) of the system as a whole.
- The optimal filter for the case in which $(x(t), \theta(t))$ are unknown is nonlinear and infinite-dimensional (in the filtering sense).
- Contrary to the linear time-invariant case, there is a fundamental limitation on the degree of robustness against linear perturbation that the H_∞ control of MJLS may offer (see [284]).

If the state space of the Markov chain is *infinitely countable*, there are even further distinctions as illustrated below:

- Mean-square stability is no longer equivalent to L^2-stability (see [148]).
- One cannot guarantee anymore that the maximal solution of the Riccati equation associated to the quadratic optimal control problem is a strong solution (see [17]).

In order to develop a theory for MJLS using an operator theoretical approach, the following steps are adopted:

(S.1) "Markovianize" the problem by considering as the state of the model the pair $(x(t), \theta(t))$.
(S.2) Establish a connection between $x(t)$ and measurable functions of the pair $(x(t), \theta(t))$.
(S.3) Devise adequate operators from (S.2) by considering the first and second moments.

These steps were followed in [152] (see also [77]), using the identities

$$x(t)=\sum_{i=1}^{N}x(t)1_{\{\theta(t)=i\}} \quad \text{and} \quad x(t)x(t)^*=\sum_{i=1}^{N}x(t)x(t)^*1_{\{\theta(t)=i\}},$$

where $1_{\{\theta(t)=i\}}$ represents the Dirac measure over the set $\{\theta(t)=i\}$ (see (2.2)), and $*$ the transpose conjugate. The key point here is to work with $x(t)1_{\{\theta(t)=i\}}$ and $x(t)x(t)^*1_{\{\theta(t)=i\}}$, which are measurable functions of $(x(t),\theta(t))$, and obtain differential equations for the first and second moments in terms of some appropriate operators (see Sect. 3.3). This approach has uncovered many new differences between the MJLS and its linear classical counterpart, and has allowed the development of several new theoretical results and applications for MJLS. For instance, it is possible to devise a spectral criterion for mean-square stability and study stability radius in the same spirit as the one found in the linear system theory. What unifies the body of results in this book, and has certainly helped us to choose its content, is this particular approach, which we will call the *analytical point of view* (from now on APV).

It is worth mentioning two other distinctive approaches, which, together with the APV, have given rise to a host of important results on various topics of MJLS: the so-called Multiple Model (MM) approach and the Hidden Markov Model (HMM) approach. In the MM approach the idea is, roughly speaking, to devise strategies that decide in an efficient way which mode is running the system and work with the linear system associated to this mode (see, e.g., [22] for a comprehensive treatment on this subject). The HMM approach focuses on what is known in the control literature as a class of partially observed stochastic dynamical systems. The basic framework for the HMM consists of a Markov process $\theta(t)$ that is not directly observed but is hidden in a noisy observation process $y(t)$. Roughly speaking, the aim is to estimate the Markov process, given the related observations, and from this estimation to derive a control strategy for the hidden Markov process (usually the transition matrix of the chain depends on the control variable $u(t)$). See, e.g., [130] for a modern treatment on this topic.

The present book can be seen, roughly speaking, as a continuation of the book [81], which dealt with the discrete-time case, to the continuous-time case. From a practical viewpoint, perhaps the most important aspect which favors the study of continuous-time systems is that many models in science are based upon specific relations between the system variables and their *instantaneous* rates of variation (some examples being Newton's second law of classical mechanics, Maxwell's nonstationary equations of electromagnetics, the law of continuity of fluid transport phenomena, Verhulst's logistic equation, and Gompertz's model of tumor growth, among many others). In these cases, the direct application of discrete-time systems theory may be a potential cause for distortion to the model in the form, e.g., of spurious effects such as artificial energy dissipation, instability, oscillations, and a potentially cumbersome dependence on the discretization parameters (which may even be plagued by the curse of dimensionality, as pointed out in [114]). Hence, it

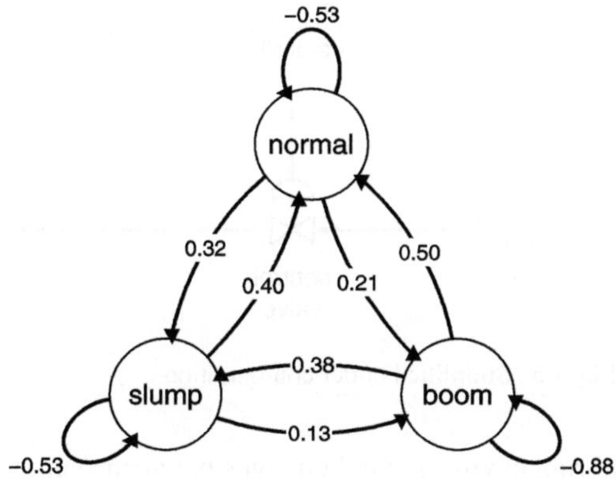

Fig. 1.4 Transitions between the different scenarios of a country's economy

is fair to say that, *prior to sampling*, a cautious study of the underlying continuous-time process should be carried out in order to ensure that such adversities would not hinder the desired specifications to be achieved in a given application.

1.2 Some Applications of MJLS

Since their inception in the early 1960s, MJLS have found many applications in a great variety of fields. These include unmanned air vehicles [266], solar power stations [275], satellite dynamics [226], economics [32, 33, 52, 108, 268], flight systems [174, 175], power systems [206, 211, 212, 290, 291], communication systems [1, 2, 208, 243], among many others. This section is devoted to a brief exposition of some selected topics regarding applications of MJLS, with special attention to those in continuous time.

The earliest application of MJLS to economics, introduced in the discrete-time scenario in [33], was based on Samuelson's multiplier–accelerator macroeconomic model (see [252, 286, 299]). The model studied in [33] consists of a very simplified relation for the dynamical evolution of a country's *national income* in terms of the *governmental expenditure*, which is weighted by two parameters (the *marginal propensity to save* and the *accelerator coefficient*). Based on the historical data obtained from the U.S. Department of Commerce from years 1929 until 1971, [33] assumed that the state of the economy could be roughly lumped in three possible operation modes ("normal", "boom", and "slump") and that the switching between them could be modeled as a homogeneous Markov chain, with the transition rates depicted in Fig. 1.4. The subsequent problem considered in [33] corresponds to an MJLS version of the optimal linear quadratic control setup. This application, which was also analyzed in [81] for the discrete-time case, will be further studied in Chap. 10.

Another practical application of MJLS, which, in the discrete-time case, has been extensively discussed in [81], is *the control of a solar power receiver*. In continuous time, this situation was initially considered in [275] with $\theta(t)$ representing

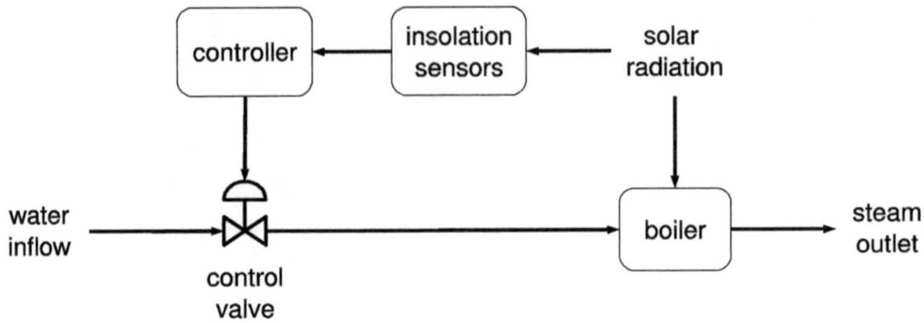

Fig. 1.5 Simplified boiler configuration

abrupt environmental changes between "sunny" and "cloudy" conditions, which are measured by sensors located on the plant. The role of control is to determine the feedwater flow rate which will enter the boiler, in such a way as to regulate the outflow temperature at the desired level. The boiler flow rate is strongly dependent upon the receiving insolation, and, as a result of this abrupt variability, several linearized models are required to characterize the evolution of the boiler when clouds interfere with the sun's rays. The control law described in [275] makes use of the state feedback and a measurement of $\theta(t)$ through the use of flux sensors on the receiver panels. In Fig. 1.5 a very simplified representation of the system is shown. Although several important blocks such as the boiler internal models for steam and metal dynamics, thermal couplings, and feedforward compensation loops are not displayed, their abstraction evidences the dependence of the control system on the abrupt variability of the received insolation.

Robotic manipulator arms are employed in a great deal of modern applications, which span areas as diverse as deep sea engineering, manufacturing processes, space technology, or teleoperated medicine, for instance. The adequate operation of such devices, however, is severely compromised by the occurrence of failures, which may be intolerable in safety-critical applications, for example. Furthermore, repairing the faulty arm may frequently turn out to be a difficult task during the course of operation (which may occur if, e.g., the robot is functioning in a hazardous environment). On this regard, a great deal of research has been carried out on the control of robotic arms with less actuators than degrees of freedom, commonly referred to as underactuated manipulators in the specialized literature [11, 12, 92]. In practical terms, a manipulator arm is said to be *underactuated* whenever the motor on at least one of its joints is in a passive state. The basic principle in the operation of these devices is then to explore the dynamic coupling that the active joints impose on the passive ones, in such a way as to drive the arm to the desired position in spite of its faulty condition.

The introduction of MJLS theory to tackle the control of underactuated robotic arms was made by Siqueira and Terra in [261], in the discrete-time setting. This approach was subsequently described in [81] and, more recently, brought to the continuous-time setting in [262, 263]. In Chap. 10 the robust control of the planar 3-link underactuated arm depicted in Fig. 1.6 is further studied by means of the results devised in this book.

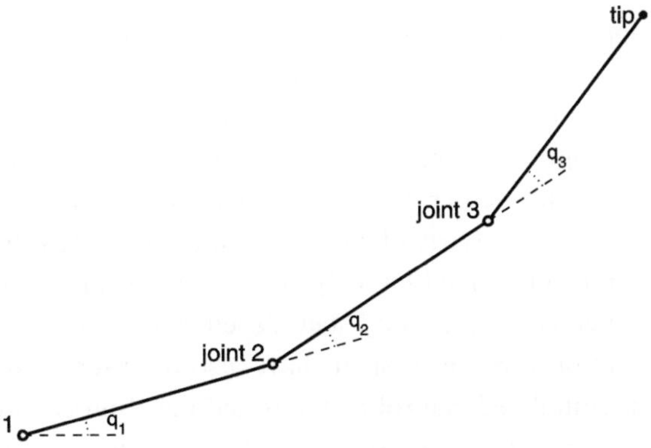

Fig. 1.6 Geometric modeling of the planar 3-link arm

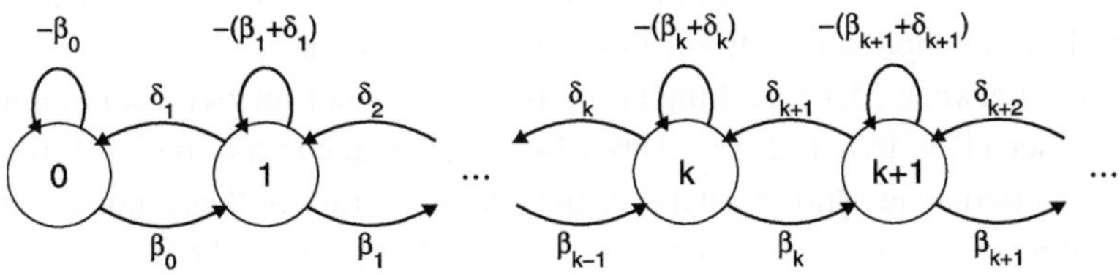

Fig. 1.7 Birth–death process considered in [175]

In yet another field, MJLS were employed in [175] for the stability analysis of *controlled flight systems affected by electromagnetic disturbances*. This problem stems from the susceptibility of electronic devices against external disturbances such as lightning, thermal noise, and radio signals, for instance, which is recognized as a potential cause for *computer upsets* in digital controllers. In order to cope with the resulting adversities, which may range from random bit errors to permanent computer failures, the strategy in [175] is to model the accumulative effect of external disturbances on the system by means of a continuous-time birth–death Markov process of the form indicated in Fig. 1.7. The rationale for this choice has been, as pointed out in [175], that by doing so the arrival of new disturbances is dictated by a Poisson process with exponentially distributed sojourn times.

An important issue in this situation is that, while the aircraft dynamics treated in [175, Sect. IV.B] evolves in continuous time, the occurrence of electromagnetic disturbances takes place in the sampled-data digital controller implementation. This leaves open an interesting conjecture of whose aspects of continuous- and discrete-time MJLS theory should be relevant to the problem at hand.

In [211, 212] the *modeling and control of power systems* subject to Markov jumps has been addressed in the continuous-time scenario. In this case the switching mechanism is used to model random changes in the load, generating unit outages, and transmission line faults, for instance. As shown in [212], intermittent couplings between electrical machines operating in a network can render the overall system unstable in a stochastic sense, a result which somewhat resembles the analysis previously carried out for (1.1) in the cases (1.3) and (1.4). In [211] the main results were applied to the problem of dynamic security assessment, which amounts to determining, with desired probability, whether certain parameters of the electrical system are guaranteed to remain within a safe region of operation at a given period. A "security measure" was defined, which corresponds to a quantitative indicator of the vulnerability of the current system state and network topology to stochastic contingency events. Two types of switchings are considered therein: *primary* ones, driven by a continuous-time Markov process taking values in a finite set, and *secondary* events, which are modeled by state-dependent (controlled) jumps. Recent advances in the control of power systems have also been reported in [206, 290, 291] by means of decentralized control methods and the S-procedure. An alternative account of this problem via robust control methods is presented in details in Sect. 10.3.

The *modeling of communication systems* via MJLS is by now another promising trend in the applications front. In discrete-time this is boldly motivated by the connection between MJLS and the Gilbert–Elliott model for burst communication channels (see [129, 167, 172, 173, 186, 244, 254]), which in its simplest form corresponds to a two-state Markov chain. A convenient feature of these models (besides their relative simplicity) is that they are capable of describing the fact that eventual packet losses typically occur during *intervals* of time (e.g., while a wireless link is obstructed). In other words, isolated packet losses are not common events, and it usually takes a while before communication, once lost, is restored. As pointed out in [167, 173, 186, 254], large packet-loss rates imply poor performance or even instability, and therefore controllers implemented within a network may provide considerably better results if their design takes into account the probabilistic nature of the network.

In the continuous-time scenario, references [1, 2] treated the problem of *dynamic routing* in mobile networks. In loose terms, this amounts to determining a route within the network topology constraints, through which information packets will travel from one given node to another. Of course, many of these routing operations will typically occur at the same time and between different nodes and directions, so that the routing algorithm must be able to provide a satisfactory quality of service for all customers (for example, by delivering packets in the shortest time, with as little lag as possible, and with a very low packet loss rate), without violating physical constraints such as link capacity, memory size (queueing length), or power availability. The approach in [1, 2] considers that sudden variations, modeled by MJLS, occur in the network due to, e.g., mobility and topological variations. The consideration of *time delay* in the underlying model is of major importance, owing to the time that packets must wait on a queue before being processed, together with the fact that the network nodes are geographically separated. Furthermore, this latter constraint motivated the consideration of a *decentralized* control scheme. The ultimate problem considered in [1, 2] was the minimization of the worst-case queueing length, with the aid of H_∞ control methods.

1.3 Prerequisites and General Remarks

As prerequisite for this book, it is desirable some knowledge on the classical linear control theory (stability results, the linear quadratic Gaussian (LQG) control problem, H_∞ control problem, and Riccati equations), some familiarity with continuous-time Markov chains and probability theory, and some basic knowledge of operator theory.

In this book we follow an approach that combines probability and operator theory to develop the results. By doing this we believe that the book provides a unified and rigorous treatment of recent results for the control theory of continuous-time MJLS. Most of the material included in the book was published after 1990. The goal is to provide a complete and unified picture on the main topics of the control theory of continuous-time MJLS such as mean-square stability, quadratic control and H_2 control for the complete and partial observations (also called partial information) cases, associated coupled differential and algebraic Riccati equations, linear filtering, and H_∞ control. The book also intends to present some design algorithms mainly based on linear matrix inequalities (LMIs) optimization tool packages.

One of the objectives of the book is to introduce, as far as possible in a friendly way, a bent of the MJLS theory that we have named here as the analytical point of view. We believe (and do hope) that experts in linear systems with Markov jump parameters will find in this book the minimal essential tools to follow this approach. Moreover, the stochastic control problems for MJLS considered in this book provide one of those few cases in the stochastic control field in which explicit solutions can be obtained, being a useful material for a course and for introducing students into an interesting and active research area. In addition, we do hope to motivate the reader, especially the graduate students, in such a way that this book could be a starting point for further developments and applications of continuous-time MJLS. From the application point of view we believe that the book provides a powerful theory with potential application in systems whose dynamics are subject to abrupt changes, as those found in safety-critical and high-integrity systems, industrial plants, economic systems, etc.

1.4 Overview of the Chapters

We next present a brief overview of the contents of the chapters of the book.

Chapter 2 is dedicated to present some background material needed throughout the book, as the notation, norms, and spaces that are appropriate for our approach. It also presents some important auxiliary results, especially related to the stability concepts to be considered in Chap. 3. A few facts on Markov chains and the bear essential of infinitesimal generator is also included. We also recall some basic facts regarding LMIs, which are useful for the design techniques and the H_∞ control problem.

Chapter 3 deals with mean-square stability for continuous-time MJLS. It is shown that mean-square stability is equivalent to the maximal real part of the eigenvalues of an augmented matrix being less than zero or to the existence of a solution of a Lyapunov equation. As aforementioned, stability of all modes of operation is neither necessary nor sufficient for global stability of the system. The criterion based on the eigenvalues of an augmented matrix reveals that a balance between the modes and the transition rate matrix is essential for mean-square stability.

It is worth pointing out that the setup of Chap. 3 is on the domain of complex matrices and vectors. By doing this we can use a useful result on the decomposition of a matrix into some positive semidefinite matrices, which is very important to prove the equivalence on stability results. But it is shown later on in this chapter that the results obtained are valid even in the setup of real matrices and vectors. In order to simplify the notation and proofs, in the remaining chapters of the book we consider just the real case.

Chapter 4 analyzes the quadratic optimal control problem for MJLS in the usual finite- and infinite-time horizon framework. We consider in this chapter that the controller has access to both the state variable $x(t)$ and the jump parameter $\theta(t)$. The case in which the controller has access only to an output $y(t)$ and $\theta(t)$ is considered in Chap. 6 and called the partial observation (or partial information) case. The solution for the quadratic optimal control problems of Chap. 4 relies, in part, on the study of a finite set of coupled differential and algebraic Riccati equations (CDRE and CARE, respectively). These equations are studied in the Appendix A.

Chapter 5 restudies the infinite-horizon quadratic optimal control for MJLS but now from another point of view, usually known in the literature of linear systems as the H_2 control. The advantage of the H_2 approach is that it allows one to consider parametric uncertainties and solve the problem using LMIs optimization tools.

Chapter 6 deals with the finite-horizon quadratic optimal control problem and the H_2 control problem of continuous-time MJLS for the partial information case. The main result shown is that the optimal control is obtained from two sets of coupled differential (for the finite-horizon case) and algebraic (for the H_2 case) Riccati equations, one set associated with the optimal control problem when the state variable is available, as analyzed in Chaps. 4 and 5, and the other set associated with the optimal filtering problem. This establishes the so-called separation principle for continuous-time MJLS.

Chapter 7 aims to derive the best linear mean square estimator of continuous-time MJLS assuming that only an output $y(t)$ is available. It is important to emphasize that in this chapter we assume that the jump parameter $\theta(t)$ is not known. The idea is to derive a filter which bears those desirable properties of the Kalman filter: *a recursive scheme suitable for computer implementation which allows some offline computation that alleviates the computational burden*. The linear filter has dimension Nn (where n denotes the dimension of the state vector, and N the number of states of the Markov chain). Both the finite-horizon and stationary cases are considered.

Chapter 8 is devoted to the H_∞ control of Markov jump linear systems, in the infinite-horizon setting. The statement of a bounded real lemma is the starting point toward a complete LMIs characterization of static state feedback, as well as of full-order dynamic output feedback stabilizing controllers that guarantee that a prescribed closed-loop H_∞ performance is attained. The main results include explicit formulas and the corresponding algorithms for designing the controllers of interest.

Chapters 9 and 10 are intended to conclude the book assembling some problems in the Markov jump context and the tools to solve them.

In the Appendix A some results on coupled differential and algebraic Riccati equations (CDRE and CARE, respectively) associated to the control problem are presented. Initially, we consider the problem of uniqueness, existence, positive definiteness, and continuity of the solution of the CDRE. After that we study the CARE, dealing essentially with conditions for the existence of solutions and asymptotic convergence, based on the concepts of mean-square stabilizability and detectability seen in Chap. 3. Regarding the existence of a solution, we are particularly interested in maximal and stabilizing solutions.

In the Appendix B we derive some auxiliary results related to an adjoint operator for MJLS used in the separation principle presented in Chap. 6.

1.5 Historical Remarks

The study of dynamical systems with *random parameters* can be traced back at least to [196, 200–202], and [142]. Due to the importance and complexity of the theme, a variety of techniques has emerged, and an extraordinary burst of publications associated with terminologies such as *multiple model*, *switching systems*, *hidden Markov models* and *Markov jump linear systems*, *inter alia*, has appeared in the specialized literature. Even for those approaches which embraced the idea of modeling the random parameter as a Markov chain, the methodologies were different. Each one of these topics have charted its own course, and the associated literature is by now huge. Therefore, it is out of the scope here to go into details on all of these topics (see, e.g., [22, 130, 232] and references therein for an account on multiple model, hidden Markov model, and switching systems, respectively). We focus instead on the MJLS case.

Regarding MJLS, the seminal papers [270] and [303] set the stage for future research in this area. In the first one, the jump linear quadratic (JLQ) control problem was considered for the finite-horizon setting, via a maximum principle approach (see also [269]). In the other one, dynamic programming was used, and the infinite-horizon case was also treated. In this case, although the objective had been carried out successfully, a technical inconvenience was related to the choice of the adopted stability criteria, which did not seem to be fully adequate. This, in turn, has entailed a great deal of challenges for future research. In the 1970s we can mention the papers [32, 248, 273, 274] and [33] dealing with the JLQ control problem, where the latter seems to be the first one that treated the discrete-time version of the optimal quadratic control problem for the finite-time horizon case (see also [20] for the MM approach).

Due, in part, to the lack of a suitable concept for stability, it took a while for the theory flourish. For instance, it elapsed more than ten years to appear some of the key papers that put mean-square stability (stochastic stability, or L_2-stability), mean square stabilizability, and mean square detectability for MJLS in a solid ground. Without any intention of being exhaustive here, we mention, for instance, [43, 48, 71, 77, 79, 118, 137, 147, 148, 150–152, 156, 164, 170, 170, 187, 189–191, 210, 221, 230, 231, 245, 278, 282, 284], for a sample of papers dealing with *stability and stabilizability* for MJLS. For the case in which the state space of the Markov chain is *countably infinite*, it was shown in [78] and [148] that mean-square and stochastic stability (L_2-stability) are no longer equivalent (see also [147]). Another critical issue was an adequate concept for the solution of the Riccati equation associated to MJLS (*coupled Riccati equations*). As far as the authors are aware of, it was in [156] where the concept of mean-square solution for the coupled Riccati equations was first proposed. For a glimpse on some issues dealing with coupled Riccati equations, see, for instance, [4–6, 17, 66, 67, 149, 156, 246]. For *control problems* (optimal, adaptive, H_∞, H_2, robust, receding horizon, singularly perturbed, partial observations, etc.), we mention, for instance, [18, 38, 44, 46, 47, 59, 60, 69, 70, 73–75, 78, 85, 87, 93, 94, 101, 104, 108, 111, 119, 120, 122, 125, 134, 143–146, 158, 159, 171, 176, 188, 190, 224, 229, 240, 258, 279, 307]. For the *filtering problem*, the reader is referred, for instance, to [9, 35–37, 65, 82–84, 126, 127, 132, 155, 160, 217, 321, 322]. The case with *delay* was treated, for instance, in [24, 44, 45, 55, 165, 209, 218, 257, 297]. The case with *uncertainty* (including uncertainty in the transition matrix of the Markov chain) was considered in [15, 26, 42, 256, 305, 306, 316–319]. *Separation principles* were derived, for instance, in [80, 89, 90, 153]. *Structural properties* such as controllability, detectability, and observability have been studied, for instance, in [61, 63, 187, 189, 222]. See [287, 288] for problems related to detection and identification for MJLS. In addition, there is by now a growing conviction that MJLS provide models of wide applicability (see, e.g., [14] and [266]). The evidence in favor of such a proposition has been amassing rapidly over the last decades. We mention [1, 2, 8, 14, 21, 28, 32, 39, 52, 57, 108, 124, 133, 141, 173–175, 184, 186, 197, 198, 208, 213, 219, 220, 225, 226, 228, 241, 243, 249, 253, 254, 266, 275] and [323], as works dealing with *applications* of this class of systems (see also the books [22, 40, 41, 81, 121, 223, 271, 309] and references therein).

Chapter 2
A Few Tools and Notations

2.1 Outline of the Chapter

This chapter consists primarily of some background material, with the selection of topics being dictated by our later needs. In Sect. 2.2, we introduce some notation and definitions that will be used throughout the book. In Sect. 2.3, we recall some definitions and properties of semigroup operators and infinitesimal generators, and in Sect. 2.4, we present some fundamental results on the existence and uniqueness of solutions of a differential equation. In Sect. 2.5 we recall some basic definitions and results on continuous-time Markov chains with finite state space. In Sect. 2.6, we introduce the spaces that are appropriate for our approach. Next, in Sect. 2.7, we show some important auxiliary results regarding the stability of some operators. In Sect. 2.8, we recall some basic facts regarding linear matrix inequalities.

2.2 Some Basic Notation and Definitions

We use throughout the book some standard definitions and results from operator theory in Banach spaces, which can be found, for instance, in [234] or [298]. For Banach spaces $\mathbb{X}$ and $\mathbb{Y}$, we set $\mathbb{B}(\mathbb{X}, \mathbb{Y})$ for the Banach space of all bounded linear operators of $\mathbb{X}$ into $\mathbb{Y}$, with the uniform induced norm denoted by $\|\cdot\|$. For simplicity, we set $\mathbb{B}(\mathbb{X}) := \mathbb{B}(\mathbb{X}, \mathbb{X})$. We denote by $\mathbb{C}^n$ the n-dimensional complex Euclidean space, by $\mathbb{R}^n$ the n-dimensional real Euclidean space, and by $\mathbb{R}^+$ the interval $[0, \infty)$.

If $\mathbb{X}$ is a Hilbert space then $\langle \cdot ; \cdot \rangle$ will stand for the inner product, and for $\mathcal{T} \in \mathbb{B}(\mathbb{X})$, $\mathcal{T}^*$ will indicate the adjoint operator of $\mathcal{T}$. As usual, $\mathcal{T} \geq 0$ ($\mathcal{T} > 0$) will mean that the operator $\mathcal{T} \in \mathbb{B}(\mathbb{X})$ is positive semidefinite (positive definite), respectively. The superscripts $^-$, $'$, and * will denote respectively the complex conjugate, transpose, and conjugate transpose of a matrix. We will use throughout the book the symbol * even for the case in which all the elements of a matrix are real so that, in these situations, * will represent just the transpose of a matrix. We denote by $\mathbb{B}(\mathbb{C}^n, \mathbb{C}^m)$ ($\mathbb{B}(\mathbb{R}^n, \mathbb{R}^m)$ respectively) the normed bounded linear space of

O.L.V. Costa et al., *Continuous-Time Markov Jump Linear Systems*,
Probability and Its Applications, DOI 10.1007/978-3-642-34100-7_2,

all $m \times n$ complex (real) matrices, with $\mathbb{B}(\mathbb{C}^n) := \mathbb{B}(\mathbb{C}^n, \mathbb{C}^n)$ ($\mathbb{B}(\mathbb{R}^n) := \mathbb{B}(\mathbb{R}^n, \mathbb{R}^n)$) and $\mathbb{B}(\mathbb{C}^n)^+ := \{L \in \mathbb{B}(\mathbb{C}^n); L = L^* \geq 0\}$ ($\mathbb{B}(\mathbb{R}^n)^+ := \{L \in \mathbb{B}(\mathbb{R}^n); L = L' \geq 0\}$). We refer to I_ℓ as the $\ell \times \ell$ identity matrix or, for simplicity, to just I. The kernel and the range of a matrix A will be denoted by $\mathcal{N}(A)$ and $\mathcal{R}(A)$, respectively, and, for a square matrix A, we set $\mathrm{Her}(A) := A + A^*$. For $L \in \mathbb{B}(\mathbb{C}^n)^+$, $L > 0$, we write $\|\cdot\|_L$ for the norm in $\mathbb{C}^n$ induced by the inner product $\langle x, y\rangle_L = x^* L y$. We will use $\|\cdot\|$ to denote the Euclidean norm in $\mathbb{C}^n$ or the spectral induced norm in $\mathbb{B}(\mathbb{C}^n)$.

We denote by $\mathrm{Re}\{z\}$ and $\mathrm{Im}\{z\}$ the real and imaginary parts of a complex number $z \in \mathbb{C}$, so that $z = \mathrm{Re}\{z\} + \sqrt{-1}\,\mathrm{Im}\{z\}$. In addition, we denote by $\sigma(\mathcal{L})$ the spectrum of the operator $\mathcal{L} \in \mathbb{B}(\mathbb{X})$ and write

$$\mathrm{Re}\{\lambda(\mathcal{L})\} := \sup\{\mathrm{Re}\{\lambda\}; \lambda \in \sigma(\mathcal{L})\}.$$

The eigenvalues of a matrix $P \in \mathbb{B}(\mathbb{C}^n)$ will be denoted by $\lambda_i(P)$.

We recall that the trace operator $\mathrm{tr} : \mathbb{B}(\mathbb{C}^n) \to \mathbb{C}$ is a linear functional with the following properties:

(i) $\mathrm{tr}(KL) = \mathrm{tr}(LK)$;
(ii) For any $M, P \in \mathbb{B}(\mathbb{C}^n)^+$ with $P > 0$, we have:

$$\Bigl(\min_{i=1,\dots,n} \lambda_i(P)\Bigr)\,\mathrm{tr}(M) \leq \mathrm{tr}(MP) \leq \Bigl(\max_{i=1,\dots,n} \lambda_i(P)\Bigr)\,\mathrm{tr}(M). \tag{2.1}$$

In a probability space $(\Omega, \mathcal{F}, P)$ the Dirac measure over a set $A \in \mathcal{F}$ is defined by $1_A(\cdot)$, meaning that

$$1_A(\omega) = \begin{cases} 1 & \text{if } \omega \in A, \\ 0 & \text{otherwise.} \end{cases} \tag{2.2}$$

We say that $x = \{x(t); t \in \mathbb{R}^+\} \in L_2^n(\Omega, \mathcal{F}, P)$ if $x(t)$ is a stochastic process taking values in $\mathbb{R}^n$ and satisfying

$$\|x\|_2^2 := \int_0^\infty E\bigl[\|x(t)\|^2\bigr]\,dt < \infty, \tag{2.3}$$

where $E[\cdot]$ denotes the mathematical expectation with respect to P. In other words, $L_2^n(\Omega, \mathcal{F}, P)$ represents the space of square-integrable stochastic processes.

2.3 Semigroup Operators and Infinitesimal Generator

In this section, we recall from [242] some basic definitions and properties of semigroup operators and infinitesimal generators that will be useful in the sequel. We recall that (see [242, Chap. 1]) in a Banach space $\mathbb{X}$ a one parameter family $\phi(s) \in \mathbb{B}(\mathbb{X})$, $s \in \mathbb{R}^+$, is called a semigroup of bounded linear operators on $\mathbb{X}$ if

$\phi(0) = I$ and the semigroup property $\phi(t+s) = \phi(t)\phi(s)$ for every $t, s \in \mathbb{R}^+$ is satisfied. The infinitesimal generator $\mathcal{L}$ of the semigroup $\phi(t)$ is defined as

$$\mathcal{L}x := \lim_{h \downarrow 0} \frac{\phi(h)x - x}{h} = \left. \frac{d^+\phi(t)x}{dt} \right|_{t=0} \tag{2.4}$$

whenever the limit in (2.4) exists. The domain of $\mathcal{L}$, denoted by $\mathfrak{D}(\mathcal{L})$, is the set of all elements $x \in \mathbb{X}$ for which the limit in (2.4) exists. A semigroup $\phi(s)$ of bounded linear operators is called uniformly continuous at $t = 0$ if $\lim_{t \downarrow 0} \|\phi(t) - I\| = 0$ and strongly continuous at $t = 0$ if $\lim_{t \downarrow 0} \phi(t)x = x$ for every $x \in \mathbb{X}$. The following theorem is proved in Chap. 1 of [242] (see in particular Theorem 1.2).

Theorem 2.1 *A linear operator $\mathcal{L}$ on a Banach space $\mathbb{X}$ is the infinitesimal generator of a uniformly continuous semigroup $\phi(t)$ if and only if $\mathcal{L} \in \mathbb{B}(\mathbb{X})$. Furthermore, associated to $\mathcal{L} \in \mathbb{B}(\mathbb{X})$, there exists a unique uniformly continuous semigroup $\phi_{\mathcal{L}}(t)$ given by*

$$\phi_{\mathcal{L}}(t) = e^{\mathcal{L}t} := \sum_{k=0}^{\infty} \frac{1}{k!} \mathcal{L}^k t^k \in \mathbb{B}(\mathbb{X}). \tag{2.5}$$

The following corollary is also presented in Chap. 1 of [242].

Corollary 2.2 *Let $\phi(t)$ be a uniformly continuous semigroup. Then*:

(a) *There exists a constant $c \geq 0$ such that $\|\phi(t)\| \leq e^{ct}$.*
(b) *There exists a unique $\mathcal{L} \in \mathbb{B}(\mathbb{X})$ such that $\phi(t) = \phi_{\mathcal{L}}(t) = e^{\mathcal{L}t}$.*
(c) *The operator $\mathcal{L} \in \mathbb{B}(\mathbb{X})$ given by* (b) *is the infinitesimal generator of the uniformly continuous semigroup $\phi(t)$, and $\phi(t)\mathcal{L} = \mathcal{L}\phi(t)$.*
(d) *We have that*

$$\frac{d\phi_{\mathcal{L}}(t)}{dt} = \mathcal{L}\phi_{\mathcal{L}}(t). \tag{2.6}$$

2.4 The Fundamental Theorem for Differential Equations

In this section we present some fundamental results on the existence and uniqueness of solutions of a differential equation that will be used throughout the book. The results of this section follow those presented in [53]. We start by presenting the definition of a real-valued function and a matrix-function of class **PC** (piecewise constant, see [53]).

Definition 2.3 We say that a real-valued function f defined on $\mathbb{R}^+$ is of class **PC** if on every bounded interval $[t_0, t_1] \subset \mathbb{R}^+$, $f(t)$ is continuous everywhere except at a finite number of discontinuity points τ_k, where the one-sided limits $f(\tau_k+)$ (the right limit) and $f(\tau_k-)$ (the left limit) are well defined and finite. We say that a matrix-valued function A on $\mathbb{R}^+$ is of class **PC** if all its elements A_{ij} are of class **PC**.

Consider now the following n-dimensional differential equation:

$$\dot{x}(t) = p\bigl(x(t), t\bigr) \tag{2.7}$$

with initial condition $x(t_0) = x_0$. Let D be a set in $\mathbb{R}^+$ which contains at most a finite number of points per unit interval. In addition, consider the following two assumptions:

(A1) For each $x \in \mathbb{R}^n$, the function $t \in \mathbb{R}^+ \setminus D \to p(x,t) \in \mathbb{R}^n$ is continuous, and for any $\tau \in D$, the left-hand and the right-hand limits $p(x, \tau-)$ and $p(x, \tau+)$, respectively, are finite vectors in $\mathbb{R}^n$.

(A2) There is a real-valued function $k : \mathbb{R}^+ \to \mathbb{R}^+$ of class **PC** such that the global Lipschitz condition is satisfied:

$$\bigl\| p(x,t) - p(y,t) \bigr\| \le k(t)\|x - y\| \quad \forall t \in \mathbb{R}^+,\ \forall x, y \in \mathbb{R}^n. \tag{2.8}$$

The next result is proved in Theorem B1.6 of [53] (p. 470).

Theorem 2.4 *Consider the differential equation* (2.7) *with initial condition* $x(t_0) = x_0$, *and suppose that Assumptions* (A1) *and* (A2) *are satisfied. Then*:

(i) *for each* $(t_0, x_0) \in \mathbb{R}^+ \times \mathbb{R}^n$, *there exists a continuous function* $\phi : \mathbb{R}^+ \to \mathbb{R}^n$ *such that* $\phi(t_0) = x_0$, *and for all* $t \in \mathbb{R}^+ \setminus D$, *we have that* $\dot{\phi}(t) = p(\phi(t), t)$;

(ii) *this function is unique.*

Consider now the linear differential equation

$$\dot{x}(t) = A(t)x(t) + B(t)u(t), \tag{2.9}$$

where A, B, and u are of class **PC**. We have that for the global Lipschitz condition,

$$\bigl\| \bigl(A(t)x + B(t)u(t)\bigr) - \bigl(A(t)y + B(t)u(t)\bigr) \bigr\| = \bigl\| A(t)(x - y) \bigr\| \le \bigl\| A(t) \bigr\| \|x - y\|,$$

and thus Assumption (A2) is satisfied with $k(t) = \|A(t)\|$. Moreover, Assumption (A1) is satisfied by taking D as the union of the sets of discontinuity points of A, B, and u. By Theorem 2.4, (2.9) has a unique continuous solution $x(t)$. Moreover, the following results can be obtained (see Fact 37 and Theorem 70 in Chap. 2 of [53]).

Theorem 2.5 *For all* $t_0 \in \mathbb{R}^+$, *there exists a unique continuous matrix function (called state transition matrix)* $\Phi(\cdot, t_0) : \mathbb{R}^+ \to \mathbb{B}(\mathbb{R}^n)$, *a solution of the homogeneous linear matrix differential equation*

$$\frac{\partial \Phi(t, t_0)}{\partial t} = A(t)\Phi(t, t_0), \quad a.a.\ t \in \mathbb{R}^+,$$

$$\Phi(t_0, t_0) = I$$

(*where a.a. t means almost all t, since for those t where A is discontinuous, the left-hand side of the differential equation is not defined in the usual sense*). *Moreover, the unique solution* $x(t)$ *of* (2.9) *is given by*

$$x(t) = \Phi(t, t_0)x_0 + \int_{t_0}^{t} \Phi(t, \tau)B(\tau)u(\tau)\,d\tau. \tag{2.10}$$

Remark 2.6 For the case in which A is constant, we have from (2.6) that the state transition matrix is given by $\Phi(t, \tau) = e^{A(t-\tau)}$.

2.5 Continuous-Time Markov Chains

For a positive integer number N, we define $\mathcal{S} := \{1, \dots, N\}$. Let $(\Omega, \mathcal{F}, P)$ be a complete probability space equipped with a filtration $\{\mathcal{F}_t; t \in \mathbb{R}^+\}$ satisfying the usual hypotheses, that is, a right-continuous filtration augmented by all null sets in the P-completion of $\mathcal{F}$. Throughout the book we will consider a homogeneous Markov chain $\{\theta(t); t \in \mathbb{R}^+\}$ adapted to the filtration $\{\mathcal{F}_t; t \in \mathbb{R}^+\}$, denoted as $\theta = \{(\theta(t), \mathcal{F}_t); t \in \mathbb{R}^+\}$, with right-continuous trajectories and taking values on the set $\mathcal{S}$. We recall that by adapted to the filtration $\{\mathcal{F}_t; t \in \mathbb{R}^+\}$ we mean that for each $t \in \mathbb{R}^+$, $\theta(t)$ is $\mathcal{F}_t$-measurable, and from the Markov property we have for $s, t \in \mathbb{R}^+$ with $t \geq s$, that $P(\theta(t) = j|\mathcal{F}_s) = P(\theta(t) = j|\theta(s))$. By homogeneous we mean that $P(\theta(t) = j|\theta(s))$ depends upon s and t only through the difference $t - s$, so that we can define

$$p_{ij}(t) := P\big(\theta(t + s) = j|\theta(s) = i\big), \quad i, j \in \mathcal{S}.$$

Define for all $t \in \mathbb{R}^+$, the $N \times N$ dimensional transition matrices $T(t)$ by letting the element in row i, column j of these matrices be $p_{ij}(t)$. Set also the vector $P(t) := [p_1(t) \ \dots \ p_N(t)]' \in \mathbb{R}^N$, where

$$p_i(t) := P\big(\theta(t) = i\big) \tag{2.11}$$

for $i \in \mathcal{S}$. The initial distribution of θ will be denoted by $\nu = \{\nu_i; i \in \mathcal{S}\}$, so that $p_i(0) = \nu_i$ in (2.11). As shown in [239], Proposition 1.2 of Chap. 7, p. 137,

$$P(t) = T(t)'P(0), \tag{2.12}$$

and the transition matrices $T(t)$ satisfy the semigroup property (also called the Chapman–Kolmogorov equation)

$$T(t + s) = T(t)T(s).$$

Furthermore, the following result is proved in [239], Theorem 2.1 and Corollary 2.2 of Chap. 7, pp. 139–140 (we recall that the notation $o(h)$ denotes a function on $h > 0$ such that $\lim_{h\downarrow 0} \frac{o(h)}{h} = 0$).

Theorem 2.7 *Let $\{T(t); t \in \mathbb{R}^+\}$ be the semigroup of transition matrices of a continuous-time Markov chain $\{\theta(t); t \in \mathbb{R}^+\}$. Then there exists an $N \times N$ matrix Π which is the infinitesimal generator of the semigroup $\{T(t); t \in \mathbb{R}^+\}$. Define as λ_{ij} the element in row i and column j of Π. Then for $h \geq 0$,*

$$P\big(\theta(t+h) = j|\theta(t) = i\big) = \begin{cases} \lambda_{ij}h + o(h), & i \neq j, \\ 1 + \lambda_{ii}h + o(h), & i = j, \end{cases} \tag{2.13}$$

with $0 \leq \lambda_{ij}$, $i \neq j$, and $0 \leq \lambda_i := -\lambda_{ii} = \sum_{\{j:\, j \neq i\}} \lambda_{ij}$ for all $i \in \mathcal{S}$. Moreover, $\{T(t); t \in \mathbb{R}^+\}$ is the unique solution of the backward Kolmogorov differential equation:

$$\frac{dT(t)}{dt} = \Pi T(t), \quad t \in \mathbb{R}^+, \qquad T(0) = I. \tag{2.14}$$

Similarly, $\{T(t); t \in \mathbb{R}^+\}$ is the unique solution of the forward Kolmogorov differential equation:

$$\frac{dT(t)}{dt} = T(t)\Pi, \quad t \in \mathbb{R}^+, \qquad T(0) = I. \tag{2.15}$$

Remark 2.8 Notice that the matrix Π is also called the stationary transition rate matrix of θ.

As seen in Remark 2.6, from (2.15) (or (2.14)) we have that

$$T(t) = e^{\Pi t}, \quad t \in \mathbb{R}^+. \tag{2.16}$$

It follows from (2.15) that $P(t)$ satisfies the forward Kolmogorov differential equation

$$\frac{dP(t)}{dt} = \Pi' P(t), \quad t \in \mathbb{R}^+, \tag{2.17}$$

whose solution, by (2.12) and (2.16), is

$$P'(t) = P'(0)e^{\Pi t}, \quad t \in \mathbb{R}^+. \tag{2.18}$$

In what follows we denote by $\mathbf{e} \in \mathbb{R}^N$ the vector formed by 1s in all its components. We recall the following definitions from [309].

Definition 2.9 Π is said to be weakly irreducible if the system of equations $x'\Pi = 0$, $x'\mathbf{e} = 1$ has a unique solution $\pi \in \mathbb{R}^N$ and $\pi \geq 0$. This solution is called the quasi-stationary distribution. Π is said to be irreducible if the system of equations $x'\Pi = 0$, $x'\mathbf{e} = 1$ has a unique solution $\pi \in \mathbb{R}^N$ and $\pi > 0$. This solution π is called the stationary distribution.

Clearly if Π is irreducible, then it is weakly irreducible, but, as shown in [309], p. 21, the reverse is not in general true. The following result is shown in [309], Lemma 4.4, p. 56.

Lemma 2.10 *Consider the forward Kolmogorov differential equations* (2.15) *and suppose that Π is weakly irreducible with quasi-stationary distribution π. Then $T(t) \to \bar{P}$ as $t \to \infty$, where $\bar{P} = \mathbf{e}\pi'$. Moreover,*

$$\left\| e^{\Pi t} - \bar{P} \right\| \leq \bar{\alpha} e^{-\beta t}, \quad t \in \mathbb{R}^+, \tag{2.19}$$

for some positive constants $\bar{\alpha} > 0$, $\beta > 0$.

For any $N \times N$ matrix V consider the norm $\|V\|_{\max} = \max_{i,j} |v_{ij}|$. From the equivalence of the norms in finite-dimensional spaces (see Theorem 5.10.6 in [234]) we have that $\|V\|_{\max} \leq c\|V\|$ for some $c > 0$. In particular, we have under the hypothesis of Lemma 2.10 and from (2.16) that

$$\max_{i,j} |p_{ij}(t) - \pi_j| = \left\| e^{\Pi t} - \bar{P} \right\|_{\max} \leq c \left\| e^{\Pi t} - \bar{P} \right\| \leq c\bar{\alpha} e^{-\beta t}, \quad t \in \mathbb{R}^+. \tag{2.20}$$

Thus, under the hypothesis of Lemma 2.10, from (2.18) and (2.20) it follows that for some positive constant $\alpha > 0$,

$$\max_{i,j} \left| p_{ij}(t) - \pi_j \right| \leq \alpha e^{-\beta t} \quad \text{and} \quad \max_{j} \left| p_j(t) - \pi_j \right| \leq \alpha e^{-\beta t}. \tag{2.21}$$

We next recall the definition of irreducibility of a continuous-time Markov chain (see, for instance, [91]).

Definition 2.11 The continuous-time Markov chain θ is said to be irreducible if for any two states i_1 and i_2 in $\mathcal{S}$, there exists $t \in \mathbb{R}^+$ such that $p_{i_1 i_2}(t) > 0$.

From (2.21) it is easy to see that if Π is irreducible, then we can find $t > 0$ such that $p_{ij}(t) > \pi_j - \alpha e^{-\beta t} > 0$ since $\pi_j > 0$. On the other hand, it is shown in [91], page 185, that for the case of finite number of states, if the continuous-time Markov chain θ is irreducible, then Π is irreducible. Along the book we will sometimes assume the hypothesis that θ is irreducible, so that the exponential convergence (2.21), also known as exponential ergodicity, will be satisfied. Notice that if the number of states is infinite countable, then some extra conditions are needed to get exponential ergodicity (see, for instance, Sect. 6.6 in [10]).

In Chap. 9, we will consider the robust linear filtering problem for the case in which Π is not exactly known but instead there are known irreducible stationary transition rate matrices Π^κ such that

$$\Pi = \sum_{\kappa=1}^{\ell} \rho^\kappa \Pi^\kappa \tag{2.22}$$

for some $0 \leq \rho^\kappa \leq 1$, $\kappa = 1, \ldots, \ell$, $\sum_{\kappa=1}^{\ell} \rho^\kappa = 1$. The next result will be useful in Chap. 9, Sect. 9.5.

Proposition 2.12 *Let Π be as in* (2.22) *with each Π^κ irreducible. Then Π is an irreducible transition rate matrix.*

Proof We have that Π is a transition rate matrix since

$$-\lambda_{ii} = -\sum_{\kappa=1}^{\ell} \rho^{\kappa} \lambda_{ii}^{\kappa} = \sum_{\kappa=1}^{\ell} \rho^{\kappa} \sum_{j \neq i} \lambda_{ij}^{\kappa} = \sum_{j \neq i} \sum_{\kappa=1}^{\ell} \rho^{\kappa} \lambda_{ij}^{\kappa} = \sum_{j \neq i} \lambda_{ij}.$$

Suppose by contradiction that Π is not irreducible. Then, according to [91], Sect. 3.10, by relabeling the states appropriately, Π and Π^{κ} can be written as

$$\Pi = \begin{pmatrix} \Pi_{11} & 0 \\ \Pi_{21} & \Pi_{22} \end{pmatrix} = \sum_{\kappa=1}^{\ell} \rho^{\kappa} \Pi^{\kappa} = \sum_{\kappa=1}^{\ell} \rho^{\kappa} \begin{pmatrix} \Pi_{11}^{\kappa} & \Pi_{12}^{\kappa} \\ \Pi_{21}^{\kappa} & \Pi_{22}^{\kappa} \end{pmatrix},$$

where Π_{11} is a square matrix. But since $\rho^{\kappa} \geq 0$ and each element of the matrix Π_{12}^{κ} is nonnegative, we must have for κ such that $\rho^{\kappa} > 0$ that $\Pi_{12}^{\kappa} = 0$, in contradiction with the hypothesis that Π^{κ} is irreducible. □

We next present a useful result that will be employed in Chap. 4 to get Dynkin's formula for the MJLS. First, we recall the following definition.

Definition 2.13 The stochastic process $M = \{M(t); t \in \mathbb{R}^+\}$ is a martingale with respect to the filtration $\{\mathcal{F}_t; t \in \mathbb{R}^+\}$ if M is adapted to $\{\mathcal{F}_t; t \in \mathbb{R}^+\}$, the random variable $M(t)$ is integrable for each $t \in \mathbb{R}^+$, and for each $s, t \in \mathbb{R}^+$ with $t \geq s$, $E(M(t)|\mathcal{F}_s) = M(s)$ a.s. (almost surely).

Define

$$\chi_i(t) := 1_{\{\theta(t)=i\}} \tag{2.23}$$

and the random vector $\chi(t) = [\chi_1(t) \ \ldots \ \chi_N(t)]'$ taking values on the set $\mathcal{S}_v = \{e_1, \ldots, e_N\}$, where $e_i \in \mathbb{R}^N$ is formed by 1 in the ith component and zero elsewhere. The following result was proved in Lemmas 1.1 and 1.5 in [130], Appendix B.

Lemma 2.14 *Let $\mathcal{F}_t$ be the right-continuous complete filtration generated by $\sigma\{\theta(r); 0 \leq r \leq t\}$ and set*

$$M(t) := \chi(t) - \chi(0) - \int_0^t \Pi' \chi(s-)\, ds. \tag{2.24}$$

Then $M = \{M(t); t \in \mathbb{R}^+\}$ is a martingale with respect to the filtration $\{\mathcal{F}_t; t \in \mathbb{R}^+\}$. Furthermore, for a differentiable function $f(t)$ taking values in $\mathbb{R}^N$, set $f(t, \chi(t)) = f(t)'\chi(t)$. Then

$$f\big(t, \chi(t)\big) = f\big(0, \chi(0)\big) + \int_0^t \frac{df(s)'}{ds} \chi(s)\, ds + \int_0^t f(s)' \Pi' \chi(s-)\, ds + \int_0^t f(s)'\, dM(s), \tag{2.25}$$

and defining

$$\mathfrak{I}(t) := \int_0^t f(s)' dM(s), \tag{2.26}$$

it follows that $\mathfrak{I} = \{\mathfrak{I}(t); t \in \mathbb{R}^+\}$ *is a martingale with respect to the filtration* $\{\mathcal{F}_t; t \in \mathbb{R}^+\}$.

Remark 2.15 The notation $g(t-)$ represents the left-hand limit of a function $g(t)$ in (2.24) and (2.25).

We conclude this section presenting a result on the strong Markov property of the augmented state $(x(t), \theta(t))$ associated to systems (1.1) and (1.2) (see also [56, Chap. 2]).

Theorem 2.16 *Let* $x(t)$ *be a solution of system* (1.1) *or* (1.2) *and set*

$$\vartheta_t = \big(x(t), \theta(t)\big). \tag{2.27}$$

The augmented state process $\vartheta = \{\vartheta_t, t \geq 0\}$ *satisfies the strong Markov property.*

Proof It follows from Corollary 2.19 in [310]. □

2.6 The Space of Sequences of N Matrices

In order to analyze the stochastic models in the next chapters, we will use the indicator function on the jump parameter taking values in $\mathcal{S}$ to *markovianize* the state. This, in turn, will decompose the matrices associated to the second moment and control problems into N matrices. We will consider throughout the book that all matrices involved in our dynamic systems will be real. However it will be convenient to consider in this chapter and Chap. 3 the complex case for reasons that will become clear in Proposition 2.23 below (see also Remark 2.24 and Proposition 3.19). Due to that, a natural and convenient space to be used is $\mathbb{H}_{\mathbb{C}}^{n,m}$, defined as the linear space made up of all sequences of N matrices $\mathbf{V} = (V_1, \dots, V_N)$ with $V_i \in \mathbb{B}(\mathbb{C}^n, \mathbb{C}^m)$. For simplicity, set $\mathbb{H}_{\mathbb{C}}^{n} := \mathbb{H}_{\mathbb{C}}^{n,n}$. For $\mathbf{V} = (V_1, \dots, V_N) \in \mathbb{H}_{\mathbb{C}}^{n,m}$, we write $\mathbf{V}^* = (V_1^*, \dots, V_N^*) \in \mathbb{H}_{\mathbb{C}}^{m,n}$ and say that $\mathbf{V} \in \mathbb{H}_{\mathbb{C}}^{n}$ is Hermitian if $\mathbf{V} = \mathbf{V}^*$. We define $\mathbb{H}_{\mathbb{C}}^{n*} := \{\mathbf{V} = (V_1, \dots, V_N) \in \mathbb{H}_{\mathbb{C}}^{n}; V_i = V_i^*, i = 1, \dots, N\}$ and $\mathbb{H}_{\mathbb{C}}^{n+} := \{\mathbf{V} = (V_1, \dots, V_N) \in \mathbb{H}_{\mathbb{C}}^{n*}; V_i \geq 0, i = 1, \dots, N\}$ and write, for $\mathbf{V} = (V_1, \dots, V_N) \in \mathbb{H}_{\mathbb{C}}^{n}$ and $\mathbf{S} = (S_1, \dots, S_N) \in \mathbb{H}_{\mathbb{C}}^{n}$, that $\mathbf{V} \geq \mathbf{S}$ if $\mathbf{V} - \mathbf{S} = (V_1 - S_1, \dots, V_N - S_N) \in \mathbb{H}_{\mathbb{C}}^{n+}$ and that $\mathbf{V} > \mathbf{S}$ if $V_i - S_i > 0$ for each $i \in \mathcal{S}$.

For $\mathbf{V} = (V_1, \dots, V_N) \in \mathbb{H}_{\mathbb{C}}^{n,m}$, we consider the following norms $\|\cdot\|_\kappa$ in $\mathbb{H}_{\mathbb{C}}^{n,m}$, $\kappa = 1, 2, \max$:

$$\|\mathbf{V}\|_1 := \sum_{i \in \mathcal{S}} \|V_i\|, \tag{2.28}$$

$$\|\mathbf{V}\|_2 := \left(\sum_{i\in\mathcal{S}} \operatorname{tr}\left(V_i^* V_i\right)\right)^{1/2}, \tag{2.29}$$

$$\|\mathbf{V}\|_{\max} := \max\left\{\|V_i\|; i \in \mathcal{S}\right\}. \tag{2.30}$$

From the equivalence of the norms in finite-dimensional spaces (see [234]), we have that all the above norms are equivalent (see also Remark 2.19 below). It is easy to see that

$$\mathbf{H} \le \mathbf{L} \quad \Rightarrow \quad \|\mathbf{H}\|_1 \le \|\mathbf{L}\|_1. \tag{2.31}$$

It is also easy to verify that $(\mathbb{H}_{\mathbb{C}}^{n,m}, \|\cdot\|_\kappa)$ are Banach spaces and, in fact, $(\mathbb{H}_{\mathbb{C}}^{n,m}, \|\cdot\|_2)$ is a Hilbert space, with inner product given, for $\mathbf{S} = (S_1, \dots, S_N)$ and $\mathbf{V} = (V_1, \dots, V_N)$ in $\mathbb{H}_{\mathbb{C}}^{n,m}$, by

$$\langle \mathbf{V}; \mathbf{S}\rangle = \sum_{i\in\mathcal{S}} \operatorname{tr}\left(V_i^* S_i\right). \tag{2.32}$$

We will restrict the above definitions to the case of real matrices by writing $\mathbb{H}$. For instance, the linear space made up of all sequences of N real matrices $\mathbf{V} = (V_1, \dots, V_N)$ with $V_i \in \mathbb{B}(\mathbb{R}^n, \mathbb{R}^m)$ will be denoted by $\mathbb{H}^{n,m}$, and so on.

For an operator $\Upsilon \in \mathbb{B}(\mathbb{H}_{\mathbb{C}}^{n,m}, \mathbb{H}_{\mathbb{C}}^{\ell,r})$, we will consider the following induced norm in $\mathbb{B}(\mathbb{H}_{\mathbb{C}}^{n,m}, \mathbb{H}_{\mathbb{C}}^{\ell,r})$:

$$\|\Upsilon\| = \sup\left\{\frac{\|\Upsilon(\mathbf{V})\|_1}{\|\mathbf{V}\|_1}; \mathbf{V} \in \mathbb{H}_{\mathbb{C}}^{n,m}, \mathbf{V} \neq 0\right\}. \tag{2.33}$$

We say that an operator $\Upsilon \in \mathbb{B}(\mathbb{H}_{\mathbb{C}}^{n})$ is positive if it maps $\mathbb{H}_{\mathbb{C}}^{n+}$ into $\mathbb{H}_{\mathbb{C}}^{n+}$, that is, $\Upsilon(\mathbf{H}) \in \mathbb{H}_{\mathbb{C}}^{n+}$ whenever $\mathbf{H} \in \mathbb{H}_{\mathbb{C}}^{n+}$.

The following result will be useful in the sequel.

Lemma 2.17 *Suppose that for $\mathbf{H}^T \in \mathbb{H}_{\mathbb{C}}^{n+}$, $T \in (0, \infty)$, we have $H_i^{T_1} \le H_i^{T_2} \le dI$ for every $T_1 < T_2$ and some constant $0 < d < \infty$ which does not depend on i, T_1, and T_2. Then there exists $\mathbf{H} \in \mathbb{H}_{\mathbb{C}}^{n+}$ such that $\mathbf{H}^T \to \mathbf{H}$ as $T \to \infty$.*

Proof The assertion follows from a standard monotonicity result concerning positive semi-definite matrices (see Lemma 3.1 in [302]). □

For $D_i \in \mathbb{B}(\mathbb{R}^n)$, $i \in \mathcal{S}$, $\operatorname{diag}(D_i)$ is an Nn square matrix where the matrices D_i are put together corner-to-corner diagonally, with all other entries being zero.

Define now the operators φ, $\hat{\varphi}_i$, and $\hat{\varphi}$ in the following way: for $\mathbf{V} = (V_1, \dots, V_N) \in \mathbb{H}_{\mathbb{C}}^{n,m}$ with $V_i = [v_{i1} \ \dots \ v_{in}] \in \mathbb{B}(\mathbb{C}^n, \mathbb{C}^m)$, $v_{ij} \in \mathbb{C}^m$,

$$\varphi(V_i) := \begin{bmatrix} v_{i1} \\ \vdots \\ v_{in} \end{bmatrix} \in \mathbb{C}^{mn},$$

$$\hat{\varphi}_i(\mathbf{V}) := \varphi(V_i) \in \mathbb{C}^{mn}, \qquad \hat{\varphi}(\mathbf{V}) := \begin{bmatrix} \varphi(V_1) \\ \vdots \\ \varphi(V_N) \end{bmatrix} \in \mathbb{C}^{Nmn}. \tag{2.34}$$

Furthermore, for

$$v = \begin{bmatrix} v_1 \\ \vdots \\ v_N \end{bmatrix} \in \mathbb{C}^{Nmn},$$

$v_i \in \mathbb{C}^{mn}$, we define $V_i \in \mathbb{B}(\mathbb{R}^n, \mathbb{R}^m)$ such that $V_i := \varphi^{-1}(v_i)$ and $\mathbf{V} = (V_1, \ldots, V_N) \in \mathbb{H}_{\mathbb{C}}^{n,m}$ as

$$\mathbf{V} := \hat{\varphi}^{-1}\left(\begin{bmatrix} v_1 \\ \vdots \\ v_N \end{bmatrix}\right) := \big(\hat{\varphi}_1^{-1}(v), \ldots, \hat{\varphi}_N^{-1}(v)\big) = \big(\varphi^{-1}(v_1), \ldots, \varphi^{-1}(v_N)\big).$$

Remark 2.18 Notice that the mapping φ stacks up the columns of the matrix from left to right and makes a long vector out of the matrix. Furthermore, it can be shown, through the mapping $\hat{\varphi}$, that $(\mathbb{H}_{\mathbb{C}}^{n,m}, \|\cdot\|_2)$ and $(\mathbb{C}^{Nmn}, \|\cdot\|_2)$ are *isometrically isomorphic* spaces (if $\mathbf{V} \in \mathbb{H}_{\mathbb{C}}^{n,m}$, then $\|\mathbf{V}\|_2 = \|\hat{\varphi}(\mathbf{V})\|_2$).

Remark 2.19 From the equivalence of the norms in finite-dimensional spaces (see [234]) we have that there exist constants $c_1 > 0$ and $c_2 > 0$ such that for any $\mathbf{H} = (H_1, \ldots, H_N) \in \mathbb{H}_{\mathbb{C}}^{n,m}$,

$$c_1 \big\|\hat{\varphi}(\mathbf{H})\big\| \le \|\mathbf{H}\|_1 \le c_2 \big\|\hat{\varphi}(\mathbf{H})\big\|. \tag{2.35}$$

From (2.35) we also have for any $y \in \mathbb{C}^{Nnm}$ that

$$\frac{1}{c_2}\big\|\hat{\varphi}^{-1}(y)\big\|_1 \le \|y\| \le \frac{1}{c_1}\big\|\hat{\varphi}^{-1}(y)\big\|_1. \tag{2.36}$$

From (2.35) it is clear that $\hat{\varphi}$ is a continuous mapping from $\mathbb{H}_{\mathbb{C}}^{n,m}$ into $\mathbb{C}^{Nmn}$, because $\|\hat{\varphi}(\mathbf{H}) - \hat{\varphi}(\mathbf{Q})\| = \|\hat{\varphi}(\mathbf{H} - \mathbf{Q})\| \le \frac{1}{c_1}\|\mathbf{H} - \mathbf{Q}\|_1$. From (2.36) similar remarks hold for $\hat{\varphi}^{-1}$.

With the Kronecker product $L \otimes K \in \mathbb{B}(\mathbb{C}^{sn}, \mathbb{C}^{rm})$ defined in the usual way for any $L \in \mathbb{B}(\mathbb{C}^s, \mathbb{C}^r)$ and $K \in \mathbb{B}(\mathbb{C}^n, \mathbb{C}^m)$, the following properties hold (see, e.g., [50]):

$$\text{(i) } (L \otimes K)^* = L^* \otimes K^* \quad \text{and} \quad \text{(ii) } \varphi(LKH) = \big(H' \otimes L\big)\varphi(K). \tag{2.37}$$

Recall also that for $L \in \mathbb{B}(\mathbb{C}^n)$ and $K \in \mathbb{B}(\mathbb{C}^m)$, the Kronecker sum is defined as

$$L \oplus K := L \otimes I_m + I_n \otimes K \in \mathbb{B}\big(\mathbb{C}^{nm}\big).$$

2.7 Auxiliary Results

In this section we present several auxiliary results that will be useful along the book. The first result presents a useful decomposition. Notice that in order to get this decomposition, we need to work on the space of complex matrices.

Lemma 2.20 *For any* $\mathbf{H} \in \mathbb{H}^n_{\mathbb{C}}$, *there exist* $\mathbf{H}^i \in \mathbb{H}^{n+}_{\mathbb{C}}$, $i = 1, 2, 3, 4$, *such that*

$$\mathbf{H} = (\mathbf{H}^1 - \mathbf{H}^2) + \sqrt{-1}(\mathbf{H}^3 - \mathbf{H}^4). \tag{2.38}$$

Proof For any $W \in \mathbb{B}(\mathbb{C}^n)$, there exist W^j, $j = 1, 2, 3, 4$, such that $W^j \geq 0$ and $\|W^j\| \leq \|W\|$ for $j = 1, 2, 3, 4$, and $W = (W^1 - W^2) + \sqrt{-1}(W^3 - W^4)$. Indeed, we can write

$$W = V^1 + \sqrt{-1}V^2,$$

where

$$V^1 = \frac{1}{2}(W^* + W),$$
$$V^2 = \frac{\sqrt{-1}}{2}(W^* - W).$$

Since V^1 and V^2 are self-adjoint (that is, $V^i = V^{i*}$, $i = 1, 2$) and every self-adjoint element in $\mathbb{B}(\mathbb{C}^n)$ can be decomposed into positive and negative parts (see [234], p. 464), we have that there exist $W^i \in \mathbb{B}(\mathbb{C}^n)^+, i = 1, 2, 3, 4$, such that

$$V^1 = W^1 - W^2,$$
$$V^2 = W^3 - W^4.$$

Therefore, for any $\mathbf{H} = (H_1, \ldots, H_N) \in \mathbb{H}^n_{\mathbb{C}}$, we can find $\mathbf{H}^j \in \mathbb{H}^{n+}_{\mathbb{C}}$, $j = 1, 2, 3, 4$, such that (2.38) holds. □

For $A \in \mathbb{B}(\mathbb{R}^n)$, we have that A is the infinitesimal generator of the uniformly continuous semigroup $\phi_A(t) = e^{At}$ (see Sect. 2.3). The next result is proved in Chap. 2 of [312], Theorem 2.3 (see also Sect. 4.2 of [242]).

Proposition 2.21 *Let* $A \in \mathbb{B}(\mathbb{R}^n)$. *The following conditions are equivalent*:

(i) $\operatorname{Re}\{\lambda(A)\} < 0$.
(ii) *There are constants* $k > 0$ *and* $b > 0$ *such that*

$$\|e^{At}\| \leq ke^{-bt} \quad \textit{for all } t \geq 0.$$

(iii) $\|e^{At}x_0\| \to 0$ *as* $t \to \infty$ *for every* $x_0 \in \mathbb{C}^n$.
(iv) $\int_0^\infty \|e^{At}x_0\|dt < \infty$ *for every* $x_0 \in \mathbb{C}^n$.

The result also holds replacing $\mathbb{C}^n$ *by* $\mathbb{R}^n$ *in* (iii) *and* (iv).

The next proposition, adapted from [242], will be useful in deriving some stability results in Chap. 3.

Proposition 2.22 *Let* $A \in \mathbb{B}(\mathbb{R}^n)$ *and* $\{f(t); t \in \mathbb{R}^+\}$ *be a continuous function in* $\mathbb{R}^n$ *such that* $\lim_{t\to\infty} f(t) = f_0$. *Consider*

$$\dot{y}(t) = Ay(t) + f(t). \tag{2.39}$$

If $\operatorname{Re}\{\lambda(A)\} < 0$, *then for any initial condition* $y(0) = y_0 \in \mathbb{R}^n$,

$$\lim_{t\to\infty} y(t) = -\mathcal{A}^{-1} f_0.$$

Let us consider now $\mathcal{L} \in \mathbb{B}(\mathbb{H}^n)$. We recall from Sect. 2.3 that $\mathcal{L}$ is the infinitesimal generator of the uniformly continuous semigroup

$$\phi_{\mathcal{L}}(t) = e^{\mathcal{L}t} = \sum_{k=0}^{\infty} \frac{1}{k!} \mathcal{L}^k t^k \in \mathbb{B}\big(\mathbb{H}^n_{\mathbb{C}}\big).$$

Moreover, from the definition of the operators $\hat{\varphi}$ and $\hat{\varphi}^{-1}$ in Sect. 2.6 we have that for any $\mathbf{V} \in \mathbb{H}^n_{\mathbb{C}}$, $\mathbf{Q} = \mathcal{L}(\mathbf{V})$ implies that $\hat{\varphi}(\mathbf{Q}) = (\hat{\varphi}\mathcal{L}\hat{\varphi}^{-1})v$, where $v = \hat{\varphi}(\mathbf{V}) \in \mathbb{C}^{Nn^2}$. Thus, from $(\hat{\varphi}\mathcal{L}\hat{\varphi}^{-1})$ we can obtain a matrix representation of the operator $\mathcal{L}$. Indeed, let $\{e_i\}$, $i = 1, \ldots, Nn^2$, be the canonical basis for $\mathbb{R}^{Nn^2}$ (that is, e_i is a vector in $\mathbb{R}^{Nn^2}$ formed by 1 at the ith position, 0 elsewhere), and $\mathcal{A}_i = (\hat{\varphi}\mathcal{L}\hat{\varphi}^{-1})e_i \in \mathbb{R}^{Nn^2}$, $i = 1, \ldots, Nn^2$. Define the matrix $\mathcal{A} \in \mathbb{B}(\mathbb{R}^{Nn^2})$ with columns given by $\mathcal{A}_i$ (that is, the ith column of $\mathcal{A}$ is $\mathcal{A}_i$). From this we have that

$$\mathcal{A} = \big(\hat{\varphi}\mathcal{L}\hat{\varphi}^{-1}\big) \quad \text{and} \quad \mathcal{L} = \big(\hat{\varphi}^{-1}\mathcal{A}\hat{\varphi}\big). \tag{2.40}$$

In particular, we have from (2.40) that the spectra of $\mathcal{A}$ and $\mathcal{L}$ are the same, that is,

$$\sigma(\mathcal{A}) = \sigma(\mathcal{L}). \tag{2.41}$$

From (2.40), (2.41), Proposition 2.21, and the decomposition of square matrices into positive semi-definite matrices as in Lemma 2.20 we have the following result.

Proposition 2.23 *Let* $\mathcal{L} \in \mathbb{B}(\mathbb{H}^n)$. *The following assertions are equivalent*:

(i) $\operatorname{Re}\{\lambda(\mathcal{L})\} < 0$.
(ii) *There are constants* $k > 0$ *and* $b > 0$ *such that*

$$\|e^{\mathcal{L}t}\| \le k e^{-bt} \quad \textit{for all } t \ge 0.$$

(iii) $\|e^{\mathcal{L}t}(\mathbf{V})\|_1 \to 0$ *as* $t \to \infty$ *for every* $\mathbf{V} \in \mathbb{H}^{n+}_{\mathbb{C}}$.
(iv) $\int_0^\infty \|e^{\mathcal{L}t}(\mathbf{V})\|_1\, dt < \infty$ *for every* $\mathbf{V} \in \mathbb{H}^{n+}_{\mathbb{C}}$.

If the operator $e^{\mathcal{L}t}$ *is positive for every* $t \in \mathbb{R}^+$, *then* (iii) *and* (iv) *can be replaced by respectively*

(iii′) $\|e^{\mathcal{L}t}(\mathbf{V})\|_1 \to 0$ *as* $t \to \infty$ *for every* $\mathbf{V} \in \mathbb{H}^{n+}$.
(iv′) $\int_0^\infty \|e^{\mathcal{L}t}(\mathbf{V})\|_1 dt < \infty$ *for every* $\mathbf{V} \in \mathbb{H}^{n+}$.

Proof By applying Proposition 2.21 with the matrix representation $\mathcal{A}$ of the operator $\mathcal{L}$ given by (2.40) we get from (2.41) and from the equivalence of the norms in $\mathbb{C}^{Nn^2}$ and $\mathbb{H}^n_{\mathbb{C}}$ (see Remark 2.19) that the following assertions are equivalent:

(i) $\mathrm{Re}\{\lambda(\mathcal{L})\} < 0$.
(ii) There are constants $k > 0$ and $b > 0$ such that

$$\|e^{\mathcal{L}t}\| \le ke^{-bt} \quad \text{for all } t \ge 0.$$

(iiia) $\|e^{\mathcal{L}t}(\mathbf{V})\|_1 \to 0$ as $t \to \infty$ for every $\mathbf{V} \in \mathbb{H}^n_{\mathbb{C}}$.
(iva) $\int_0^\infty \|e^{\mathcal{L}t}(\mathbf{V})\|_1 \, dt < \infty$ for every $\mathbf{V} \in \mathbb{H}^n_{\mathbb{C}}$.

Let us now show that (ii) is equivalent to (iii). It is easy to see that (ii) implies (iii). Suppose that (iii) holds. From the decomposition (2.38), for any $\mathbf{H} \in \mathbb{H}^n_{\mathbb{C}}$, we can find $\mathbf{H}^i \in \mathbb{H}^{n+}_{\mathbb{C}}$, $i = 1, 2, 3, 4$, such that $\mathbf{H} = (\mathbf{H}^1 - \mathbf{H}^2) + \sqrt{-1}(\mathbf{H}^3 - \mathbf{H}^4)$. From the linearity of the semigroup $e^{\mathcal{L}t}$ we have from (iii) that, as $t \to \infty$,

$$\|e^{\mathcal{L}t}(\mathbf{H})\|_1 \le \sum_{i=1}^4 \|e^{\mathcal{L}t}(\mathbf{H}^i)\|_1 \to 0.$$

Thus, we have that (iiia) holds, and since (iiia) implies (ii), we have that (iii) implies (ii). Therefore, we have that (i), (ii), and (iii) are all equivalent. Using similar reasoning, we can show that (ii) is equivalent to (iv).

Suppose now that the operator $e^{\mathcal{L}t}$ is positive for every $t \in \mathbb{R}^+$. Clearly, (iii) implies (iii′) since $\mathbb{H}^{n+} \subset \mathbb{H}^{n+}_{\mathbb{C}}$. Suppose now that (iii′) holds. For any $\mathbf{H} \in \mathbb{H}^{n+}_{\mathbb{C}}$, define $\mathbb{I}(\mathbf{H}) \in \mathbb{H}^{n+}$ as follows:

$$\mathbb{I}(\mathbf{H}) := \big(\|H_1\| I_n, \ldots, \|H_N\| I_n\big).$$

Clearly, $\mathbf{H} \le \mathbb{I}(\mathbf{H})$. From the fact that $e^{\mathcal{L}t}$ is a positive operator we get that $e^{\mathcal{L}t}(\mathbf{H}) \le e^{\mathcal{L}t}(\mathbb{I}(\mathbf{H}))$, and thus $\|e^{\mathcal{L}t}(\mathbf{H})\|_1 \le \|e^{\mathcal{L}t}(\mathbb{I}(\mathbf{H}))\|_1$. From (iii′) we have that

$$\|e^{\mathcal{L}t}(\mathbf{H})\|_1 \le \|e^{\mathcal{L}t}(\mathbb{I}(\mathbf{H}))\|_1 \to 0$$

as $t \to \infty$. Therefore, we have shown that (i), (ii), (iii), and (iii′) are all equivalent. Using similar reasoning, we can show that (i) (or (ii) or (iii′)) is equivalent to (iv′), completing the proof. □

Remark 2.24 It should be noticed that to show Proposition 2.23, we needed the decomposition in Lemma 2.20, which in turn required to work with the space of complex matrices. As mentioned before, we will be mainly interested in the case in which all matrices and initial conditions involved in the dynamic equations are real. It will be shown in Chap. 3 that the operators associated to the second moments of the MJLS are positive, so that the equivalence results in Proposition 2.23 considering real matrices $\mathbf{V} \in \mathbb{H}^{n+}$ can be applied. This is the reasoning behind the proof of Proposition 3.19, which shows the equivalence between the concepts of stochastic stability considering only real initial conditions.

2.8 Linear Matrix Inequalities

Some miscellaneous definitions and results involving matrices and matrix equations are presented in this section. These results will be used throughout the book, especially those related with the concept of linear matrix inequalities (or in short LMIs), which will play a very important role in the next chapters.

Definition 2.25 (Generalized inverse) The generalized inverse (Moore–Penrose inverse) of a matrix $A \in \mathbb{B}(\mathbb{R}^n, \mathbb{R}^m)$ is the unique matrix $A^\dagger \in \mathbb{B}(\mathbb{R}^m, \mathbb{R}^n)$ such that

(i) $AA^\dagger A = A$,
(ii) $A^\dagger A A^\dagger = A^\dagger$,
(iii) $(AA^\dagger)^* = AA^\dagger$,
(iv) $(A^\dagger A)^* = A^\dagger A$.

For more on this subject, see [54]. The Schur complements presented below are used to convert quadratic equations into larger dimension linear ones and vice versa (see [251], p. 13).

Lemma 2.26 (Schur complements) *Consider a symmetric matrix Q such that*

$$Q = \begin{bmatrix} Q_{11} & Q_{12} \\ Q_{12}^* & Q_{22} \end{bmatrix}.$$

(i) $Q > 0$ *if and only if*

$$\begin{cases} Q_{22} > 0, \\ Q_{11} - Q_{12} Q_{22}^{-1} Q_{12}^* > 0, \end{cases}$$

or

$$\begin{cases} Q_{11} > 0, \\ Q_{22} - Q_{12}^* Q_{11}^{-1} Q_{12} > 0. \end{cases}$$

(ii) $Q \geq 0$ *if and only if*

$$\begin{cases} Q_{22} \geq 0, \\ Q_{12} = Q_{12} Q_{22}^{\dagger} Q_{22}, \\ Q_{11} - Q_{12} Q_{22}^{\dagger} Q_{12}^{*} \geq 0, \end{cases}$$

or

$$\begin{cases} Q_{11} \geq 0, \\ Q_{12} = Q_{11} Q_{11}^{\dagger} Q_{12}, \\ Q_{22} - Q_{12}^{*} Q_{11}^{\dagger} Q_{12} \geq 0. \end{cases}$$

Next, we present the definition of a linear matrix inequality.

Definition 2.27 A linear matrix inequality is any constraint that can be written or converted to

$$F(x) = F_0 + x_1 F_1 + x_2 F_2 + \cdots + x_m F_m < 0, \tag{2.42}$$

where x_i are the variables, and the symmetric matrices $F_i \in \mathbb{B}(\mathbb{R}^n)$ for $i = 1, \dots, m$ are known.

The linear matrix inequality (2.42) is referred to as a strict linear matrix inequality. Also of interest is the nonstrict linear matrix inequality, where $F(x) \leq 0$. From the practical point of view, LMIs are usually presented as

$$f(X_1, \dots, X_N) < g(X_1, \dots, X_N),$$

where f and g are affine functions of the unknown matrices $X_1, \dots, X_N$. Quadratic forms can usually be converted to affine ones using the Schur complements. Therefore, we will make no distinctions between quadratic and affine forms, or between a set of LMIs or a single one, and will refer to all of them as simply LMIs. For more on LMIs, the reader is referred to [49, 293] or any of the many works on the subject.

The next result, from [167], Lemma 7.3, will be useful in Chap. 9, Sect. 9.5, which deals with the robust linear filtering problem via LMIs.

Lemma 2.28 *Let real $m \times m$ symmetric matrices Z and Y be such that*

$$\begin{bmatrix} Y & I \\ I & Z \end{bmatrix} > 0.$$

Then for an arbitrary nonsingular $m \times m$ matrix U, there exist $m \times m$ symmetric matrices $\widehat{Z}$, $\widehat{Y}$ and an $m \times m$ matrix V such that

$$\begin{bmatrix} Z & U \\ U^* & \widehat{Z} \end{bmatrix}^{-1} = \begin{bmatrix} Y & V \\ V^* & \widehat{Y} \end{bmatrix} > 0.$$

Moreover,

$$
\begin{aligned}
V &= (I - YZ)\left(U^{-1}\right)^*,\\
\widehat{Y} &= U^{-1}Z\left(Y - Z^{-1}\right)Z\left(U^{-1}\right)^* > 0,\\
\widehat{Z} &= U^*\left(Z - Y^{-1}\right)^{-1}U > 0.
\end{aligned}
$$

Moreover,

Chapter 3
Mean-Square Stability

3.1 Outline of the Chapter

This chapter deals with mean-square stability (MSS) for continuous-time MJLS. We follow an operator-theoretical approach to deal with this subject, trying as much as possible to trace a parallel with the stability theory results for continuous-time linear systems. In this way the MSS of MJLS is studied via the spectrum of an augmented matrix, which captures the idea that a balance between the modes and the transition probability matrix is essential for MSS, or via the existence of a positive-definite solution for a set of coupled Lyapunov equations. The organization of the chapter is as follows. The model and problem statements are described in Sect. 3.2, while the main operators and some auxiliary results are presented in Sect. 3.3. Conditions for mean-square stability for the homogeneous case written in terms of an augmented matrix or a set of coupled Lyapunov equations are presented in Sect. 3.4. For the case of one mode of operation (no jumps in the parameters), these criteria reconcile to well-known stability results for continuous-time linear systems. Furthermore, it is proved that the Lyapunov equation can be written down in two equivalent forms, each providing an easy-to-check sufficient condition. We consider in Sect. 3.5 two scenarios regarding additive disturbances: the one in which the disturbances are characterized via a Wiener process and the one characterized by any function in $L_2^m(\Omega, \mathcal{F}, P)$. For the first case, it is shown that MSS is equivalent to asymptotic wide-sense stationarity (AWSS), while for the second case, it is shown that the state variable belongs to $L_2^n(\Omega, \mathcal{F}, P)$ whenever the disturbance belongs to $L_2^m(\Omega, \mathcal{F}, P)$. Section 3.6 deals with the concepts of mean-square stabilizability and detectability.

3.2 The Models and Problem Statement

In what follows we recall from Chap. 2 that for a positive integer N, we define $\mathcal{S} := \{1, \dots, N\}$. Let $(\Omega, \mathcal{F}, P)$ be a complete probability space equipped with a filtration

O.L.V. Costa et al., *Continuous-Time Markov Jump Linear Systems*,
Probability and Its Applications, DOI 10.1007/978-3-642-34100-7_3,

$\{\mathcal{F}_t, t \in \mathbb{R}^+\}$ satisfying the usual hypotheses, that is, a right-continuous filtration augmented by all null sets in the P-completion of $\mathcal{F}$. Throughout the book we will consider a homogeneous Markov chain $\theta = \{(\theta(t), \mathcal{F}_t); t \in \mathbb{R}^+\}$ taking values in $\mathcal{S}$ as presented in Sect. 2.5.

Let $\mathbf{A} := (A_1, \dots, A_N) \in \mathbb{H}^n$ and $\mathbf{J} := (J_1, \dots, J_N) \in \mathbb{H}^{r,n}$. We deal in this chapter with three types of linear systems with Markov jump parameters. First, we consider the homogeneous system

$$\begin{aligned} &\dot{x}(t) = A_{\theta(t)} x(t), \quad t \in \mathbb{R}^+ \\ &\theta(0) = \theta_0, \qquad x(0) = x_0, \qquad P(\theta_0 = i) = \nu_i. \end{aligned} \tag{3.1}$$

We consider next the class of dynamical systems modeled by the following stochastic differential equation:

$$\begin{aligned} &\dot{x}(t) = A_{\theta(t)} x(t) + J_{\theta(t)} w(t), \quad t \in \mathbb{R}^+ \\ &\theta(0) = \theta_0, \qquad x(0) = x_0, \qquad P(\theta_0 = i) = \nu_i, \end{aligned} \tag{3.2}$$

where $\{w(t); t \in \mathbb{R}^+\}$ is any $L_2^r(\Omega, \mathcal{F}, P)$-function, which is the usual scenario for the H_∞ approach, to be considered in Chap. 8.

The third model to be considered in this chapter is the class of dynamical systems modeled by the following Itô stochastic differential equation:

$$\begin{aligned} &dx(t) = A_{\theta(t)} x(t)\,dt + J_{\theta(t)}\,dw(t), \quad t \in \mathbb{R}^+ \\ &\theta(0) = \theta_0, \qquad x(0) = x_0, \qquad P(\theta_0 = i) = \nu_i, \end{aligned} \tag{3.3}$$

where $W = \{(w(t), \mathcal{F}_t); t \in \mathbb{R}^+\}$ is an r-dimensional Wiener process with incremental covariance operator $I\,dt$, independent of the Markov chain θ and initial condition x_0. In addition, we assume for this case that $\{\theta(t); t \in \mathbb{R}^+\}$ is an irreducible Markov chain. We recall that, as seen in Sect. 2.5 (see (2.21)), in this case there exists a limiting probability $\{\pi_i;\ i \in \mathcal{S}\}$ that does not depend on the initial distribution, with $\{\sum_{i\in\mathcal{S}} \pi_i = 1\}$, and satisfies, for some positive constants $\alpha > 0$ and $\beta > 0$,

$$\max_j \left| p_j(t) - \pi_j \right| \le \alpha e^{-\beta t}. \tag{3.4}$$

As in (2.27), we will use along this book the notation

$$\vartheta_t = \bigl(x(t), \theta(t)\bigr),$$

and when we say arbitrary initial condition $\vartheta_0 = (x_0, \theta_0)$, we mean any distribution $\nu = \{\nu_i; i \in \mathcal{S}\}$ for θ_0 ($P(\theta_0 = i) = \nu_i$, $i \in \mathcal{S}$) and any distribution for x_0 satisfying $E(\|x_0\|^2) < \infty$.

In this chapter it will be convenient to consider that $x_0 \in \mathbb{C}^n$ (see the proof of Theorem 3.15), so that, although the matrices A_i and J_i are real, $x(t)$ will take values in $\mathbb{C}^n$. In Proposition 3.19 we show that the stability results derived for $x_0 \in \mathbb{C}^n$ also hold when we restrict $x_0 \in \mathbb{R}^n$.

In what follows, we will be mainly interested in deriving convergence results to the first and second moments of the state variable $x(t)$. These moments are defined next for each $t \in \mathbb{R}^+$:

$$q(t) := E\big(x(t)\big) \in \mathbb{C}^n, \tag{3.5}$$

$$Q(\tau, t) := E\big(x(t+\tau)x(t)^*\big) \in \mathbb{B}\big(\mathbb{C}^n\big), \tag{3.6}$$

$$Q(t) := Q(0, t) \in \mathbb{B}\big(\mathbb{C}^n\big)^+. \tag{3.7}$$

Recalling that $1_{\{.\}}$ stands for the Dirac measure (see (2.2)), we have that

$$x(t) = \sum_{i \in \mathcal{S}} x(t) 1_{\{\theta(t)=i\}},$$

$$x(t+s)x(t)^* = \sum_{i \in \mathcal{S}} x(t+s)x(t)^* 1_{\{\theta(t)=i\}}, \quad s \geq 0.$$

We will obtain in Sect. 3.3 differential equations for the first and second moments of $x(t)1_{\{\theta(t)=i\}}$. For that, we define

$$q_i(t) := E\big(x(t)1_{\{\theta(t)=i\}}\big) \in \mathbb{C}^n, \tag{3.8}$$

$$Q_i(t) := E\big(x(t)x(t)^* 1_{\{\theta(t)=i\}}\big) \in \mathbb{B}\big(\mathbb{C}^n\big)^+, \tag{3.9}$$

$$Q_i(s, t) := E\big(x(t+s)x(t)^* 1_{\{\theta(t+s)=i\}}\big) \in \mathbb{B}\big(\mathbb{C}^n\big). \tag{3.10}$$

Set also

$$\hat{q}(t) := \begin{bmatrix} q_1(t) \\ \vdots \\ q_N(t) \end{bmatrix}, \tag{3.11}$$

$$\mathbf{Q}(t) := \big(Q_1(t), \ldots, Q_N(t)\big), \tag{3.12}$$

$$\mathbf{Q}(s, t) := \big(Q_1(s, t), \ldots, Q_N(s, t)\big). \tag{3.13}$$

Notice that

$$\|\mathbf{Q}(t)\|_1 = \sum_{i \in \mathcal{S}} \|Q_i(t)\| \leq \sum_{i \in \mathcal{S}} E\big[\|x(t)\|^2 1_{\{\theta(t)=i\}}\big] = E\big[\|x(t)\|^2\big], \tag{3.14}$$

$$\begin{aligned} E\big(\|x(t)\|^2\big) &= \sum_{j \in \mathcal{S}} \operatorname{tr}\big(E\big(x(t)x(t)^* 1_{\{\theta(t)=j\}}\big)\big) \\ &= \sum_{j \in \mathcal{S}} \operatorname{tr}\big(Q_j(t)\big) \leq n \sum_{j \in \mathcal{S}} \|Q_j(t)\| = n\|\mathbf{Q}(t)\|_1, \end{aligned} \tag{3.15}$$

and that $\mathbf{Q}(t) \in \mathbb{H}_{\mathbb{C}}^{n+}$. Similarly we have that $\hat{q}(t) \in \mathbb{C}^{Nn}$ and $\mathbf{Q}(s, t) \in \mathbb{H}_{\mathbb{C}}^{n}$.

In the next three definitions we present some stability concepts that are often found in the literature for MJLS.

Definition 3.1 A linear system with Markov jump parameters is stochastically stable (StS) if for arbitrary initial condition ϑ_0, we have

$$\int_0^\infty E\big[\|x(t)\|^2\big]\,dt < \infty.$$

Definition 3.2 A linear system with Markov jump parameters is mean-square stable (MSS) if there exist $q \in \mathbb{C}^n$ and $Q \in \mathbb{B}(\mathbb{C}^n)^+$ such that for arbitrary initial condition ϑ_0, we have:

(a) $\|q(t) - q\| \to 0$ as $t \to \infty$,
(b) $\|Q(t) - Q\| \to 0$ as $t \to \infty$.

Definition 3.3 A linear system with Markov jump parameters is asymptotically wide-sense stationary (AWSS) if there exist $q \in \mathbb{C}^n$ and $\mathcal{Q}(\tau) \in \mathbb{B}(\mathbb{C}^n)^+$ such that for arbitrary initial condition ϑ_0, we have:

(a) $\|q(t) - q\| \to 0$ as $t \to \infty$,
(b) $\|Q(\tau, t) - \mathcal{Q}(\tau)\| \to 0$ as $t \to \infty$.

Remark 3.4 In the case of systems (3.2), (3.3), we also assume that q, Q, and $\mathcal{Q}(\tau)$ are independent of $w(t)$.

3.3 Main Operators and Auxiliary Results

We first consider the homogeneous equation (3.1), restated here for convenience as follows:

$$\begin{aligned} dx(t) &= A_{\theta(t)}x(t)dt, \quad t \in \mathbb{R}^+, \\ x(t_0) &= x_0, \qquad P(\theta_0 = i) = \nu_i. \end{aligned} \tag{3.16}$$

In addition, let T_k denote the kth jump time of the Markov process $\{\theta(t); t \geq 0\}$ and define

$$\varUpsilon := \Big\{\omega \in \varOmega;\ \lim_{k\to\infty} T_k(\omega) \to \infty\Big\}. \tag{3.17}$$

We recall that $P(\varUpsilon) = 1$. For each realization of the Markov process $\{\theta(t); t \geq 0\}$ in $\varUpsilon$, we have that $\{A_{\theta(t)}; t \geq 0\}$ are matrix-valued functions on $\mathbb{R}^+$ of the class **PC** (piecewise continuous, see Definition 2.3 or [53], p. 411), and therefore, according to Theorem 2.5 (see also [53], p. 11), there exists a unique continuous solution $\varPhi(\cdot, t_0)$ from $\mathbb{R}^+$ to $\mathbb{B}(\mathbb{R}^n)$ of the homogeneous linear matrix differential equation

$$\begin{aligned} \frac{\partial \varPhi(t, t_0)}{\partial t} &= A_{\theta(t)}\varPhi(t, t_0), \\ \varPhi(t_0, t_0) &= I \end{aligned} \tag{3.18}$$

for almost all $t \in \mathbb{R}^+$. Moreover, the solution of (3.16) is given by

$$x(t) = \Phi(t, t_0)x_0. \tag{3.19}$$

In order to obtain differential equations for the first and second moments of $x(t)1_{\{\theta(t)=i\}}$, we define the following linear operators $\mathcal{L} \in \mathbb{B}(\mathbb{H}^n_{\mathbb{C}})$, $\mathcal{F} \in \mathbb{B}(\mathbb{H}^n_{\mathbb{C}})$, and $\mathcal{T} \in \mathbb{B}(\mathbb{H}^n_{\mathbb{C}})$:

$$\begin{aligned} &\mathcal{F}(\cdot) = \big(\mathcal{F}_1(\cdot), \dots, \mathcal{F}_N(\cdot)\big); \\ &\mathcal{L}(\cdot) = \big(\mathcal{L}_1(\cdot), \dots, \mathcal{L}_N(\cdot)\big); \qquad \mathcal{T}(\cdot) = \big(\mathcal{T}_1(\cdot), \dots, \mathcal{T}_N(\cdot)\big), \end{aligned} \tag{3.20}$$

where, for $\mathbf{P} = (P_1, \dots, P_N) \in \mathbb{H}^n_{\mathbb{C}}$ and $i \in \mathcal{S}$,

$$\begin{aligned} \mathcal{F}_i(\mathbf{P}) &:= A_i P_i + \sum_{j\in\mathcal{S}} \lambda_{ji} P_j, \\ \mathcal{L}_i(\mathbf{P}) &:= A_i P_i + P_i A_i^* + \sum_{j\in\mathcal{S}} \lambda_{ji} P_j, \\ \mathcal{T}_i(\mathbf{P}) &:= A_i^* P_i + P_i A_i + \sum_{j\in\mathcal{S}} \lambda_{ij} P_j. \end{aligned} \tag{3.21}$$

The next result establishes a link between $\mathcal{T}$ and $\mathcal{L}$.

Lemma 3.5 *With the inner product as defined in* (2.32), $\mathcal{T}^* = \mathcal{L}$, *i.e.*, $\mathcal{T}$ *is the adjoint operator of* $\mathcal{L}$ *in the Hilbert space* $(\mathbb{H}^n_{\mathbb{C}}, \|\cdot\|_2)$.

Proof For any $\mathbf{P}, \mathbf{V} \in \mathbb{H}^n_{\mathbb{C}}$, we have from (2.32) that

$$\begin{aligned} \big\langle \mathcal{L}(\mathbf{P}); \mathbf{V}\big\rangle &= \sum_{j\in\mathcal{S}} \operatorname{tr}\big(\mathcal{L}_j(\mathbf{P})^* V_j\big) \\ &= \sum_{j\in\mathcal{S}} \operatorname{tr}\bigg(\Big(A_j P_j + P_j A_j^* + \sum_{i\in\mathcal{S}} \lambda_{ij} P_i\Big)^* V_j\bigg) \\ &= \sum_{i\in\mathcal{S}} \operatorname{tr}\bigg(P_i^* \Big(A_i^* V_i + V_i A_i + \sum_{j\in\mathcal{S}} \lambda_{ij} V_j\Big)\bigg) \\ &= \big\langle \mathbf{P}; \mathcal{T}(\mathbf{V})\big\rangle, \end{aligned}$$

showing the result. □

Before deriving differential equations, to compute the first and second moments of the state variable of (3.1), we need the following auxiliary result.

Lemma 3.6 *Consider a stochastic process* $\{f(t)\}$ *such that* $f(t)$ *is* $\mathcal{F}_t$*-measurable and* $E(f(t)1_{\{\theta(t)=i\}}) := f_i(t)$ *exists. Then*

$$E\big(f(t)(1_{\{\theta(t+h)=i\}} - 1_{\{\theta(t)=i\}})\big) = \sum_{j\in\mathcal{S}} \lambda_{ji} f_j(t)h + o(h). \tag{3.22}$$

Proof We have from (2.13) that

$$\begin{aligned} &E\big(f(t)(1_{\{\theta(t+h)=i\}} - 1_{\{\theta(t)=i\}})\big) \\ &= \sum_{j\in\mathcal{S}} E\big(E\big(f(t)1_{\{\theta(t+h)=i\}}1_{\{\theta(t)=j\}}|\mathcal{F}_t\big)\big) - E\big(f(t)1_{\{\theta(t)=i\}}\big) \\ &= \sum_{j\in\mathcal{S}} P\big(\theta(t+h)=i|\theta(t)=j\big)f_j(t) - f_i(t) \\ &= \sum_{j\in\mathcal{S}} \lambda_{ji} f_j(t)h + o(h), \end{aligned}$$

showing the desired result. □

For notational simplicity, we will represent, from now on, *the infinitesimal variation equation* in (3.22) by the more compact differential equation

$$E\big(f(t)d(1_{\{\theta(t)=i\}})\big) = \sum_{j\in\mathcal{S}} \lambda_{ji} f_j(t)\,dt, \tag{3.23}$$

and throughout the book we will adopt, whenever necessary, similar notation.

We will apply the Kronecker product (see Chap. 2) to the operator $\mathcal{L}$ in order to write differential equations for the second moments of $x(t)1_{\{\theta(t)=i\}}$ in a matrix form. Bearing this in mind, we introduce the following notation:

$$F := \Pi' \otimes I_n + \operatorname{diag}(A_i); \qquad V := \Pi' \otimes I_{n^2}; \qquad G := \operatorname{diag}(I_n \oplus A_i); \tag{3.24}$$

$$H := \operatorname{diag}(\bar{A}_i \oplus A_i); \qquad \mathcal{A} := V + H; \qquad \mathcal{B} := V + G. \tag{3.25}$$

The next proposition provides differential equations to compute the first and second moments of the state variable of (3.1).

Proposition 3.7 *For* $t \in \mathbb{R}^+$, *we have for* (3.1) *that*

$$\dot{\hat{q}}(t) = F\hat{q}(t), \tag{3.26}$$

$$\dot{\mathbf{Q}}(t) = \mathcal{L}\big(\mathbf{Q}(t)\big). \tag{3.27}$$

Proof From Lemma 3.6 and applying Itô's rule to (3.8), we have from (3.1) (bearing in mind the notation in (3.22) and (3.23)) that

$$dq_j(t) = E\big[dx(t)1_{\{\theta(t)=j\}} + x(t)\,d1_{\{\theta(t)=j\}}\big]$$

$$= A_j E\big[x(t)1_{\{\theta(t)=j\}}\big]\,dt + \sum_{i\in\mathcal{S}} \lambda_{ij} q_i(t)\,dt$$
$$= A_j q_j(t)\,dt + \sum_{i\in\mathcal{S}} \lambda_{ij} q_i(t)\,dt,$$

and (3.26) follows. Similarly, for (3.9), we have

$$\begin{aligned} dQ_j(t) &= E\big[dx(t)x(t)^* 1_{\{\theta(t)=j\}} + x(t)\,dx(t)^* 1_{\{\theta(t)=j\}} + x(t)x(t)^*\,d(1_{\{\theta(t)=j\}})\big] \\ &= A_j E\big[x(t)x(t)^* 1_{\{\theta(t)=j\}}\big]\,dt + E\big[x(t)x(t)^* 1_{\{\theta(t)=j\}}\big]A_j^*\,dt \\ &\quad + \sum_{i\in\mathcal{S}} \lambda_{ij} E\big[x(t)x(t)^* 1_{\{\theta(t)=i\}}\big]\,dt \\ &= \mathcal{L}_j\big(\mathbf{Q}(t)\big)\,dt, \end{aligned}$$

showing (3.27). □

The following result gives the matrix representation for the operators $\mathcal{L}$, $\mathcal{T}$, $\mathcal{F}$ in terms of the matrices $\mathcal{A}$, $\mathcal{A}^*$, and $\mathcal{B}$, respectively (see also (2.40) for the general case).

Proposition 3.8 *For $\mathcal{A} \in \mathbb{B}(\mathbb{R}^{Nn^2})$, $\mathcal{B} \in \mathbb{B}(\mathbb{R}^{Nn^2})$, and $\mathcal{A}^* \in \mathbb{B}(\mathbb{R}^{Nn^2})$ defined as in* (3.24)–(3.25) *we have, for any $\mathbf{Q} \in \mathbb{H}_{\mathbb{C}}^n$, that*:

(a) $\hat{\varphi}(\mathcal{L}(\mathbf{Q})) = \mathcal{A}\hat{\varphi}(\mathbf{Q})$,
(b) $\hat{\varphi}(\mathcal{T}(\mathbf{Q})) = \mathcal{A}^*\hat{\varphi}(\mathbf{Q})$,
(c) $\hat{\varphi}(\mathcal{F}(\mathbf{Q})) = \mathcal{B}\hat{\varphi}(\mathbf{Q})$.

Proof It follows from the definition of $\hat{\varphi}$ in Sect. 2.6, in conjunction with (2.37), bearing in mind the definition of the operators $\mathcal{F}$, $\mathcal{L}$, and $\mathcal{T}$ in (3.20)–(3.21) and matrices $\mathcal{A} \in \mathbb{B}(\mathbb{R}^{Nn^2})$, $\mathcal{B} \in \mathbb{B}(\mathbb{R}^{Nn})$, and $\mathcal{A}^* \in \mathbb{B}(\mathbb{R}^{Nn^2})$ defined as in (3.24)–(3.25). □

As a consequence of the previous proposition, we have the following lemma, which presents a matrix representation for the differential equation (3.27).

Lemma 3.9 *Let $\mathcal{A} \in \mathbb{B}(\mathbb{R}^{Nn^2})$ be defined as in* (3.24)–(3.25), *and consider the homogeneous system $\dot{y}(t) = \mathcal{A}y(t)$, $t \in \mathbb{R}^+$, with initial condition $y(0) = \hat{\varphi}(\mathbf{Q})$, $\mathbf{Q} \in \mathbb{H}_{\mathbb{C}}^n$. Then,*

$$y(t) = e^{\mathcal{A}t}y(0) = \hat{\varphi}\big(e^{\mathcal{L}t}(\mathbf{Q})\big). \tag{3.28}$$

The result also holds replacing $\mathcal{A}$ and $\mathcal{L}$ by $\mathcal{A}^*$ and $\mathcal{T}$, respectively.

Proof We begin by noticing that the solution of the above differential equation is given by $y(t) = e^{\mathcal{A}t}y(0)$. Consider any $y \in \mathbb{C}^{Nn^2}$ and take $\mathbf{Y} = \hat{\varphi}^{-1}(y) \in \mathbb{H}_{\mathbb{C}}^n$. From

Proposition 3.8 it follows that $\hat{\varphi}(\mathcal{L}^k(\mathbf{Y})) = \mathcal{A}^k\hat{\varphi}(\mathbf{Y}) = \mathcal{A}^k y$. From the continuity of the operator $\hat{\varphi}(\cdot)$ (see Remark 2.19) we have that

$$\begin{aligned} e^{\mathcal{A}t}y &= \sum_{k=0}^{\infty}\frac{t^k}{k!}\mathcal{A}^k y = \sum_{k=0}^{\infty}\frac{t^k}{k!}\mathcal{A}^k\hat{\varphi}(\mathbf{Y}) = \lim_{r\to\infty}\sum_{k=0}^{r}\frac{t^k}{k!}\hat{\varphi}\big(\mathcal{L}^k(\mathbf{Y})\big) \\ &= \lim_{r\to\infty}\hat{\varphi}\left(\sum_{k=0}^{r}\frac{t^k}{k!}\mathcal{L}^k(\mathbf{Y})\right) = \hat{\varphi}\left(\lim_{r\to\infty}\sum_{k=0}^{r}\frac{t^k}{k!}\mathcal{L}^k(\mathbf{Y})\right) \\ &= \hat{\varphi}\big(e^{\mathcal{L}t}(\mathbf{Y})\big). \end{aligned} \tag{3.29}$$

Thus, from (3.29) we have that

$$y(t) = e^{\mathcal{A}t}\hat{\varphi}(\mathbf{Q}) = \hat{\varphi}\big(e^{\mathcal{L}t}(\mathbf{Q})\big),$$

showing the result. □

The next result shows the positiveness of the operators $\mathcal{L}$ and $\mathcal{T}$.

Lemma 3.10 *For $\mathcal{L}$ defined as in* (3.20) *and* (3.21), *and $y(t)$ given as in* (3.28), *we have that*:

(a) *The operator $e^{\mathcal{L}t}$ is positive, that is, $\mathbf{Q} \in \mathbb{H}_{\mathbb{C}}^{n+}$ implies $e^{\mathcal{L}t}(\mathbf{Q}) \in \mathbb{H}_{\mathbb{C}}^{n+}$, for every $t \in \mathbb{R}^+$.*

(b) *For $\mathbf{Q} \in \mathbb{H}_{\mathbb{C}}^{n+}$, $\hat{\varphi}^{-1}(y(t)) \in \mathbb{H}_{\mathbb{C}}^{n+}$ and, consequently, $\varphi^{-1}(y(t)) \in \mathbb{B}(\mathbb{C}^n)^+$ for all $j \in \mathcal{S}$ and $t \in \mathbb{R}^+$.*

Items (a) *and* (b) *also hold replacing $\mathcal{L}$ by $\mathcal{T}$.*

Proof The proof of (a) follows the same steps as in the proof of Lemma A.2. To prove this result, it suffices to show that $Y_i(t) \in \mathbb{B}(\mathbb{C}^n)^+$ for $i \in \mathcal{S}$ and any $t \in \mathbb{R}^+$, where $\mathbf{Y}(t) := (Y_1(t), \dots, Y_N(t))$, with $\mathbf{Y}(0) = \mathbf{Q} \in \mathbb{H}_{\mathbb{C}}^{n+}$, satisfies: $\dot{\mathbf{Y}}(t) = \mathcal{L}(\mathbf{Y}(t))$, or $\dot{Y}_i(t) = \mathcal{L}_i(\mathbf{Y}(t))$. Notice that the unique solution to this equation is $\mathbf{Y}(t) = e^{\mathcal{L}t}(\mathbf{Q})$. From (3.21), defining $\tilde{A}_i = A_i + \frac{1}{2}\lambda_{ii}I$, we have that

$$\dot{Y}_i(t) = \tilde{A}_i Y_i(t) + Y_i(t)\tilde{A}_i^* + \sum_{\{j\neq i\}}\lambda_{ji}Y_j(t).$$

Furthermore,

$$Y_i(t) = e^{\tilde{A}_i t}Q_i e^{\tilde{A}_i^* t} + \int_0^t e^{\tilde{A}_i(t-s)}\left(\sum_{\{j\neq i\}}\lambda_{ji}Y_j(s)\right)e^{\tilde{A}_i^*(t-s)}\,ds. \tag{3.30}$$

The above equation has a unique integrable solution $Y_i(t)$ that can be found by successive approximations as follows. Consider the sequence $\{Y_i^k(t) : k = 0, 1, \dots,$

$t \in \mathbb{R}^+\}$ for $i \in \mathcal{S}$ obtained recursively as

$$Y_i^{k+1}(t) = e^{\tilde{A}_i t} Q_i e^{\tilde{A}_i^* t} + \int_0^t e^{\tilde{A}_i (t-s)} \left(\sum_{\{j \neq i\}} \lambda_{ji} Y_j^k(s) \right) e^{\tilde{A}_i^* (t-s)} \, ds, \tag{3.31}$$
$$Y_i^0(t) = 0, \quad i \in \mathcal{S}.$$

Bearing in mind that $\lambda_{ij} \geq 0$ for $i \neq j$, $i \in \mathcal{S}$, it is easy to see that $Y_i^{k+1}(t) \geq Y_i^k(t) \geq 0$ for $k = 0, 1, \ldots$ and any $t \in \mathbb{R}^+$. Next, we prove that, for $i \in \mathcal{S}$, $\|Y_i^k(t)\| \leq \ell_i(t)$ for every $k = 0, 1, \ldots$ and any $t \in \mathbb{R}^+$, where

$$\dot{\ell}_i(t) = 2\|\tilde{A}_i\|\ell_i(t) + \sum_{\{j \neq i\}} \lambda_{ji} \ell_j(t),$$
$$\ell_i(0) = \|Y_i(0)\|.$$

This is carried out by induction as follows. First notice that the assertion above is obviously true for $k = 0$. Assuming now that it holds for some k, i.e., $\|Y_i^k(t)\| \leq \ell_i(t)$ for $i \in \mathcal{S}$ and $t \in \mathbb{R}^+$, we have that

$$\begin{aligned}
\ell_i(t) &= e^{2\|\tilde{A}_i\|t} \ell_i(0) + \int_0^t e^{2\|\tilde{A}_i\|(t-s)} \left(\sum_{\{j \neq i\}} \lambda_{ji} \ell_j(s) \right) ds \\
&\geq \left\| e^{\tilde{A}_i t} Q_i e^{\tilde{A}_i^* t} + \int_0^t e^{\tilde{A}_i \, (t-s)} \left(\sum_{\{j \neq i\}} \lambda_{ji} Y_j^k \right) e^{\tilde{A}_i^* (t-s)} \, ds \right\| \\
&= \left\| Y_i^{k+1}(t) \right\|,
\end{aligned}$$

and the assertion follows. Finally, using Lemma 2.17, we get that for each $t \in \mathbb{R}^+$, $\lim_{k \to \infty} Y_i^k(t) = \hat{Y}_i(t) \geq 0$ for some $\hat{Y}_i(t) \in \mathbb{B}(\mathbb{C}^n)^+$, which is Lebesgue measurable in t (since each $Y_i^k(t)$ is continuous and differentiable in t from (3.31)). Therefore, taking the limit as $k \to \infty$ in (3.31) and applying the bounded convergence theorem (see [29], Sect. 16), we get that

$$\begin{aligned}
\hat{Y}_i(t) &= \lim_{k \to \infty} Y_i^{k+1}(t) \\
&= e^{\tilde{A}_i t} Q_i e^{\tilde{A}_i^* t} + \lim_{k \to \infty} \int_0^t e^{\tilde{A}_i \, (t-s)} \left(\sum_{\{j \neq i\}} \lambda_{ji} Y_j^k(s) \right) e^{\tilde{A}_i^* (t-s)} \, ds \\
&= e^{\tilde{A}_i t} Q_i e^{\tilde{A}_i^* t} + \int_0^t e^{\tilde{A}_i (t-s)} \left(\sum_{\{j \neq i\}} \lambda_{ji} \lim_{k \to \infty} Y_j^k(s) \right) e^{\tilde{A}_i^* (t-s)} \, ds \\
&= e^{\tilde{A}_i t} Q_i e^{\tilde{A}_i^* t} + \int_0^t e^{\tilde{A}_i \, (t-s)} \left(\sum_{\{j \neq i\}} \lambda_{ji} \hat{Y}_j(s) \right) e^{\tilde{A}_i^* (t-s)} \, ds.
\end{aligned} \tag{3.32}$$

From (3.32) we get that $\hat{Y}_i(t)$ is continuous and differentiable, and from the uniqueness of the solution $\mathbf{Y}(t)$ in (3.30) we get that $Y_i(t) = \hat{Y}_i(t) \geq 0$ for $i \in \mathcal{S}$ and $t \in \mathbb{R}^+$, i.e., $Y_i(t) \in \mathbb{B}(\mathbb{C}^n)^+$ for $i \in \mathcal{S}$ and $t \in \mathbb{R}^+$, and part (a) follows. Part (b) follows from part (a) and the definitions of $\hat{\varphi}$ and $\hat{\varphi}_j{}^{-1}$. □

The next result will be useful to write the mean-square stability results in several equivalent ways.

Proposition 3.11 *The following assertions are equivalent*:

(i) $\mathrm{Re}\{\lambda(\mathcal{A})\} < 0$.
(ii) $\mathrm{Re}\{\lambda(\mathcal{L})\} < 0$.
(iii) $\mathrm{Re}\{\lambda(\mathcal{A}^*)\} < 0$.
(iv) $\mathrm{Re}\{\lambda(\mathcal{T})\} < 0$.

Proof Clearly, we have that (i) and (iii) are equivalent and similarly (ii) and (iv) are equivalent (recall from Lemma 3.5 that $\mathcal{L} = \mathcal{T}^*$). For the equivalence between (i) and (ii), let us first suppose that (ii) holds. From the equivalence between the norms $\|\cdot\|$ and $\|\cdot\|_1$ (see Remark 2.19) we have from (3.29) and (2.35) that

$$\left\|e^{\mathcal{A}t}y\right\| = \left\|\hat{\varphi}\left(e^{\mathcal{L}t}(\mathbf{Y})\right)\right\| \leq \frac{1}{c_1}\left\|e^{\mathcal{L}t}(\mathbf{Y})\right\|_1 \leq \frac{1}{c_1}\left\|e^{\mathcal{L}t}\right\|\|\mathbf{Y}\|_1 \to 0 \quad \text{as } t \to \infty.$$

From Proposition 2.21 we get that (i) holds. Similarly, if (i) holds, then from (3.29) and (2.36) we get for any $\mathbf{Y} \in \mathbb{H}^n_{\mathbb{C}}$ by setting $y = \hat{\varphi}(\mathbf{Y}) \in \mathbb{C}^{Nn^2}$ that

$$\left\|e^{\mathcal{L}t}(\mathbf{Y})\right\|_1 = \left\|\hat{\varphi}^{-1}\left(e^{\mathcal{A}t}y\right)\right\|_1 \leq c_2\left\|e^{\mathcal{A}t}y\right\| \leq c_2\left\|e^{\mathcal{A}t}\right\|\|y\| \to 0 \quad \text{as } t \to \infty.$$

From Proposition 2.23 we get that (ii) holds. □

We need also the following auxiliary results.

Lemma 3.12 *For any second-order random variable z taking values in $\mathbb{C}^n$, and for $t, \tau \in \mathbb{R}^+$, $t \geq \tau \geq 0$, and $\Phi(t, \tau)$ as in* (3.18),

$$\left\|\Phi(t,\tau)z\right\|_2^2 \leq n\left\|e^{\mathcal{L}(t-\tau)}\right\|\|z\|_2^2. \tag{3.33}$$

Proof Consider (3.16) with initial time $t_0 = \tau$ and initial condition $x(t_0) = z$. It follows from (3.19) that over the set Υ (see (3.17)),

$$x(t) = \Phi(t, \tau)z.$$

Recalling that $P(\Upsilon) = 1$, $Q_i(t) = E(x(t)x(t)^* 1_{\{\theta(t)=i\}})$, and that $Q_i(\tau) = E(zz^* \times 1_{\{\theta(\tau)=i\}})$, it follows that

$$\left\|x(t)\right\|_2^2 = \left\|\Phi(t,\tau)z\right\|_2^2 = \sum_{i\in\mathcal{S}}\mathrm{tr}\left(Q_i(t)\right) \leq n\sum_{i\in\mathcal{S}}\left\|Q_i(t)\right\| = n\left\|\mathbf{Q}(t)\right\|_1. \tag{3.34}$$

From (3.27) in Proposition 3.7,

$$\mathbf{Q}(t) = e^{\mathcal{L}(t-\tau)}\mathbf{Q}(\tau). \tag{3.35}$$

Plugging (3.35) into (3.34) leads to

$$\|\Phi(t,\tau)z\|_2^2 \le n\|e^{\mathcal{L}(t-\tau)}\mathbf{Q}(\tau)\|_1 \le n\|e^{\mathcal{L}(t-\tau)}\|\|\mathbf{Q}(\tau)\|_1. \tag{3.36}$$

The result follows from (3.36) after noticing that

$$\|\mathbf{Q}(\tau)\|_1 = \sum_{i\in\mathcal{S}}\|Q_i(\tau)\| \le \sum_{i\in\mathcal{S}}\operatorname{tr}(Q_i(\tau)) = \|z\|_2^2.$$

□

Proposition 3.13 *If* $\operatorname{Re}\{\lambda(\mathcal{A})\} < 0$, *then* $\operatorname{Re}\{\lambda(F)\} < 0$.

Proof Since $\operatorname{Re}\{\lambda(\mathcal{A})\} < 0$, it follows from Proposition 2.23 and Proposition 3.11 that $\|e^{\mathcal{L}t}\| \le \mu e^{-bt}$ for some $\mu > 0$ and $b > 0$. For the homogeneous system

$$\dot{x}(t) = A_{\theta(t)}x(t), \quad t \in \mathbb{R}^+,$$

we have from (3.27) in Proposition 3.7 and (3.14), (3.15) that

$$\begin{aligned} E(\|x(t)\|^2) &= \sum_{j\in\mathcal{S}}\operatorname{tr}\big(E\big(x(t)x(t)^* 1_{\{\theta(t)=j\}}\big)\big) \le n\|\mathbf{Q}(t)\|_1 \\ &= \|e^{\mathcal{L}t}(\mathbf{Q}(0))\|_1 \le n\mu e^{-bt}\|\mathbf{Q}(0)\|_1 \le n\mu e^{-bt}E(\|x(0)\|^2) \end{aligned}$$

and from (3.26) in Proposition 3.7 that

$$\begin{aligned} \|e^{Ft}\hat{q}(0)\|_1 &= \|\hat{q}(t)\|_1 = \sum_{j\in\mathcal{S}}\|q_j(t)\| \le \sum_{j\in\mathcal{S}}E\big(\|x(t)\|1_{\{\theta(t)=j\}}\big) = E(\|x(t)\|) \\ &\le \big(E(\|x(t)\|^2)\big)^{1/2} \le (n\mu)^{1/2}e^{-\frac{b}{2}t}\big(E(\|x(0)\|^2)\big)^{1/2}. \end{aligned} \tag{3.37}$$

From Proposition 2.21 and (3.37) we get that $\operatorname{Re}\{\lambda(F)\} < 0$. □

Remark 3.14 It is not difficult to see that $\operatorname{Re}\{\lambda(F)\} < 0$ does not imply $\operatorname{Re}\{\lambda(\mathcal{A})\} < 0$. Indeed, consider, for instance, $n = 1$, $\mathcal{S} = \{1, 2\}$, $\lambda_{11} = \lambda_{22} = -1$, $A_1 = \frac{1}{2}$, and $A_2 = -5$. It is then straightforward to show that

$$\lambda_1(F) = \frac{-6.5 - \sqrt{34.25}}{2} < 0, \qquad \lambda_2(F) = \frac{-6.5 + \sqrt{34.25}}{2} < 0$$

and that

$$\lambda_1(\mathcal{A}) = \frac{-11 + \sqrt{125}}{2} > 0, \qquad \lambda_2(\mathcal{A}) = \frac{-11 - \sqrt{125}}{2} < 0.$$

3.4 Mean-Square Stability for the Homogeneous Case

In this section necessary and sufficient conditions for mean-square stability (or, as we are going to see in Sect. 3.4.1, equivalently stochastic stability) of the homogeneous case are established. It is required that either the real part of all the elements in the spectrum of an augmented matrix be less than zero or that there exists a unique solution of a Lyapunov equation. It is proved that the Lyapunov equation can be written down in two equivalent forms, each providing an easy-to-check sufficient condition. Moreover, it is shown that for real matrices A_i, the system is MSS for the complex state space if and only if the system is MSS for the real state space.

We start in Sect. 3.4.1 by showing the equivalence between mean square stability, stochastic stability, and the real part of all the elements in the spectrum of an augmented matrix being less than zero. In Sect. 3.4.2 we show the equivalence between mean-square stability and the existence of a unique solution for a set of coupled Lyapunov equations. In Sect. 3.4.3 we summarize the main results of the section.

3.4.1 MSS, StS, and the Spectrum of an Augmented Matrix

The goal of this subsection is to show the equivalence between mean-square stability, stochastic stability, and the real part of all the elements in the spectrum of an augmented matrix being less than zero, in the spirit of the classical linear case. This result is proved in the next theorem.

Theorem 3.15 *The following assertions are equivalent:*

(i) *System* (3.1) *is StS according to Definition* 3.1.
(ii) $\mathrm{Re}\{\lambda(\mathcal{L})\} < 0$.
(iii) $\mathrm{Re}\{\lambda(\mathcal{A})\} < 0$.
(iv) *System* (3.1) *is MSS according to Definition* 3.2 *with* $q = 0$ *and* $Q = 0$.
(v) *There exist* $b > 0$ *and* $a > 0$ *such that for each* $t \in \mathbb{R}^+$,

$$E\left(\|x(t)\|^2\right) \leq a e^{-bt} E\left(\|x_0\|^2\right). \tag{3.38}$$

Proof From (3.27) in Proposition 3.7 we have that $\dot{\mathbf{Q}}(t) = \mathcal{L}(\mathbf{Q}(t))$. Therefore, from (3.14),

$$\int_0^\infty \left\|e^{\mathcal{L}t}(\mathbf{Q}(0))\right\|_1 dt = \int_0^\infty \left\|\mathbf{Q}(t)\right\|_1 dt \leq \int_0^\infty E\left(\|x(t)\|^2\right) dt \tag{3.39}$$

for arbitrary initial condition ϑ_0. Suppose now that (i) holds, so that, for arbitrary initial condition ϑ_0, we have that $\int_0^\infty E(\|x(t)\|^2)\,dt < \infty$, and consider any $\mathbf{H} = (H_1, \ldots, H_N) \in \mathbb{H}_{\mathbb{C}}^{n+}$. From the spectral decomposition of H_i (see, for instance,

[234], Chap. 6) we have that

$$H_i = \sum_{k=1}^{n} \lambda_k(H_i) e_k(H_i) e_k^*(H_i),$$

where $\lambda_k(H_i) \geq 0$ is the kth eigenvalue of H_i, and $e_k(H_i)$ the corresponding eigenvector. We take independent variables x_{0i} and θ_0 with the following distribution: $x_{0i} = \sqrt{Nn\lambda_k(H_i)} e_k(H_i)$ with probability $\frac{1}{n}$ for $k = 1, \ldots, n$ and $\theta(0) = i$ with probability $\frac{1}{N}$ for $i = 1, \ldots, N$. Consider now $x_0 = \sum_{i \in \mathcal{S}} x_{0i} 1_{\{\theta_0 = i\}}$. We get that

$$\begin{aligned} E\big(x_0 x_0^* 1_{\{\theta_0=i\}}\big) &= E\big(x_{0i} x_{0i}^* 1_{\{\theta_0=i\}}\big) \\ &= E\big(x_{0i} x_{0i}^*\big) P(\theta_0 = i) \\ &= E\big(x_{0i} x_{0i}^*\big) \frac{1}{N} \\ &= \frac{1}{N} \sum_{k=1}^{n} Nn\lambda_k(H_i) e_k(H_i) e_k^*(H_i) \frac{1}{n} \\ &= \sum_{k=1}^{n} \lambda_k(H_i) e_k(H_i) e_k^*(H_i) = H_i. \end{aligned}$$

Thus, we have obtained an initial condition x_0 and ν such that $Q_i(0) = H_i$. It follows from (3.39) that for any $\mathbf{H} = (H_1, \ldots, H_N) \in \mathbb{H}_{\mathbb{C}}^{n+}$, $\int_0^\infty \|e^{\mathcal{L}t}(\mathbf{H})\|_1 \, dt < \infty$, which implies, by Proposition 2.23, that (ii) holds. From Propositions 2.23 and 3.7 it is immediate that if (ii) holds, then (i) holds. The equivalence between (ii) and (iii) follows from Proposition 3.11. If (ii) holds, we have from Proposition 3.13 that $\mathrm{Re}\{\lambda(F)\} < 0$. Thus, from (3.26) in Proposition 3.7 and Proposition 2.21 it follows that $\hat{q}(t) \to 0$ as $t \to \infty$, and since $q(t) = \sum_{i \in \mathcal{S}} q_i(t)$, we have that $q(t) \to 0$ as $t \to \infty$. From Proposition 2.23 and (3.15), (3.33),

$$\begin{aligned} \|x(t)\|_2^2 &\leq n \|\mathbf{Q}(t)\|_1 = n \|e^{\mathcal{L}t} \mathbf{Q}(0)\|_1 \\ &\leq n \|e^{\mathcal{L}t}\| \|\mathbf{Q}(0)\|_1 \leq a e^{-bt} E\big(\|x_0\|^2\big) \end{aligned} \tag{3.40}$$

for some $a > 0$ and $b > 0$. This shows that (ii) implies (iv) and (v). It is immediate from (3.40) and Propositions 2.23 and 3.13 that (v) implies (ii) and (iv). □

The next examples, borrowed from [223], illustrate the cases in which

(1) each mode is unstable, but the overall system is stable, and
(2) each mode is stable, but the overall system is unstable.

These examples illustrate that it is necessary to combine the transition probability of the Markov chain with the eigenvalues of the matrices A_i, as in matrix $\mathcal{A}$, in order to get an adequate criterion for mean-square stability of system (3.1).

Example 3.16 (Each mode is unstable, but the overall system is stable) Consider an MJLS with

$$A_1 = \begin{bmatrix} \frac{1}{2} & -1 \\ 0 & -2 \end{bmatrix}, \qquad A_2 = \begin{bmatrix} -2 & -1 \\ 0 & \frac{1}{2} \end{bmatrix}, \qquad \Pi = \begin{bmatrix} -\beta & \beta \\ \beta & -\beta \end{bmatrix},$$

so that *each mode is unstable*. However, depending on the value of β, the overall system will be mean-square stable. In fact, it may be checked that

$$\det(\lambda I - \mathcal{A}) = \left(\lambda^2 + (2\beta+3)\lambda + 3\beta - 4\right)^2 \left(\lambda^2 + (2\beta+3)\lambda + 3\beta + \frac{9}{4}\right)^2,$$

and thus, from the Routh–Hurwitz criterion, the system is MSS if and only if $\beta > 4/3$, i.e., $\operatorname{Re}\{\lambda(\mathcal{A})\} < 0$ if and only if $\beta > 4/3$. This shows that as the number of jumps per unit of time increases, the effect of switching between the unstable modes makes the overall system mean-square stable.

Example 3.17 (Each mode is stable, but the overall system is unstable) Consider now an MJLS with

$$A_1 = \begin{bmatrix} -1 & 10 \\ 0 & -1 \end{bmatrix}, \qquad A_2 = \begin{bmatrix} -1 & 0 \\ 10 & -1 \end{bmatrix}, \qquad \Pi = \begin{bmatrix} -\beta & \beta \\ \beta & -\beta \end{bmatrix},$$

whose *modes are both individually stable*, so that $\operatorname{Re}\{\lambda(\mathcal{A})\} < 0$ for $\beta = 0$. However, as proven next, the overall system will be unstable in the mean-square sense if $\beta \geq 1/24$. In fact, in the general case of $\beta > 0$ it is straightforward to check that $\mathcal{A}$ is a Metzler matrix (i.e., a matrix whose off-diagonal entries are nonnegative), so that $\operatorname{Re}\{\lambda(\mathcal{A})\} \in \sigma(\mathcal{A})$ is easily obtained from a suitable version of the Perron–Frobenius theorem [215, Chap. 6]. Therefore, due to the continuity of the spectrum, the smallest $\beta > 0$ such that mean-square stability is lost must be such that $\operatorname{Re}\{\lambda(\mathcal{A})\} = 0$, which in this case is equivalent to $\det(\mathcal{A}) = 0$. Finally, it is a routine exercise to verify that

$$\det(\mathcal{A}) = -\frac{32}{3}\left(\beta - \frac{1}{24}\right)(\beta+1)\left(\beta^2 + 27\beta + 1\right),$$

which, indeed, has a root at $\beta = 1/24$. This shows that as the number of jumps per unit of time increases, the effect of switching between the stable modes makes the overall system mean-square unstable.

We conclude this subsection by showing that the results obtained in this chapter considering the complex case also hold for the real case, that is, the situation in which the initial condition x_0 in (3.16) is real. We state below Definition 2.1 presented in [137].

Definition 3.18 [137] System (3.1) is stochastically stable if for any $x_0 \in \mathbb{R}^n$ and any probability distribution ν for θ_0,

$$\int_0^\infty E\left(\|x(t)\|^2\right) dt < \infty.$$

Notice that in Definition 3.18, we consider only real initial conditions for x_0, while in Definition 3.1 complex conditions for x_0 are allowed (see also Remark 2.24). We next show that in fact Definitions 3.1 and 3.18 are equivalent.

Proposition 3.19 *System* (3.1) *is stochastically stable according to Definition* 3.18 *if and only if it is StS according to Definition* 3.1.

Proof Clearly, if system (3.1) is StS according to Definition 3.1, it is stochastically stable for the real state space case as in Definition 3.18. Suppose now that system (3.1) is stochastically stable for the real state space case as in Definition 3.18. For any $\mathbf{H} \in \mathbb{H}_{\mathbb{C}}^{n+}$, define, as in Proposition 2.23, $\mathbb{I}(\mathbf{H}) \in \mathbb{H}^{n+}$ as follows:

$$\mathbb{I}(\mathbf{H}) := \left(\|H_1\| I_n, \ldots, \|H_N\| I_n\right).$$

Clearly $\mathbf{H} \leq \mathbb{I}(\mathbf{H})$. We consider real initial conditions x_0 and ν such that $Q_i(0) = \|H_i\| I_n$, so that $\mathbf{Q}(0) = \mathbb{I}(\mathbf{H})$ (for instance, x_0 and θ_0 independent with $E(x_0 x_0^*) = \|\mathbf{H}\|_1 I_n$ and $\nu_i = \frac{\|H_i\|}{\|\mathbf{H}\|_1}$). Let $x_{\mathbf{H}}(t)$ denote the trajectory for this initial condition. From Lemma 3.10 we have that for all $t \in \mathbb{R}^+$,

$$\mathbf{0} \leq e^{\mathcal{L}t}(\mathbf{H}) \leq e^{\mathcal{L}t}\left(\mathbb{I}(\mathbf{H})\right). \tag{3.41}$$

From (3.14) and (3.41),

$$\int_0^\infty \left\|e^{\mathcal{L}t}(\mathbf{H})\right\|_1 dt \leq \int_0^\infty \left\|e^{\mathcal{L}t}\left(\mathbb{I}(\mathbf{H})\right)\right\|_1 dt \leq \int_0^\infty E\left(\|x_{\mathbf{H}}(t)\|^2\right) dt < \infty$$

for all $\mathbf{H} \in \mathbb{H}_{\mathbb{C}}^{n+}$, and thus, by Propositions 2.23 and 3.15, system (3.1) is StS according to Definition 3.1. □

3.4.2 Coupled Lyapunov Equations

In this subsection we will be interested in obtaining equivalence results between mean-square stability and the unique solution of a set of coupled Lyapunov equations. Some easy-to-check conditions will also be established. We start by showing that mean-square stability implies the uniqueness of solution for a set of coupled Lyapunov equations, which can be written in two equivalent ways (in what follows, recall the definition of $\hat{\varphi}$ in (2.34)).

Proposition 3.20 *If* $\mathrm{Re}\{\lambda(\mathcal{A})\} < 0$, *then for every* $\mathbf{S} = (S_1, \dots, S_N) \in \mathbb{H}^n$, *there exists a unique* $\mathbf{G} = (G_1, \dots, G_N) \in \mathbb{H}^n$ *such that*

$$\mathcal{L}(\mathbf{G}) + \mathbf{S} = 0. \tag{3.42}$$

Moreover,

(a) $G_i = -\hat{\varphi}_i^{-1}(\mathcal{A}^{-1}\hat{\varphi}(\mathbf{S}))$;
(b) $\hat{\varphi}(\mathbf{G}) = \int_0^\infty e^{\mathcal{A}t}\hat{\varphi}(\mathbf{S})\,dt$;
(c) $\mathbf{S} \in \mathbb{H}^{n*}$ *iff* $\mathbf{G} \in \mathbb{H}^{n*}$;
(d) $\mathbf{S} \in \mathbb{H}^{n+}$ *implies* $\mathbf{G} \in \mathbb{H}^{n+}$.

These results also hold replacing $\mathcal{L}$ *by* $\mathcal{T}$ *and* $\mathcal{A}$ *by* $\mathcal{A}^*$, *and, in this case,* (3.42) *reads as*

$$\mathcal{T}(\mathbf{G}) + \mathbf{S} = 0. \tag{3.43}$$

Proof (a) From Proposition 3.8(a) we have that

$$\hat{\varphi}\big(\mathcal{L}(\mathbf{G})\big) = \mathcal{A}\hat{\varphi}(\mathbf{G}),$$

and therefore (3.42) is equivalent to

$$\mathcal{A}\hat{\varphi}(\mathbf{G}) = -\hat{\varphi}(\mathbf{S}). \tag{3.44}$$

The expression for G_i follows immediately from the assumption on $\mathcal{A}$ and the definition of $\hat{\varphi}$. Assume now that there exists $\bar{\mathbf{G}} = (\bar{G}_1, \dots, \bar{G}_N) \in \mathbb{H}^n$ such that $\mathcal{L}_i(\bar{\mathbf{G}}) + S_i = 0$. Then, bearing in mind (3.44), we have $\mathcal{A}\hat{\varphi}(\bar{\mathbf{G}} - \mathbf{G}) = 0$, or $\hat{\varphi}(\bar{\mathbf{G}} - \mathbf{G}) = 0$, which implies that $\bar{\mathbf{G}} - \mathbf{G} = 0$, and the uniqueness follows.

(b) It follows from (3.44), bearing in mind that $\mathrm{Re}\{\lambda(\mathcal{A})\} < 0$, Proposition 2.21, and that

$$\begin{aligned}\mathcal{A}\int_0^\infty e^{\mathcal{A}t}\hat{\varphi}(\mathbf{S})\,dt &= \int_0^\infty \mathcal{A}e^{\mathcal{A}t}\hat{\varphi}(\mathbf{S})\,dt = \int_0^\infty \frac{de^{\mathcal{A}t}\hat{\varphi}(\mathbf{S})}{dt}\,dt\\ &= e^{\mathcal{A}t}\hat{\varphi}(\mathbf{S})\big|_0^\infty = -\hat{\varphi}(\mathbf{S}).\end{aligned}$$

(c) From (3.43) and Lemma 3.10 we have that $\mathcal{L}(\mathbf{G}^*) + \mathbf{S}^* = 0$, and the result follows from the fact that $\mathcal{L}(\mathbf{G}^* - \mathbf{G}) + (\mathbf{S}^* - \mathbf{S}) = 0$.

(d) is a consequence of $e^{\mathcal{A}t}\hat{\varphi}(\mathbf{Q}) = \hat{\varphi}(e^{\mathcal{L}t}(\mathbf{Q}))$, bearing in mind Lemma 3.10. □

We next present equivalent forms of Lyapunov equation and Lyapunov inequality for mean-square stability of system (3.1).

Theorem 3.21 *The following assertions are equivalent to mean-square stability of system* (3.1):

(a) $\mathrm{Re}\{\lambda(\mathcal{A})\} < 0$.
(b) *For some* $G_j > 0$ *in* $\mathbb{B}(\mathbb{C}^n)$, $j \in \mathcal{S}$, *we have* $\mathcal{L}_i(\mathbf{G}) < 0$, $i \in \mathcal{S}$.

(c) *For any* $S_i > 0$ *in* $\mathbb{B}(\mathbb{C}^n)$, $i \in \mathcal{S}$, *there is a unique* $\mathbf{G} = (G_1, \ldots, G_N)$, $G_i > 0$ *in* $\mathbb{B}(\mathbb{C}^n)$, $i \in \mathcal{S}$, *such that*

$$\mathcal{L}(\mathbf{G}) + \mathbf{S} = 0. \tag{3.45}$$

Moreover,

$$G_i = \hat{\varphi}_i^{-1}\big(-\mathcal{A}^{-1}\hat{\varphi}(\mathbf{S})\big), \quad i \in \mathcal{S}.$$

Furthermore, the above results also hold if we replace $\mathcal{L}$ *by* $\mathcal{T}$ *and* $\mathcal{A}$ *by* $\mathcal{A}^*$, *or* $\mathbb{C}$ *by* $\mathbb{R}$.

Proof Clearly (c) implies (b). Suppose now that (b) holds. We consider the homogeneous system

$$\dot{y}(t) = \mathcal{A}^* y(t), \quad t \in \mathbb{R}^+, \qquad y(0) \in \hat{\varphi}\big(\mathbb{H}^{n+}\big), \tag{3.46}$$

where

$$\hat{\varphi}\big(\mathbb{H}^{n+}\big) = \big\{y \in \mathbb{C}^{Nn^2}; y = \hat{\varphi}(\mathbf{Q}), \mathbf{Q} \in \mathbb{H}^{n+}\big\}.$$

From Proposition 3.8 we have that

$$\hat{\varphi}_j^{-1}\big(\dot{y}(t)\big) = \mathcal{T}_j\big(\hat{\varphi}_1^{-1}\big(y(t)\big), \ldots, \hat{\varphi}_N^{-1}\big(y(t)\big)\big), \quad j \in \mathcal{S}, \tag{3.47}$$

with $\hat{\varphi}_j^{-1}(y(0)) \in \mathbb{B}(\mathbb{C}^n)^+$. It follows from Lemma 3.10 that $\hat{\varphi}_j^{-1}(y(t)) \in \mathbb{B}(\mathbb{C}^n)^+$ for all $j \in \mathcal{S}$ and all $t \in \mathbb{R}^+$, and thus $y(t) \subset \hat{\varphi}(\mathbb{H}^{n+})$, $t \in \mathbb{R}^+$. Define now the function $\phi : \hat{\varphi}(\mathbb{H}^{n+}) \to \mathbb{R}$ as

$$\phi_j(y) := \operatorname{tr}\big(\hat{\varphi}_j^{-1}(y)G_j\big) = \operatorname{tr}\big(G_j^{1/2}\hat{\varphi}_j^{-1}(y)G_j^{1/2}\big) \geq 0, \quad j \in \mathcal{S},$$

$$\phi(y) := \sum_{j=1}^{N} \phi_j(y) \geq 0.$$

In order to prove that ϕ is a Lyapunov function for system (3.46), we need to show that:

(i) $\phi(y) \to \infty$ whenever $\|y\| \to \infty$ and $y \in \hat{\varphi}(\mathbb{H}^{n+})$;
(ii) $\phi(0) = 0$;
(iii) $\phi(y) > 0$ for all $y \in \hat{\varphi}(\mathbb{H}^{n+})$, $y \neq 0$;
(iv) ϕ is continuous;
(v) $\dot{\phi}(y(t)) < 0$ whenever $y(t) \in \hat{\varphi}(\mathbb{H}^{n+})$, $y(t) \neq 0$.

Now, for $y \in \hat{\varphi}(\mathbb{H}^{n+})$, let $\lambda_{ij}(y) \geq 0$ denote the ith eigenvalue of $\hat{\varphi}_j^{-1}(y)$, and $\lambda_i(G_j) > 0$ the ith eigenvalue of G_j. Define

$$c_0 := \min_{1 \leq i \leq n, 1 \leq j \leq N} \lambda_i(G_j) > 0$$

(since $G_j > 0$) and

$$c_1 := \max_{1\le i\le n, 1\le j\le N} \lambda_i(G_j) > 0.$$

From (2.1(ii)) we have that

$$c_0\left(\sum_{j=1}^{N}\sum_{i=1}^{n}\lambda_{ji}(y)\right) \le \phi(y) \le c_1\left(\sum_{j=1}^{N}\sum_{i=1}^{n}\lambda_{ji}(y)\right). \tag{3.48}$$

Note that

$$\|y\|^2 = \operatorname{tr}(yy^*) = \sum_{j=1}^{N} \operatorname{tr}\big((\hat{\varphi}_j^{-1}(y))^2\big) = \sum_{j=1}^{N}\sum_{i=1}^{n}\big(\lambda_{ji}(y)\big)^2,$$

and bearing in mind the positiveness of $\lambda_{ji}(y)$, we get that $\|y\| \to \infty$ iff

$$\sum_{j=1}^{N}\sum_{i=1}^{n}\big(\lambda_{ji}(y)\big) \to \infty,$$

and $y = 0$ iff $\lambda_{ji}(y) = 0$, $i = 1, \ldots, n$, $j \in \mathcal{S}$. Thus, from these results and (3.48) we get (i)–(iii). Since the continuity of ϕ is easily verified, it remains only to show (v). Now, from the definition of ϕ, (3.47), Lemma 3.10, and Lemma 3.5 we have:

$$\begin{aligned}
\dot{\phi}\big(y(t)\big) &= \sum_{j=1}^{N}\dot{\phi}_j\big(y(t)\big) = \sum_{j=1}^{N}\operatorname{tr}\big(\hat{\varphi}_j^{-1}\big(\dot{y}(t)\big)G_j\big)\\
&= \sum_{j=1}^{N}\operatorname{tr}\big(\mathcal{T}_j\big(\hat{\varphi}_1^{-1}\big(y(t)\big), \ldots, \hat{\varphi}_N^{-1}\big(y(t)\big)\big)G_j\big)\\
&= \big\langle\mathcal{T}\big(\hat{\varphi}_1^{-1}\big(y(t)\big), \ldots, \hat{\varphi}_N^{-1}\big(y(t)\big)\big)^*; \mathbf{G}\big\rangle\\
&= \big\langle\mathcal{T}\big(\big(\hat{\varphi}^{-1}\big(y(t)\big)\big)^*\big); \mathbf{G}\big\rangle\\
&= \big\langle\hat{\varphi}^{-1}\big(y(t)\big)^*; \mathcal{L}(\mathbf{G})\big\rangle < 0
\end{aligned}$$

whenever $y(t) \neq 0 \in \hat{\varphi}(\mathbb{H}^{n+})$. Therefore, we have shown that (3.46) is asymptotically stable (cf. [183]), and thus $\|\exp(\mathcal{A}^* t)y\| \to 0$ as $t \to \infty$ for all $y \in \hat{\varphi}(\mathbb{H}^{n+})$, which yields from Proposition 2.23 that $\operatorname{Re}\{\lambda(\mathcal{A}^*)\} = \operatorname{Re}\{\lambda(\mathcal{A})\} < 0$.

Finally, from Proposition 3.20 we have that (a) implies (c) and from Theorem 3.15 that (a) is equivalent to mean-square stability. The fact that $\mathbb{R}$ and $\mathbb{C}$ may be interchanged is proven as follows. Obviously, the existence of $G_j \in \mathbb{B}(\mathbb{R}^n)$ in (b) is sufficient for $G_j \in \mathbb{B}(\mathbb{C}^n)$. The necessity is due to $A_i \in \mathbb{B}(\mathbb{R}^n)$ for all $i \in \mathcal{S}$. In this case, whenever (b) is true, we have $\mathbf{G} = \mathbf{G}^R + \sqrt{-1}\mathbf{G}^I$ for some $\mathbf{G}^R = (G_1^R, \ldots, G_N^R)$ and $\mathbf{G}^I = (G_1^I, \ldots, G_N^I)$ in $\mathbb{H}^n$. Thus, by taking the conjugate transpose we have (bearing in mind that the entries of A_i are real)

$$\mathcal{L}_i(\mathbf{G}^R) + \sqrt{-1}\mathcal{L}_i(\mathbf{G}^I) < 0, \quad \mathcal{L}_i(\mathbf{G}^R) - \sqrt{-1}\mathcal{L}_i(\mathbf{G}^I) < 0, \quad i \in \mathcal{S},$$

so that, summing both expressions, we obtain the real version of (b). The proof for (c) is analogous if we keep in mind that the entries of $\mathcal{A}$ are real. □

Remark 3.22 Note that the Lyapunov operator $\mathcal{L}$ works on a Hilbert space of dimension $\frac{Nn(n+1)}{2}$ rather than Nn^2, the dimension of the matrix $\mathcal{A}$. This information can then be used to write up the Lyapunov operator $\mathcal{L}$ as a square matrix of dimension $\frac{Nn(n+1)}{2}$. Once this is done, it would be more advantageous to check if $\mathrm{Re}\{\lambda(\mathcal{L})\} < 0$ by looking at the eigenvalues of this reduced-order matrix.

We show now that from Theorem 3.21 we can derive some *easy-to-check conditions* for mean-square stability of (3.1).

Corollary 3.23 *Conditions* (i) *and* (ii) *below are equivalent*:

(i) $\exists \alpha_i > 0$, $i \in \mathcal{S}$, *such that for each* $i \in \mathcal{S}$,

$$\alpha_i \lambda_{\max}(A_i + A_i^*) + \sum_{j\in\mathcal{S}} \lambda_{ij}\alpha_j < 0,$$

(ii) $\exists \alpha_i > 0$, $i \in \mathcal{S}$, *such that for each* $i \in \mathcal{S}$,

$$\alpha_i \lambda_{\max}(A_i + A_i^*) + \sum_{j\in\mathcal{S}} \lambda_{ji}\alpha_j < 0,$$

where $\lambda_{\max}(\mathcal{T}) := \max\{\lambda : \lambda$ *is an eigenvalue of the operator* $\mathcal{T}\}$. *Moreover, if the above conditions* (*one of them*) *are satisfied, then system* (3.1) *is MSS.*

Proof Consider the homogeneous scalar system

$$\dot{\tilde{x}}(t) = \tilde{a}_{\theta(t)}\tilde{x}(t), \quad t \in \mathbb{R}^+, \tag{3.49}$$

where $\tilde{a}_i := \frac{1}{2}\lambda_{\max}(A_i + A_i^*)$, $i \in \mathcal{S}$. Then by applying Theorem 3.21 to system (3.49) we obtain that conditions (i) and (ii) above are equivalent. Suppose now that condition (i) is satisfied and set $G_j = \alpha_j I_n > 0$, $j \in \mathcal{S}$. Since

$$\begin{aligned} A_i^* G_i + G_i A_i + \sum_{j\in\mathcal{S}} \lambda_{ij} G_j &= \alpha_i(A_i + A_i^*) + \sum_{j\in\mathcal{S}} \lambda_{ij}\alpha_j I_n \\ &\le \Big(\alpha_i \lambda_{\max}(A_i + A_i^*) + \sum_{j\in\mathcal{S}} \lambda_{ij}\alpha_j\Big) I_n < 0, \end{aligned}$$

we get from Theorem 3.21(a) that system (3.1) is MSS. □

Corollary 3.24 *Suppose that for some real numbers* $\delta_i > 0$, $i \in \mathcal{S}$, *one of the following conditions is satisfied*:

(1) $\lambda_{\max}[A_i + A_i^* + \frac{1}{\delta_i}(\sum_{\{j\in\mathcal{S};\, j\neq i\}} \lambda_{ij}\delta_j)I_n] < -\lambda_{ii}$,
(2) $\lambda_{\max}[A_i + A_i^* + \frac{1}{\delta_i}(\sum_{\{j\in\mathcal{S};\, j\neq i\}} \lambda_{ji}\delta_j)I_n] < -\lambda_{ii}$.

Then system (3.1) *is MSS. Moreover, these conditions are equivalent to those in Corollary* 3.23.

Proof Immediate from Corollary 3.23 after noticing that for any symmetric matrix U, $\lambda_{\max}(U + rI) = \lambda_{\max}(U) + r$. □

3.4.3 Summary

Finally, we conclude this section by summarizing the main equivalence results for the homogeneous case.

Theorem 3.25 *The assertions below are equivalent*:

(a) *System* (3.1) *is MSS.*
(b) *System* (3.1) *is StS.*
(c) $\mathrm{Re}\{\lambda(\mathcal{L})\} < 0$.
(d) $\mathrm{Re}\{\lambda(\mathcal{A})\} < 0$.
(e) (*Exponential MSS*) *There exist* $b > 0$ *and* $a > 0$ *such that for each* $t \geq 0$, (3.38) *holds.*
(f) (*Coupled Lyapunov equations*) *Given any* $\mathbf{S} = (S_1, \dots, S_N) > 0$ *in* $\mathbb{H}^{n+}$, *there exists* $\mathbf{P} = (P_1, \dots, P_N) > 0$ *in* $\mathbb{H}^{n+}$ *satisfying* $\mathcal{T}(\mathbf{P}) + \mathbf{S} = 0$.
(g) (*Adjoint coupled Lyapunov equations*) *Given any* $\mathbf{S} = (S_1, \dots, S_N) > 0$ *in* $\mathbb{H}^{n+}$, *there exists* $\mathbf{P} = (P_1, \dots, P_N) > 0$ *in* $\mathbb{H}^{n+}$ *satisfying* $\mathcal{L}(\mathbf{P}) + \mathbf{S} = 0$.

Moreover, if $\mathrm{Re}\{\lambda(\mathcal{A})\} < 0$, *then, for any* $\mathbf{S} \in \mathbb{H}^n$, *we have that*:

(i) *there exists a unique* $\mathbf{P} \in \mathbb{H}^n$ *such that* $\mathcal{T}(\mathbf{P}) + \mathbf{S} = 0$.
(ii) *if* $\mathbf{S} \geq \mathbf{T}$ ($\mathbf{S} > \mathbf{T}$) *and* $\mathcal{T}(\mathbf{P}) + \mathbf{S} = 0$, $\mathcal{T}(\mathbf{L}) + \mathbf{T} = 0$, *then* $\mathbf{P} \geq \mathbf{L}$ ($\mathbf{P} > \mathbf{L}$).

These results also hold if $\mathcal{T}$ *is replaced by* $\mathcal{L}$.

Proof It follows from Theorem 3.15, Proposition 3.20, and Theorem 3.21. □

3.5 The $L_2^r(\Omega, \mathcal{F}, P)$ and Jump Diffusion Cases

We consider in this section two scenarios regarding the additive disturbance. In Sect. 3.5.1, we consider the one as in (3.2), characterized by functions in

$L_2^r(\Omega, \mathcal{F}, P)$. In Sect. 3.5.2, we consider the scenario as in (3.3), a jump diffusion where the noise is characterized via a Wiener process. For both cases, under suitable conditions, it is shown that mean-square stability is equivalent to asymptotic wide-sense stationarity (AWSS). In addition, it is shown that the state $\{x(t); t \in \mathbb{R}^+\} \in L_2^n(\Omega, \mathcal{F}, P)$ for $L_2^r(\Omega, \mathcal{F}, P)$-disturbances. In Sect. 3.5.3, we summarize the main results of this section.

3.5.1 The $L_2^r(\Omega, \mathcal{F}, P)$ Disturbance Case

We consider in this subsection the class of dynamical systems modeled by the stochastic equation (3.2), restated here just for the sake of convenience:

$$\begin{aligned} \dot{x}(t) &= A_{\theta(t)}x(t) + J_{\theta(t)}w(t), \quad t \in \mathbb{R}^+, \\ \vartheta_0 &= (x_0, \theta_0), \qquad P(\theta_0 = i) = \nu_i, \end{aligned} \tag{3.50}$$

where the additive disturbance $\{w(t); t \in \mathbb{R}^+\}$ is any $L_2^r(\Omega, \mathcal{F}, P)$-function. We need the following result.

Lemma 3.26 *Let $\{w(t); t \in \mathbb{R}^+\} \in L_2^r(\Omega, \mathcal{F}, P)$ and define for $\lambda > 0$,*

$$\chi(t) = \int_0^t e^{-\lambda(t-\tau)} \|w(\tau)\|_2 \, d\tau. \tag{3.51}$$

Then $\chi(t) \to 0$ as $t \to \infty$.

Proof The proof follows the same arguments as in [242], pp. 119–120. Given any $\epsilon > 0$, consider $t_\epsilon > 0$ such that $\int_{t_\epsilon}^\infty \|w(\tau)\|_2^2 \, d\tau \le \epsilon^2$. Then, for $t > t_\epsilon$,

$$\chi(t) = e^{-\lambda(t-t_\epsilon)} \int_0^{t_\epsilon} e^{-\lambda(t_\epsilon-\tau)} \|w(\tau)\|_2 \, d\tau + \int_{t_\epsilon}^t e^{-\lambda(t-\tau)} \|w(\tau)\|_2 \, d\tau. \tag{3.52}$$

We have from the Schwarz inequality that

$$\begin{aligned} \left(\int_0^{t_\epsilon} e^{-\lambda(t_\epsilon-\tau)} \|w(\tau)\|_2 \, d\tau \right)^2 &\le \int_0^{t_\epsilon} e^{-2\lambda(t_\epsilon-\tau)} \, d\tau \int_0^{t_\epsilon} \|w(\tau)\|_2^2 \, d\tau \\ &\le \frac{1}{2\lambda} \|w\|_2^2 \end{aligned} \tag{3.53}$$

and

$$\begin{aligned} \left(\int_{t_\epsilon}^{t} e^{-\lambda(t-\tau)} \|w(\tau)\|_2 \, d\tau \right)^2 &\le \int_{t_\epsilon}^{t} e^{-2\lambda(t-\tau)} \, d\tau \int_{t_\epsilon}^{\infty} \|w(\tau)\|_2^2 \, d\tau \\ &\le \frac{1}{2\lambda} \epsilon^2. \end{aligned} \tag{3.54}$$

Taking the limit as $t \to \infty$ in (3.52), from (3.53) and (3.54) we obtain that $0 \le \lim_{t\to\infty} \chi(t) \le \frac{\epsilon}{(2\lambda)^{1/2}}$, showing the desired result. □

We prove now the following result.

Theorem 3.27 *The following assertions are equivalent*:

(i) $\mathrm{Re}\{\lambda(\mathcal{A})\} < 0$.
(ii) $\{x(t); t \in \mathbb{R}^+\} \in L_2^n(\Omega, \mathcal{F}, P)$ *for any* $\{w(t); t \in \mathbb{R}^+\} \in L_2^r(\Omega, \mathcal{F}, P)$ *and initial conditions* ϑ_0.

Moreover, if (i) *or* (ii) *is satisfied, then the following condition holds*:

(iii) $\lim_{t\to\infty} E(x(t+s)^* x(t)) = 0$ *for any* $s \ge 0$, $\{w(t); t \in \mathbb{R}^+\} \in L_2^r(\Omega, \mathcal{F}, P)$, *and initial conditions* ϑ_0.

Proof (i) ⇒ (ii): First, notice that over the set Υ (see (3.17)) we have from (3.50) and Theorem 2.5 (see also Theorem 2.1.70 of [53], p. 17) that

$$x(t) = \Phi(t,0)x(0) + \int_0^t \Phi(t,\tau) J_{\theta(\tau)} w(\tau)\, d\tau. \tag{3.55}$$

By the triangular inequality, recalling that $P(\Upsilon) = 1$, it follows from (3.55) that

$$\|x(t)\|_2 \le \|\Phi(t,0)x(0)\|_2 + \int_0^t \|\Phi(t,\tau) J_{\theta(\tau)} w(\tau)\|_2\, d\tau. \tag{3.56}$$

From (3.33),

$$\|\Phi(t,0)x(0)\|_2^2 \le n\|e^{\mathcal{L}t}\|\,\|x(0)\|_2^2. \tag{3.57}$$

From (3.33) again, with $z = J_{\theta(\tau)} w(\tau)$, we have that

$$\begin{aligned} \|\Phi(t,\tau) J_{\theta(\tau)} w(\tau)\|_2^2 &\le n\|e^{\mathcal{L}(t-\tau)}\|\,\|J_{\theta(\tau)} w(\tau)\|_2^2 \\ &\le n\|\mathbf{J}\|_{\max}^2 \|e^{\mathcal{L}(t-\tau)}\|\,\|w(\tau)\|_2^2. \end{aligned} \tag{3.58}$$

From (3.56), (3.57), and (3.58) we have that for some $\lambda > 0$ and $a > 0$,

$$\|x(t)\|_2 \le a\left(e^{-\lambda t}\|x(0)\|_2 + \int_0^t e^{-\lambda(t-\tau)}\|w(\tau)\|_2\, d\tau\right). \tag{3.59}$$

Consider $\chi(t)$ as in (3.51). If we define

$$f(t) = \begin{cases} e^{-\lambda t}, & t \ge 0, \\ 0, & t < 0, \end{cases}$$

$$g(t) = \begin{cases} \|w(t)\|_2, & t \ge 0, \\ 0, & t < 0, \end{cases}$$

we have that χ can be written as the convolution (denoted here by $*$) of f and g, that is, $\chi(t) = (f * g)(t)$. Since $f \in L_1(\mathbb{R}^+)$[1] and $g \in L_2(\mathbb{R}^+)$, it follows that the convolution $f * g \in L_2(\mathbb{R}^+)$ and moreover, for some $b > 0$,

$$\int_0^\infty \chi(t)^2\, dt \le b^2 \int_0^\infty f(t)\, dt \int_0^\infty g(t)^2\, dt = \frac{b^2}{\lambda} \|w\|_2^2 \tag{3.60}$$

(cf. [250]). Taking (3.59) into square, we get

$$\|x(t)\|_2^2 \le 2a^2\big(e^{-2\lambda t}\|x(0)\|_2^2 + \chi(t)^2\big), \tag{3.61}$$

and from (3.60) and (3.61) it follows that for some $c > 0$,

$$\|x\|_2^2 = \int_0^\infty \|x(t)\|_2^2 dt \le c\big(\|x(0)\|_2^2 + \|w\|_2^2\big), \tag{3.62}$$

showing the desired result.

(ii) $\Rightarrow$ (i): Take $w(t) = 0$ for all $t \in \mathbb{R}^+$. From (ii) we have that for arbitrary initial condition ϑ_0, $\|x\|_2^2 < \infty$, that is,

$$\|x\|_2^2 = \int_0^\infty \|x(t)\|_2^2 dt < \infty.$$

Thus, system (3.1) is StS as in Definition 3.1, and the result follows from Proposition 3.15.

(i) $\Rightarrow$ (iii): From Lemma 3.26 and (3.61) it follows that $E(\|x(t)\|^2) \to 0$ as $t \to \infty$. The result follows since

$$\begin{aligned}\big|E\big(x(t+s)^* x(t)\big)\big| &\le E\big(\|x(t+s)\|\,\|x(t)\|\big)\\ &\le \big(E\big(\|x(t+s)\|^2\big)E\big(\|x(t)\|^2\big)\big)^{1/2} \to 0\end{aligned}$$

as $t \to \infty$. □

3.5.2 The Jump Diffusion Case

In this subsection we deal with mean-square stability issues for the class of systems described in Sect. 3.2 by (3.3), i.e.,

$$\begin{aligned} &dx(t) = A_{\theta(t)} x(t)\, dt + J_{\theta(t)}\, dw(t), \quad t \in \mathbb{R}^+,\\ &\vartheta_0 = (x_0, \theta_0), \qquad \theta_0 \text{ with distribution } \nu. \end{aligned} \tag{3.63}$$

[1] For $1 \le p < \infty$, $L_p(\mathbb{R}^+)$ is the space of Lebesgue-measurable functions f from $\mathbb{R}^+$ to $\mathbb{R}$ such that $\int_0^\infty |f(t)|^p\, dt < \infty$.

Let $\mathbf{R}(t) := (R_1(t), \dots, R_N(t)) \in \mathbb{H}_{\mathbb{C}}^{n+}$ with $R_i(t) := J_i J_i^* p_i(t)$, where we recall that $p_i(t) = P(\theta(t) = i)$. The next proposition provides differential equations to compute the first and second moments of the state variable and follows the same steps as in the proof of Proposition 3.7.

Proposition 3.28 *For $t \in \mathbb{R}^+$, we have*:

$$\dot{\hat{q}}(t) = F\hat{q}(t), \tag{3.64}$$

$$\dot{\mathbf{Q}}(t) = \mathcal{L}\big(\mathbf{Q}(t)\big) + \mathbf{R}(t), \tag{3.65}$$

$$\dot{\mathbf{Q}}(s,t) = \mathcal{F}\big(\mathbf{Q}(s,t)\big). \tag{3.66}$$

Proof By Lemma 3.6, applying Itô's rule to (3.8) and recalling that $w(t)$ and $\theta(t)$ are independent and $E[dw(t)] = 0$, we have from (3.63) that

$$\begin{aligned} dq_j(t) &= E\big(dx(t)1_{\{\theta(t)=j\}} + x(t)d1_{\{\theta(t)=j\}}\big) \\ &= A_j E\big(x(t)1_{\{\theta(t)=j\}}\big)\,dt + J_j E\big(dw(t)\big)p_j(t) + \sum_{i\in\mathcal{S}} \lambda_{ij} q_i(t)\,dt \\ &= A_j q_j(t)\,dt + \sum_{i\in\mathcal{S}} \lambda_{ij} q_i(t)\,dt, \end{aligned}$$

and (3.64) follows. Similarly, noticing that

$$E\big(dw(t)\,dw(t)^* 1_{\{\theta(t)=j\}}\big) = p_j(t)\,dt \tag{3.67}$$

we have for (3.9) that

$$\begin{aligned} dQ_j(t) &= E\big(dx(t)x(t)^* 1_{\{\theta(t)=j\}} + x(t)\,dx(t)^* 1_{\{\theta(t)=j\}} + x(t)x(t)^*\,d(1_{\{\theta(t)=j\}}) \\ &\quad + dx(t)\,dx(t)^* 1_{\{\theta(t)=j\}}\big) \\ &= A_j E\big(x(t)x(t)^* 1_{\{\theta(t)=j\}}\big)\,dt + E\big(x(t)x(t)^* 1_{\{\theta(t)=j\}}\big)A_j^*\,dt \\ &\quad + \sum_{i\in\mathcal{S}} \lambda_{ij} E\big(x(t)x(t)^* 1_{\{\theta(t)=i\}}\big)\,dt + J_j J_j^* p_j(t)\,dt \\ &= \mathcal{L}_j\big(\mathbf{Q}(t)\big)\,dt + R_j(t)\,dt, \end{aligned}$$

showing (3.65). Finally, from (3.10),

$$\begin{aligned} dQ_j(s,t) &= E\big(dx(t+s)x(t)^* 1_{\{\theta(t+s)=j\}} + x(t+s)x(t)^*\,d(1_{\{\theta(t+s)=j\}})\big) \\ &= A_j E\big(x(t+s)x(t)^* 1_{\{\theta(t+s)=j\}}\big)\,ds \\ &\quad + \sum_{i\in\mathcal{S}} \lambda_{ij} E\big(x(t+s)x(t)^* 1_{\{\theta(t+s)=i\}}\big)\,ds \\ &= \mathcal{F}_j\big(\mathbf{Q}(s,t)\big)\,ds, \end{aligned}$$

showing (3.66). □

From this we have the following result.

Proposition 3.29 *If* $\mathrm{Re}\{\lambda(\mathcal{A})\} < 0$, *then system* (3.3) *is MSS according to Definition* 3.2 *and AWSS according to Definition* 3.3, *with* $q = 0$ *and*

$$Q = \sum_{i \in \mathcal{S}} \hat{\varphi}_i^{-1}\big(-\mathcal{A}^{-1}\hat{\varphi}(\mathbf{R})\big), \tag{3.68}$$

$$\mathcal{Q}(s) = \sum_{i \in \mathcal{S}} \hat{\varphi}_i^{-1}\big(e^{\mathcal{B}s}\mathcal{A}^{-1}\hat{\varphi}(\mathbf{R})\big), \tag{3.69}$$

where $\mathbf{R} = (R_1, \dots, R_N) \in \mathbb{H}_{\mathbb{C}}^{n+}$, $R_i := \pi_i J_i J_i^*$, $i \in \mathcal{S}$.

Proof First, notice that since $\mathrm{Re}\{\lambda(\mathcal{A})\} < 0$, we have from Proposition 3.13 that $\mathrm{Re}\{\lambda(F)\} < 0$. Thus, from (3.26) in Proposition 3.7 and Proposition 2.23 it follows that $\hat{q}(t) \to 0$ as $t \to \infty$, and since $q(t) = \sum_{i \in \mathcal{S}} q_i(t)$, we have that $q(t) \to 0$ as $t \to \infty$. Now, from (3.65) and Proposition 3.8(a) we have that

$$\hat{\varphi}\big(\dot{\mathbf{Q}}(t)\big) = \mathcal{A}\hat{\varphi}\big(\mathbf{Q}(t)\big) + \hat{\varphi}\big(\mathbf{R}(t)\big).$$

If we define $y(t) := \hat{\varphi}(\mathbf{Q}(t))$ and $f(t) := \hat{\varphi}(\mathbf{R}(t))$, we get that

$$\dot{y}(t) = \mathcal{A}y(t) + f(t).$$

From the forward differential equation to the Markov chain (Sect. 3.2), bearing in mind the definition of $\mathbf{R}(t)$, it follows that $f(t)$ is continuous. Furthermore, recalling that θ has limiting probabilities $\{\pi_i;\ i \in \mathcal{S}\}$ satisfying expression (3.4), we get that

$$\big\|\mathbf{R}(t) - \mathbf{R}\big\|_1 = \sum_{i \in \mathcal{S}} \big\|R_i(t) - R_i\big\| \le \|\mathbf{J}\|_2^2 \max_{j \in \mathcal{S}}\big\{\big|p_j(t) - \pi_j\big|\big\}, \tag{3.70}$$

and therefore $\lim_{t\to\infty}\mathbf{R}(t) = \mathbf{R}$. Now by Proposition 2.22 it follows that $\hat{\varphi}(\mathbf{Q}(t)) \to -\mathcal{A}^{-1}\hat{\varphi}(\mathbf{R})$ as $t \to \infty$. By noting that $Q(t) = \sum_{i \in \mathcal{S}} Q_i(t)$ it follows that $Q(t) \to Q$ as $t \to \infty$, with Q as in (3.68). From Proposition 3.8(c) and Proposition 3.28 we have that

$$\hat{\varphi}\big(\dot{\mathbf{Q}}(s,t)\big) = \mathcal{B}\hat{\varphi}\big(\mathbf{Q}(s,t)\big),$$

and therefore $\hat{\varphi}(\mathbf{Q}(s,t)) = e^{\mathcal{B}s}(\mathbf{Q}(0,t))$. Moreover, as seen above, we have that $\hat{\varphi}(\mathbf{Q}(t)) \to \mathcal{A}^{-1}\hat{\varphi}(\mathbf{R})$ as $t \to \infty$. It follows that $\hat{\varphi}(\mathbf{Q}(s,t)) \to e^{\mathcal{B}s}\mathcal{A}^{-1}\hat{\varphi}(\mathbf{R})$ as $t \to \infty$, and since $Q(s,t) = \sum_{i \in \mathcal{S}} Q_i(s,t)$, we have that

$$Q(t,s) \to \sum_{i \in \mathcal{S}} \hat{\varphi}_i^{-1}\big(e^{\mathcal{B}s}\mathcal{A}^{-1}\hat{\varphi}(\mathbf{R})\big) = \mathcal{Q}(s),$$

completing the proof. □

Remark 3.30 From Proposition 3.29 we have that the L_2^r-result of Theorem 3.27 does not apply for the Wiener disturbance setting.

The following corollary is immediate from the previous results.

Corollary 3.31 *For system* (3.3), *we have that*

$$\dot{\mathbf{Q}}(t) = \mathcal{L}(\mathbf{Q}(t)) + \mathbf{R}(t). \tag{3.71}$$

If $\operatorname{Re}\{\lambda(\mathcal{L})\} < 0$, *then* $\mathbf{Q}(t) \to -\mathcal{L}^{-1}(\mathbf{R}) \geq 0$ *as* $t \to \infty$, *and moreover,* $-\mathcal{L}^{-1}(\mathbf{R})$ *is the unique solution of the following equation in* $\mathbf{Z} \in \mathbb{H}_{\mathbb{C}}^n$:

$$\mathcal{L}(\mathbf{Z}) + \mathbf{R} = 0. \tag{3.72}$$

We have the following result:

Theorem 3.32 *The following statements are equivalent*:

(a) $\operatorname{Re}\{\lambda(\mathcal{A})\} < 0$.
(b) *System* (3.3) *is MSS according to Definition* 3.2(b).
(c) *System* (3.3) *is AWSS according to Definition* 3.3.

Proof From Proposition 3.29 we have that (a) implies (b) and (c). In addition, it is obvious that AWSS implies MSS. It remains to prove that (b) implies (a). First, we have that $Q(t) = \sum_{i \in \mathcal{S}} Q_i(t)$ and from (3.27) in Proposition 3.7 and Proposition 3.8(a) that

$$\hat{\varphi}(\dot{\mathbf{Q}}(t)) = \mathcal{A}\hat{\varphi}(\mathbf{Q}(t)) + \hat{\varphi}(\mathbf{R}(t)).$$

Therefore,

$$\hat{\varphi}(\mathbf{Q}(t)) = e^{\mathcal{A}t}\hat{\varphi}(\mathbf{Q}(0)) + \int_0^t e^{\mathcal{A}(t-s)}\hat{\varphi}(\mathbf{R}(s))\,ds$$

and

$$Q(t) = \sum_{i \in \mathcal{S}} \hat{\varphi}_i^{-1}\left(e^{\mathcal{A}t}\hat{\varphi}(\mathbf{Q}(0))\right) + \sum_{i \in \mathcal{S}} \hat{\varphi}_i^{-1}\left(\int_0^t e^{\mathcal{A}(t-s)}\hat{\varphi}(\mathbf{R}(s))\,ds\right). \tag{3.73}$$

Now, by hypothesis, there exists $Q \in \mathbb{B}(\mathbb{C}^n)^+$ such that $Q(t) \to Q$ as $t \to \infty$ for any $Q(0) = E(x_0 x_0^*)$ and Q does not depend on x_0. Furthermore, notice that for $x_0 = 0$, we have that the second term on the right-hand side of (3.73) converges to Q as $t \to \infty$, and thus the first term goes to zero for any x_0 and ν. The rest of the proof follows as in the proof of Theorem 3.15. □

3.5.3 Summary

We summarize the main results of this section in the next theorem.

Theorem 3.33 *The following assertions are equivalent*:

(a) $\mathrm{Re}\{\lambda(\mathcal{A})\} < 0$.
(b) *For system* (3.2), *we have that* $\{x(t); t \in \mathbb{R}^+\} \in L_2^n(\Omega, \mathcal{F}, P)$ *for any* $\{w(t); t \in \mathbb{R}^+\} \in L_2^r(\Omega, \mathcal{F}, P)$ *and initial conditions* ϑ_0.
(c) *System* (3.3) *is MSS according to Definition* 3.2(b).
(d) *System* (3.3) *is AWSS according to Definition* 3.3.

Proof It follows from Theorems 3.27 and 3.32. □

3.6 Mean-Square Stabilizability and Detectability

In this section we will deal with the concepts of mean-square stabilizability and mean-square detectability, tracing a parallel with similar concepts found in the literature of linear systems. As seen in Theorem 3.25, mean-square stability and stochastic stability are equivalent, so we will use the term mean-square stabilizability (detectability) and stochastic stabilizability (detectability, respectively) meaning the same thing. Using the two versions of the coupled Lyapunov equations seen in Theorem 3.25, we will present in Sect. 3.6.1 some LMIs test conditions to verify if an MJLS is mean-square stabilizable/detectable or not. In Sect. 3.6.2 we will provide necessary and sufficient conditions for mean-square stabilizability subject to partial information on the jump variable. We will assume that the Markov jump parameter is not exactly known, but instead an estimate of it is available to the controller. Under some additional assumptions, a solution via LMIs will also be provided. Section 3.6.3 deal with the dynamic output mean-square stabilizability case. Necessary and sufficient LMIs test conditions are presented to verify if an MJLS is mean-square stabilizable through a dynamic output feedback system.

3.6.1 Definitions and LMIs Conditions

Consider the following class of differential equations:

$$\dot{x}(t) = A_{\theta(t)}x(t) + B_{\theta(t)}u(t), \tag{3.74}$$

$$\vartheta_0 = (x_0, \theta_0), \tag{3.75}$$

where $\mathbf{A} = (A_1, \dots, A_N) \in \mathbb{H}^n$, $\mathbf{B} = (B_1, \dots, B_N) \in \mathbb{H}^{m,n}$, and, as before, $x(t) \in \mathbb{C}^n$ denotes the state vector, and $u(t) \in \mathbb{C}^m$ the control input, and the homogeneous Markov process $\theta = \{(\theta(t), \mathcal{F}_t), t \in \mathbb{R}^+\}$ is as defined in Sect. 3.2.

Definition 3.34 (Mean-square stabilizability) We say that the system $(\mathbf{A}, \mathbf{B}, \Pi)$ is mean square (or stochastically) stabilizable (SS) if there exists $\mathbf{K} = (K_1, \dots, K_N) \in \mathbb{H}^{n,m}$ such that for arbitrary initial condition ϑ_0, we have that

$$\int_0^\infty E(\|x(t)\|^2)\, dt < \infty, \tag{3.76}$$

where $x(t)$ is given by (3.74) with $t \in \mathbb{R}^+$ and $u(t) = -K_{\theta(t)}x(t)$, i.e.,

$$\dot{x}(t) = \widetilde{A}_{\theta(t)}x(t), \quad t \in \mathbb{R}^+, \tag{3.77}$$

with $\widetilde{A}_{\theta(t)} = A_{\theta(t)} - B_{\theta(t)}K_{\theta(t)}$. In this case we say that $\mathbf{K}$ stabilizes $(\mathbf{A}, \mathbf{B}, \Pi)$. We set $\mathbb{K} := \{\mathbf{K} \in \mathbb{H}^{n,m}; \mathbf{K}$ stabilizes $(\mathbf{A}, \mathbf{B}, \Pi)$ in the mean-square sense$\}$.

Definition 3.35 (Mean-square detectability) Consider $\mathbf{C} = (C_1, \ldots, C_N) \in \mathbb{H}^{n,r}$. We say that the system $(\mathbf{C}, \mathbf{A}, \Pi)$ is mean-square (or stochastically) detectable (SD) if there exists $\mathbf{G} = (G_1, \ldots, G_N) \in \mathbb{H}^{r,n}$ such that for arbitrary initial condition ϑ_0, we have that

$$\int_0^\infty E\bigl(\|x(t)\|^2\bigr)\,dt < \infty, \tag{3.78}$$

where $x(t)$ is given by

$$\dot{x}(t) = \widetilde{A}_{\theta(t)}x(t), \quad t \in \mathbb{R}^+, \tag{3.79}$$

with $\widetilde{A}_{\theta(t)} = A_{\theta(t)} - G_{\theta(t)}C_{\theta(t)}$. In this case we say that $\mathbf{G}$ stabilizes $(\mathbf{C}, \mathbf{A}, \Pi)$.

Remark 3.36 The system $(\mathbf{C}, \mathbf{A}, \Pi)$ refers to a filter of the form

$$\dot{\widehat{x}}(t) = A_{\theta(t)}\widehat{x}(t) + G_{\theta(t)}\bigl(y(t) - C_{\theta(t)}\widehat{x}(t)\bigr), \tag{3.80}$$

where

$$\begin{cases} \dot{x}(t) = A_{\theta(t)}x(t), \\ y(t) = C_{\theta(t)}x(t). \end{cases} \tag{3.81}$$

Setting $\widetilde{x}(t) = x(t) - \widehat{x}(t)$, it is easy to see from (3.80) and (3.81) that the estimation error equation verifies

$$\dot{\widetilde{x}}(t) = \widetilde{A}_{\theta(t)}\widetilde{x}(t), \quad \widetilde{A}_{\theta(t)} = A_{\theta(t)} - G_{\theta(t)}C_{\theta(t)},$$

and thus if $\mathbf{G}$ stabilizes $(\mathbf{C}, \mathbf{A}, \Pi)$, then the mean-square estimation error tends to zero as t goes to infinity.

For $\widetilde{\mathbf{A}} = (\widetilde{A}_1, \ldots, \widetilde{A}_N) \in \mathbb{H}^n$, set the operators $\mathcal{L}$ and $\mathcal{T}$ as in (3.21) replacing A_i by $\widetilde{A}_i$, that is, for $\mathbf{H} = (H_1, \ldots, H_N) \in \mathbb{H}^n_{\mathbb{C}}$,

$$\mathcal{L}_i(\mathbf{H}) = \widetilde{A}_i H_i + H_i \widetilde{A}_i^* + \sum_{j \in \mathcal{S}} \lambda_{ji} H_j, \quad i \in \mathcal{S}, \tag{3.82}$$

$$\mathcal{T}_i(\mathbf{H}) = \widetilde{A}_i^* H_i + H_i \widetilde{A}_i + \sum_{j \in \mathcal{S}} \lambda_{ij} H_j, \quad i \in \mathcal{S}. \tag{3.83}$$

We have the following equivalence lemma.

Lemma 3.37 *Consider the operators $\mathcal{L}$ and $\mathcal{T}$ given by* (3.82) *and* (3.83), *respectively. For SS, the following assertions are equivalent*:

(SS1) *The system* $(\mathbf{A}, \mathbf{B}, \Pi)$ *is SS.*
(SS2) *There exists (a stabilizing)* $\mathbf{K} = (K_1, \ldots, K_N) \in \mathbb{H}^{n,m}$ *such that*

$$\operatorname{Re}\{\lambda(\mathcal{L})\} < 0 \quad \textit{with } \widetilde{A}_i = A_i - B_i K_i,\ i \in \mathcal{S}. \tag{3.84}$$

(SS3) (LMIs test) *There exists* $\mathbf{P} > 0$ *in* $\mathbb{H}^{n+}$ *and* $\mathbf{L} \in \mathbb{H}^{n,m}$ *such that for each* $i \in \mathcal{S}$,

$$A_i P_i + B_i L_i + P_i A_i^* + L_i^* B_i^* + \sum_{j \in \mathcal{S}} \lambda_{ji} P_j < 0. \tag{3.85}$$

Moreover, in this case, with $K_i = -L_i P_i^{-1}$, $\mathbf{K} = (K_1, \ldots, K_N) \in \mathbb{H}^{n,m}$ *stabilizes* $(\mathbf{A}, \mathbf{B}, \Pi)$.

Similarly, for SD, the following assertions are equivalent:

(SD1) *The system* $(\mathbf{C}, \mathbf{A}, \Pi)$ *is SD.*
(SD2) *There exists (a stabilizing)* $\mathbf{G} = (G_1, \ldots, G_N) \in \mathbb{H}^{r,n}$ *such that*

$$\operatorname{Re}\{\lambda(\mathcal{T})\} < 0 \quad \textit{with } \widetilde{A}_i = A_i - G_i C_i, i \in \mathcal{S}. \tag{3.86}$$

(SD3) (LMIs test) *There exist* $\mathbf{P} > 0$ *in* $\mathbb{H}^{n+}$ *and* $\mathbf{L} \in \mathbb{H}^{n,m}$ *such that*

$$A_i^* P_i + C_i^* L_i^* + P_i A_i + L_i C_i + \sum_{j \in \mathcal{S}} \lambda_{ij} P_j < 0. \tag{3.87}$$

Moreover, in this case, with $G_i = -P_i^{-1} L_i$, $\mathbf{G} = (G_1, \ldots, G_N) \in \mathbb{H}^{n,m}$ *stabilizes* $(\mathbf{C}, \mathbf{A}, \Pi)$.

Proof This is an immediate application of Theorem 3.25. □

Remark 3.38 If we specialize our framework to the nonjump case, the definitions of SS and SD recast the definitions of stabilizability and detectability of the standard linear deterministic case.

3.6.2 Mean-Square Stabilizability with $\theta(t)$ Partially Known

In this subsection we shall be concerned with the mean-square stabilizability of (3.74)–(3.75) under partial information on the jumping parameter $\theta(t)$. The goal is to find $\mathbf{K} = (K_1, \ldots, K_N) \in \mathbb{H}^{n,m}$ such that the system

$$dx(t) = (A_{\theta(t)} + B_{\theta(t)} K_{\hat{\theta}(t)}) x(t)\, dt, \quad \vartheta_0 = (x_0, \theta_0) \tag{3.88}$$

is MSS, where $\hat{\theta}(t)$ is an estimate for $\theta(t)$. We assume that

$$P(\hat{\theta}(t) = j \mid \mathcal{F}_t) = \alpha_{\theta(t) j} \tag{3.89}$$

and

$$\Psi := \begin{bmatrix} \alpha_{11} & \dots & \alpha_{1N} \\ \vdots & \ddots & \vdots \\ \alpha_{N1} & \dots & \alpha_{NN} \end{bmatrix}$$

with $\sum_{j=1}^{N} \alpha_{ij} = 1$. Notice that the closer α_{ii} is to 1, the better the estimator $\hat{\theta}(t)$ will be. Furthermore, noticing that

$$P\big(\theta(t+h) = j, \hat{\theta}(t) = \nu | \mathcal{F}_t\big) = \alpha_{\theta(t)\nu} P\big(\theta(t+h) = j | \theta(t)\big) \tag{3.90}$$

and defining

$$\mathcal{H}_j(\mathbf{K}) = \sum_{\nu=1}^{N} \alpha_{j\nu} K_\nu, \tag{3.91}$$

we have, recalling (3.9), the following result.

Proposition 3.39 *For $t \in \mathbb{R}^+$ and $j \in \mathcal{S}$, we have that*

$$\dot{Q}_j(t) = \big(A_j + B_j \mathcal{H}_j(\mathbf{K})\big) Q_j(t) + Q_j(t)\big(A_j + B_j \mathcal{H}_j(\mathbf{K})\big)^* + \sum_{i \in \mathcal{S}} \lambda_{ij} Q_i(t). \tag{3.92}$$

Proof We have that

$$x(t+h) - x(t) = (A_{\theta(t)} + B_{\theta(t)} K_{\hat{\theta}(t)}) x(t) h + o(h) \tag{3.93}$$

and thus

$$\begin{aligned}
&E\big(x(t+h)x(t+h)^* 1_{\{\theta(t+h)=j\}}\big) \\
&\quad = \sum_{i \in \mathcal{S}} \sum_{\nu \in \mathcal{S}} E\big(x(t+h)x(t+h)^* 1_{\{\theta(t+h)=j\}} 1_{\{\theta(t)=i\}} 1_{\{\hat{\theta}(t)=\nu\}}\big) \\
&\quad = \sum_{i \in \mathcal{S}} \sum_{\nu \in \mathcal{S}} E\big(\big(\big(x(t+h) - x(t)\big)\big(x(t+h) - x(t)\big)^* + x(t)\big(x(t+h) - x(t)\big)^* \\
&\qquad + \big(x(t+h) - x(t)\big)x(t)^* + x(t)x(t)^*\big) 1_{\{\theta(t+h)=j\}} 1_{\{\theta(t)=i\}} 1_{\{\hat{\theta}(t)=\nu\}}\big) \\
&\quad = \sum_{i \in \mathcal{S}} \sum_{\nu \in \mathcal{S}} E\big(E\big(\big[x(t)x(t)^* 1_{\{\theta(t)=i\}} (A_i + B_i K_\nu)^* h \\
&\qquad + (A_i + B_i K_\nu) x(t)x(t)^* 1_{\{\theta(t)=i\}} h \\
&\qquad + x(t)x(t)^* 1_{\{\theta(t)=i\}}\big] 1_{\{\theta(t+h)=j\}} 1_{\{\hat{\theta}(t)=\nu\}} | \mathcal{F}_t\big)\big) + o(h) \\
&\quad = \sum_{\nu \in \mathcal{S}} \Big(\alpha_{j\nu} (1 + \lambda_{jj} h) Q_j(t) (A_j + B_j K_\nu)^* h + (A_j + B_j K_\nu) Q_j(t) h + Q_j(t)
\end{aligned}$$

$$+\sum_{i\neq j}\alpha_{i\nu}\lambda_{ij}h\big(Q_i(t)(A_i+B_iK_\nu)^*h+(A_i+B_iK_\nu)Q_i(t)h+Q_i(t)\big)\Bigg)+o(h)$$

$$=\sum_{\nu\in\mathcal{S}}\alpha_{j\nu}\big(Q_j(t)(A_j+B_jK_\nu)^*+(A_j+B_jK_\nu)Q_j(t)+\lambda_{jj}Q_j(t)\big)h$$

$$+\sum_{i\neq j}\sum_{\nu\in\mathcal{S}}\alpha_{i\nu}\lambda_{ij}Q_i(t)h+Q_j(t)+o(h)$$

$$=\Bigg(Q_j(t)\big(A_j+B_j\mathcal{H}_j(\mathbf{K})\big)^*+\big(A_j+B_j\mathcal{H}_j(\mathbf{K})\big)Q_j(t)$$

$$+\sum_{i\in\mathcal{S}}\lambda_{ij}Q_i(t)\Bigg)h+Q_j(t)+o(h).$$

Therefore, taking the limit as $h\downarrow 0$, we obtain the desired result. □

Suppose now that $\operatorname{rank}(\Psi)=\ell\leq N$, and let $\{v_{N-\ell+1},\dots,v_N\}$ be an orthonormal basis for $\mathcal{N}(\Psi')$. Write

$$v_i=\begin{pmatrix}v_{i1}\\ \vdots\\ v_{iN}\end{pmatrix}. \tag{3.94}$$

The next theorem provides a necessary and sufficient condition to get that $\mathbf{K}=(K_1,\dots,K_N)$ stabilizes system (3.88) in the mean-square sense.

Theorem 3.40 *There exists $\mathbf{K}=(K_1,\dots,K_N)\in\mathbb{H}^{n,m}$ that stabilizes system* (3.88) *in the mean square sense if and only if there exist $\mathbf{P}\in\mathbb{H}^{n+}$, $\mathbf{P}=(P_1,\dots,P_N)>0$, and $\mathbf{F}=(F_1,\dots,F_N)\in\mathbb{H}^{n,m}$ such that*

$$A_jP_j+B_jF_j+P_jA_j^*+F_j^*B_j^*+\sum_{i\in\mathcal{S}}\lambda_{ij}P_i<0,\quad j\in\mathcal{S}, \tag{3.95}$$

$$\sum_{k=1}^{N}v_{ik}F_kP_k^{-1}=0,\quad i=N-\ell+1,\dots,N. \tag{3.96}$$

Proof From Proposition 3.39, equation (3.27), and Theorem 3.15 we have that the above system is MSS if and only if there exists $\mathbf{P}=(P_1,\dots,P_N)>0$ such that

$$\big(A_j+B_j\mathcal{H}_j(\mathbf{K})\big)P_j+P_j\big(A_j+B_j\mathcal{H}_j(\mathbf{K})\big)^*+\sum_{i\in\mathcal{S}}\lambda_{ij}P_i<0,\quad j\in\mathcal{S}, \tag{3.97}$$

that is, if and only if for some $\mathbf{P}=(P_1,\dots,P_N)>0$ and $\mathbf{F}=(F_1,\dots,F_N)$,

$$A_jP_j+B_jF_j+P_jA_j^*+F_j^*B_j^*+\sum_{i\in\mathcal{S}}\lambda_{ij}P_i<0,\quad j\in\mathcal{S}, \tag{3.98}$$

and

$$\begin{pmatrix} F_1 P_1^{-1} \\ \vdots \\ F_N P_N^{-1} \end{pmatrix} = (\Psi \otimes I_m) \begin{pmatrix} K_1 \\ \vdots \\ K_N \end{pmatrix}, \tag{3.99}$$

where the last equation has a solution if and only if

$$\sum_{k=1}^{N} v_{ik} F_k P_k^{-1} = 0, \quad i = N - \ell + 1, \dots, N. \tag{3.100}$$

□

The following result is immediate from the above theorem and solves the problem through an LMIs approach.

Corollary 3.41 *If* $\operatorname{rank}(\Psi) = N$, *then there exists* $\mathbf{K} = (K_1, \dots, K_N) \in \mathbb{H}^{n,m}$ *that stabilizes system* (3.88) *in the mean square sense if and only if there exist* $\mathbf{P} \in \mathbb{H}^{n+}$, $\mathbf{P} = (P_1, \dots, P_N) > 0$, *and* $\mathbf{F} = (F_1, \dots, F_N) \in \mathbb{H}^{n,m}$ *such that*

$$A_j P_j + B_j F_j + P_j A_j^* + F_j^* B_j^* + \sum_{i \in \mathcal{S}} \lambda_{ij} P_i < 0, \quad j \in \mathcal{S}, \tag{3.101}$$

and for $\mathbf{P}$ *and* $\mathbf{F}$ *as above*, $\mathbf{K} = (K_1, \dots, K_N)$ *is given by*

$$\begin{pmatrix} K_1 \\ \vdots \\ K_N \end{pmatrix} = (\Psi^{-1} \otimes I_m) \begin{pmatrix} F_1 P_1^{-1} \\ \vdots \\ F_N P_N^{-1} \end{pmatrix}. \tag{3.102}$$

3.6.3 Dynamic Output Mean-Square Stabilizability

The goal of this subsection is to obtain necessary and sufficient LMIs test conditions in order to verify if an MJLS is mean-square stabilizable through a dynamic output feedback system. In this case we consider the control system

$$\mathcal{G}_u = \begin{cases} \dot{x}(t) = A_{\theta(t)} x(t) + B_{\theta(t)} u(t), \\ y(t) = H_{\theta(t)} x(t), \end{cases} \tag{3.103}$$

and the following dynamic output feedback system $\mathcal{K}$:

$$\mathcal{K} = \begin{cases} \dot{\widehat{x}}(t) = \widehat{A}_{\theta(t)} \widehat{x}(t) + \widehat{B}_{\theta(t)} y(t), \\ u(t) = \widehat{C}_{\theta(t)} \widehat{x}(t), \end{cases} \tag{3.104}$$

whose order we shall assume as equal to that of (3.103), that is, $\widehat{x} \in \mathbb{C}^n$. Combining (3.103) and (3.104) and letting $\mathbf{v}(t) := [x(t)^* \ \widehat{x}(t)^*]^*$, we get that the closed-loop system can be written as

$$\dot{\mathbf{v}}(t) = \Gamma_{\theta(t)}\mathbf{v}(t) \tag{3.105}$$

with

$$\Gamma_i = \begin{bmatrix} A_i & B_i\widehat{C}_i \\ \widehat{B}_i H_i & \widehat{A}_i \end{bmatrix}, \quad i \in \mathcal{S}. \tag{3.106}$$

Definition 3.42 We say that system $\mathcal{G}_u$ is mean-square stabilizable through a dynamic output feedback system $\mathcal{K}$ if there exist $\widehat{\mathbf{A}} = (\widehat{A}_1, \dots, \widehat{A}_N)$, $\widehat{\mathbf{B}} = (\widehat{B}_1, \dots, \widehat{B}_N)$, and $\widehat{\mathbf{C}} = (\widehat{C}_1, \dots, \widehat{C}_N)$ such that system (3.105) is MSS.

Notice that, by Theorem 3.25, system (3.105) is MSS if and only if there exists $\mathbf{P} = (P_1, \dots, P_N) > 0$ such that

$$\Gamma_i^* P_i + P_i \Gamma_i + \sum_{j\in\mathcal{S}} \lambda_{ij} P_j < 0, \quad i \in \mathcal{S}. \tag{3.107}$$

The following theorem ([101], Theorem 3.1) provides necessary and sufficient conditions for the existence of a dynamic output feedback mean-square stabilizing system $\mathcal{K}$, based on some LMIs conditions. We recall that diag$(\cdot)$ indicates a block diagonal matrix with entries given by $(\cdot)$.

Theorem 3.43 *There exist* $\mathbf{P} = (P_1, \dots, P_N) > 0$, $\widehat{\mathbf{A}} = (\widehat{A}_1, \dots, \widehat{A}_N)$, $\widehat{\mathbf{B}} = (\widehat{B}_1, \dots, \widehat{B}_N)$, *and* $\widehat{\mathbf{C}} = (\widehat{C}_1, \dots, \widehat{C}_N)$ *such that* (3.107) *holds if and only if the following LMIs* (3.108)–(3.110) *have a feasible solution* $\mathbf{X} = (X_1, \dots, X_N) > 0$, $\mathbf{Y} = (Y_1, \dots, Y_N) > 0$, $\mathbf{L} = (L_1, \dots, L_N)$, $\mathbf{F} = (F_1, \dots, F_N)$ *for* $i \in \mathcal{S}$:

$$\begin{bmatrix} A_iY_i + Y_iA_i^* + B_iF_i + F_i^*B_i^* + \lambda_{ii}Y_i & \mathcal{R}_i(\mathbf{Y}) \\ \mathcal{R}_i^*(\mathbf{Y}) & -\mathcal{D}_i(\mathbf{Y}) \end{bmatrix} < 0, \tag{3.108}$$

$$A_i^*X_i + X_iA_i + L_iH_i + H_i^*L_i^* + \sum_{j\in\mathcal{S}} \lambda_{ij}X_j < 0, \tag{3.109}$$

$$\begin{bmatrix} Y_i & I \\ I & X_i \end{bmatrix} > 0 \tag{3.110}$$

with

$$\mathcal{R}_i(\mathbf{Y}) = \left[\sqrt{\lambda_{i1}}Y_i \ \dots \ \sqrt{\lambda_{i(i-1)}}Y_i \ \sqrt{\lambda_{i(i+1)}}Y_i \ \dots \ \sqrt{\lambda_{iN}}Y_i\right], \tag{3.111}$$

$$\mathcal{D}_i(\mathbf{Y}) = \mathrm{diag}(Y_1, \dots, Y_{i-1}, Y_{i+1}, \dots, Y_N). \tag{3.112}$$

Moreover, if $\mathbf{X} > 0$, $\mathbf{Y} > 0$, $\mathbf{L}$, $\mathbf{F}$ *are feasible for the LMIs* (3.108)–(3.110), *then, by setting* $\mathbf{P} = (P_1, \dots, P_N) > 0$, $\widehat{\mathbf{A}} = (\widehat{A}_1, \dots, \widehat{A}_N)$, $\widehat{\mathbf{B}} = (\widehat{B}_1, \dots, \widehat{B}_N)$, *and* $\widehat{\mathbf{C}} =$

$(\widehat{C}_1, \ldots, \widehat{C}_N)$, *for each* $i \in \mathcal{S}$, *as*

$$\widehat{A}_i = (X_i - Y_i^{-1})^{-1}\Big(A_i^* + X_i A_i Y_i + X_i B_i F_i + L_i H_i Y_i + \sum_{j \in \mathcal{S}} \lambda_{ij} Y_j^{-1} Y_i\Big) Y_i^{-1}, \tag{3.113}$$

$$\widehat{B}_i = (Y_i^{-1} - X_i)^{-1} L_i, \tag{3.114}$$

$$\widehat{C}_i = F_i Y_i^{-1}, \tag{3.115}$$

$$P_i = \begin{bmatrix} X_i & Y_i^{-1} - X_i \\ Y_i^{-1} - X_i & X_i - Y_i^{-1} \end{bmatrix} > 0, \tag{3.116}$$

we have that (3.107) *holds.*

Proof Suppose first that there exist $\mathbf{P} > 0$, $\widehat{\mathbf{A}}$, $\widehat{\mathbf{B}}$, and $\widehat{\mathbf{C}}$ such that (3.107) holds, and consider the following partition for P_i:

$$P_i = \begin{bmatrix} Z_i & U_i \\ U_i^* & \widehat{Z}_i \end{bmatrix}. \tag{3.117}$$

Without loss of generality, suppose further that U_i is nonsingular (if not, re-define U_i as $U_i + \epsilon I$ so that it is nonsingular and $\epsilon > 0$ is small enough so that (3.107) still holds) for each $i \in \mathcal{S}$. Defining the matrices

$$\mathcal{Y}_i = (Z_i - U_i \widehat{Z}_i^{-1} U_i^*)^{-1} > 0, \tag{3.118}$$

$$T_i = \begin{bmatrix} \mathcal{Y}_i & I \\ \mathcal{Y}_i & 0 \end{bmatrix}, \tag{3.119}$$

$$M_i = \begin{bmatrix} I & 0 \\ 0 & -\widehat{Z}_i^{-1} U_i^* \end{bmatrix}, \tag{3.120}$$

and multiplying (3.107) to the left by $T_i^* M_i^*$ and to the right by $M_i T_i$, we get, after performing some calculations, that

$$\begin{bmatrix} A_i \mathcal{Y}_i + \mathcal{Y}_i A_i^* + B_i \mathcal{F}_i + \mathcal{F}_i^* B_i^* & \mathcal{M}_i^* \\ \mathcal{M}_i & A_i^* Z_i + Z_i A_i + \mathcal{L}_i H_i + H_i^* \mathcal{L}_i^* \end{bmatrix} + \sum_{j \in \mathcal{S}} \lambda_{ij} \begin{bmatrix} \mathcal{Y}_i [\mathcal{Y}_j^{-1} + (U_i \widehat{Z}_i^{-1} \widehat{Z}_j - U_j) \widehat{Z}_j^{-1} (U_i \widehat{Z}_i^{-1} \widehat{Z}_j - U_j)^*] \mathcal{Y}_i & 0 \\ 0 & Z_j \end{bmatrix} < 0, \tag{3.121}$$

where

$$
\begin{aligned}
\mathcal{F}_i &= -\widehat{C}_i \widehat{Z}_i^{-1} U_i^* \mathcal{Y}_i, \\
\mathcal{L}_i &= U_i \widehat{B}_i, \\
\mathcal{M}_i &= A_i^* + Z_i A_i \mathcal{Y}_i + Z_i B_i \mathcal{F}_i + \mathcal{L}_i H_i \mathcal{Y}_i - U_i \widehat{A}_i \widehat{Z}_i^{-1} U_i^* \mathcal{Y}_i \\
&\quad + \sum_{j=1}^{N} \lambda_{ij} \big(Z_j - U_j \widehat{Z}_i^{-1} U_i^*\big) \mathcal{Y}_i.
\end{aligned} \tag{3.122}
$$

Besides, the conditions $P_i > 0$, $i \in \mathcal{S}$, are equivalent to

$$
T_i^* M_i^* P_i M_i T_i = \begin{bmatrix} \mathcal{Y}_i & I \\ I & Z_i \end{bmatrix} > 0. \tag{3.123}
$$

Noticing that

$$
\begin{aligned}
&\sum_{j\in\mathcal{S}} \lambda_{ij} \big(U_i \widehat{Z}_i^{-1} \widehat{Z}_j - U_j\big) \widehat{Z}_j^{-1} \big(U_i \widehat{Z}_i^{-1} \widehat{Z}_j - U_j\big)^* \\
&\quad = \sum_{j=1, j\neq i}^{N} \lambda_{ij} \big(U_i \widehat{Z}_i^{-1} \widehat{Z}_j - U_j\big) \widehat{Z}_j^{-1} \big(U_i \widehat{Z}_i^{-1} \widehat{Z}_j - U_j\big)^* \geq 0,
\end{aligned} \tag{3.124}
$$

it follows that (3.108)–(3.110) hold for $Y_i = \mathcal{Y}_i$, $X_i = Z_i$, $F_i = \mathcal{F}_i$, and $L_i = \mathcal{L}_i$, $i \in \mathcal{S}$.

For the sufficiency part, let us suppose that the LMIs (3.108)–(3.110) have a feasible solution $\mathbf{X} > 0$, $\mathbf{Y} > 0$, $\mathbf{L}$, $\mathbf{F}$, and set $\widehat{\mathbf{A}}$, $\widehat{\mathbf{B}}$, $\widehat{\mathbf{C}}$, $\mathbf{P}$ as in (3.113)–(3.116). Define, for $i \in \mathcal{S}$,

$$
\begin{aligned}
T_i &= \begin{bmatrix} Y_i & I \\ Y_i & 0 \end{bmatrix}, \\
\mathcal{J}_i &= A_i Y_i + Y_i A_i^* + B_i F_i + F_i^* B_i^* + \sum_{j\in\mathcal{S}} \lambda_{ij} Y_i Y_j^{-1} Y_i, \\
\mathcal{H}_i &= A_i^* X_i + X_i A_i + L_i H_i + H_i^* L_i^* + \sum_{j\in\mathcal{S}} \lambda_{ij} X_j.
\end{aligned}
$$

From (3.108) we have that $\mathcal{J}_i < 0$, and from (3.109) that $\mathcal{H}_i < 0$. It follows that

$$
T_i^* \Big(\Gamma_i^* P_i + P_i \Gamma_i + \sum_{j\in\mathcal{S}} \lambda_{ij} P_j \Big) T_i = \begin{bmatrix} \mathcal{J}_i & 0 \\ 0 & \mathcal{H}_i \end{bmatrix} < 0,
$$

showing that (3.107) holds and completing the proof. □

Remark 3.44 Notice that the choice in (3.113)–(3.116) for $\widehat{\mathbf{A}}$, $\widehat{\mathbf{B}}$, $\widehat{\mathbf{C}}$, $\mathbf{P}$ corresponds to choosing in (3.117) the trivial solution $U_i = -\widehat{Z}_i = Y_i^{-1} - X_i$ and, in (3.122), $\mathcal{M}_i = 0$.

3.7 Historical Remarks

Stability theory forms an important bedrock of control theory. It provides, inter alia, the sturdy theoretical foundation to analyze the long-run behavior of dynamical systems. From an application point of view, stability of a dynamical system is one of the primary concerns in the design and synthesis of control systems. Since the path-breaking work of J.C. Maxwell on the stability of Watt's flyball governor, together with the ground-breaking work of E.J. Routh, A. Hurwitz, and A.M. Lyapunov, stability issues gained increasing significance and have given a decisive impetus for the development of modern control theory. As a consequence of their work, a flurry of interest in this subject has given rise to an extraordinary burst of publications in the specialized literature.

Regarding MJLS, a fairly amount of techniques has been successfully developed in the past two decades or so. As far as the authors are aware of, the study of stability issues for MJLS can be traced back at least to [196]. In addition, to the best of our knowledge, [303] seems to be the first to use a stability criterion to treat the infinite-horizon control problem, considering stability for each operation mode of the system. It was soon clear that this criterion was not fully adequate to deal with the many nuances of the MJLS class. In order to flesh out an adequate theory, it was necessary to devise structural criteria right from stability concepts based on the state of the system as, for instance, mean-square stability (MSS). We mention here, for instance, [77, 137, 187, 189–191, 223] as key works in the unfolding of mean-square stability theory for MJLS. These works have put mean-square stability (including mean-square stabilizability and mean-square detectability) into a solid grounding, which has given rise to a host of new important developments in various directions and has made the theory flourish. By now, the MSS theory for MJLS is a full fledged theory that provides systematic tools for the analysis of this class of systems. Without any intention of being exhaustive, we mention [24, 42, 43, 48, 71, 112, 150, 151, 156, 221, 222, 224, 231, 245, 297] as a small sample of related works of more recent vintage (see also [81]). These papers consider the case in which the Markov process takes values in a finite state space. Some important issues regarding mean-square stability for the case in which the state space of the Markov chain is *infinite countable* can be found in [78, 147, 148, 152].

Almost sure stability for MJLS is studied, for instance, in [78, 112, 135, 136, 207, 221]. For robust stability (including the case with delay), the readers are referred, for instance, to [24, 42, 43, 71, 245, 297]. A new approach to detectability (weak detectability) is considered in [61–63] (see [64] for the infinite countable case). Stability and stabilizability issues for the case with *partial observations* are treated, for instance. in [78] and [151].

For those wishing to delve into the main ideas of stochastic stability, we mention the classical [204] and [179] as excellent sources for stability of stochastic dynamical systems (see also [199] and [210]). For a book of a more recent vintage, see [227], which contains a comprehensive treatment on the stochastic stability of Markov chains.

This chapter is, essentially, based on [101, 150–152].

Chapter 4
Quadratic Optimal Control with Complete Observations

4.1 Outline of the Chapter

This chapter deals with the quadratic optimal control problem for continuous-time MJLS in the usual finite- and infinite-horizon framework. It is assumed that both the state variable $x(t)$ and jump variable $\theta(t)$ are available to the controller. The setup adopted in this chapter is based on Dynkin's formula for the resulting Markov process obtained from the state $x(t)$ and Markov chain $\theta(t)$. Under this approach, we consider the class of admissible controllers as those in a feedback form (on $x(t)$ and $\theta(t)$) satisfying a Lipschitz condition. It is shown that the solution for the problems rely, in part, on the study of a finite set of coupled differential and algebraic Riccati equations (CDRE and CARE, respectively). These equations are studied in the Appendix A. Tracing a parallel with the classical literature on quadratic optimal control problems, it is shown that the optimal controller can be written in a linear state feedback form. The organization of the chapter is as follows. Some notation and the problem formulation are presented in Sect. 4.2, while Dynkin's formula for the resulting Markov process obtained from the state $x(t)$ and Markov chain $\theta(t)$ is characterized in Sect. 4.3. The solution of the finite-horizon and infinite-horizon quadratic optimal control problems are presented in Sects. 4.4 and 4.5, respectively.

4.2 Notation and Problem Formulation

Consider the jump controlled system $\mathcal{G}$ described as follows:

$$\mathcal{G} = \begin{cases} \dot{x}(t) = A_{\theta(t)}(t)x(t) + B_{\theta(t)}(t)u(t), \\ z(t) = C_{\theta(t)}(t)x(t) + D_{\theta(t)}(t)u(t), \end{cases} \tag{4.1}$$

where $\mathbf{A}(t) = (A_1(t), \dots, A_N(t)) \in \mathbb{H}^n$, $\mathbf{B}(t) = (B_1(t), \dots, B_N(t)) \in \mathbb{H}^{m,n}$, $\mathbf{C}(t) = (C_1(t), \dots, C_N(t)) \in \mathbb{H}^{n,p}$, and $\mathbf{D}(t) = (D_1(t), \dots, D_N(t)) \in \mathbb{H}^{m,p}$ are matrices of class **PC** (piecewise continuous, see Definition 2.3). It is also assumed that

O.L.V. Costa et al., *Continuous-Time Markov Jump Linear Systems*, Probability and Its Applications, DOI 10.1007/978-3-642-34100-7_4,

$D_i^*(t)D_i(t) > 0$ and $C_i^*(t)D_i(t) = 0$ for all $i \in \mathcal{S}$ and $t \in \mathbb{R}^+$ and that the matrices $(D_i^*(t)D_i(t))^{-1}$ are of class **PC**. We recall from (2.27) that $\vartheta_t = (x(t), \theta(t))$ and write the conditional expectation with respect to $(x(t), \theta(t))$ as

$$E_{\vartheta_t}(\cdot) = E\big(\cdot\,|x(t), \theta(t)\big).$$

We will consider in this chapter the finite-horizon and infinite-horizon quadratic optimal control problems. For the finite-horizon case, we assume that the class of admissible control policies is defined as follows.

Definition 4.1 For arbitrary $T \in \mathbb{R}^+$, the class of admissible control policies for the finite-horizon quadratic optimal control problem is denoted by $\mathcal{U}^T$ and is formed by all functions $u(t) = \bar{u}(t, x(t), \theta(t))$ in which $\bar{u} : [0, T] \times \mathbb{R}^n \times \mathcal{S} \to \mathbb{R}^m$ satisfies the following properties:

(C1) For each $x \in \mathbb{R}^n$ and $i \in \mathcal{S}$ fixed, the function

$$\bar{u}(t, x, i) \text{ is of class } \mathbf{PC}, \tag{4.2}$$

(C2) For each $i \in \mathcal{S}$, there is a real-valued function $c_i : \mathbb{R}^+ \to \mathbb{R}^+$ of class **PC** such that the Lipschitz condition is satisfied:

$$\big\|\bar{u}(t, z, i) - \bar{u}(t, y, i)\big\| \le c_i(t)\|z - y\| \tag{4.3}$$

for every $z, y \in \mathbb{R}^n$, $t \in \mathbb{R}^+$. Note that c_i may depend on $\bar{u}$.

Under conditions (C1) and (C2), we have that with probability one there exists a unique solution to (4.1). Indeed, setting

$$p\big(x, \theta(t), t\big) = A_{\theta(t)}(t)x + B_{\theta(t)}(t)\bar{u}\big(t, x, \theta(t)\big),$$

we have that, for each realization of the Markov process $\{\theta(t); t \ge 0\}$ in Υ (see (3.17)), $p(x, \theta(\cdot), \cdot)$ is of class **PC** and the global Lipschitz condition (2.8) is satisfied since for all $t \in \mathbb{R}^+$ and $x, y \in \mathbb{R}^n$,

$$\big\|p\big(x, \theta(t), t\big) - p\big(y, \theta(t), t\big)\big\| \le \big(\big\|A_{\theta(t)}(t)\big\| + \big\|B_{\theta(t)}(t)\big\|c_{\theta(t)}(t)\big)\|x - y\|.$$

Thus, by Theorem 2.4 (or the Fundamental Theorem B1.6 presented in [53], p. 470) there exists a unique solution to (4.1) for every $\omega \in \Upsilon$, and since $P(\Upsilon) = 1$, we have that with probability one there exists a unique solution to (4.1).

Remark 4.2 In order to keep the Markov property of $\{x(t), \theta(t)\}$ in (4.1), we consider control policies of the form $u(t) = \bar{u}(t, \vartheta_t)$ instead of the more expanded class consisting of the policies of the form $u(t) = \bar{u}(t, \{\vartheta_s, s \le t\})$.

For a starting time $0 \le s < T$, the terminal cost condition $\mathbf{L} = (L_1, \ldots, L_N) \in \mathbb{H}^{n+}$, and for each policy $u \in \mathcal{U}^T$, define the cost functional

$$\begin{aligned}\mathcal{J}_{[s,T],\mathbf{L}}(\vartheta_s, u) &:= E_{\vartheta_s}\left(\int_s^T \|z(t)\|^2 dt + x(T)^* L_{\theta(T)} x(T)\right) \\ &= E_{\vartheta_s}\left(\int_s^T \left(\|C_{\theta(t)}(t)x(t)\|^2 + \|D_{\theta(t)}(t)u(t)\|^2\right) dt \right. \\ &\quad \left. + x(T)^* L_{\theta(T)} x(T)\right), \end{aligned} \tag{4.4}$$

where $x(t)$ is given by (4.1). The finite-horizon optimal control problem consists of finding $\hat{u} \in \mathcal{U}^T$ which minimizes $\mathcal{J}_{[s,T],\mathbf{L}}(\vartheta_s, u)$. For $s = 0$, we set

$$\mathcal{J}_{T,\mathbf{L}}(u) := E\left(\int_0^T \|z(t)\|^2 dt + x(T)^* L_{\theta(T)} x(T)\right). \tag{4.5}$$

For the infinite-horizon case, we assume that the matrices A_i, B_i, C_i, and D_i are time-invariant, that is, they do not depend on time t. The definition of the set of admissible controls in this case is as follows.

Definition 4.3 The class of admissible control policies for the infinite-horizon quadratic optimal control problem is denoted by $\mathcal{U}$ and consists of the functions $u(t) = \bar{u}(t, x(t), \theta(t))$ in which $\bar{u} : \mathbb{R}^+ \times \mathbb{R}^n \times \mathcal{S} \to \mathbb{R}^m$ satisfy (4.2) and (4.3) for every $t \in \mathbb{R}^+$, and the following additional condition:

(C3) System (4.1) with $t \in \mathbb{R}^+$ is mean-square stable, i.e.,

$$E\left(\|x(t)\|^2\right) \to 0$$

as $t \to \infty$ for arbitrary initial distribution of ϑ_0.

For each policy $u \in \mathcal{U}$, define the cost functional

$$\begin{aligned}\mathcal{J}(u) &:= E\left(\int_0^\infty \|z(t)\|^2 dt\right) \\ &= E\left(\int_0^\infty \left(\|C_{\theta(t)}x(t)\|^2 + \|D_{\theta(t)}u(t)\|^2\right) dt\right). \end{aligned} \tag{4.6}$$

For the infinite-horizon quadratic optimal control problem the goal is to derive an optimal control policy $\hat{u}$, within the class $\mathcal{U}$, that minimizes (4.6), that is, such that

$$\mathcal{J}(\hat{u}) := \inf_{u \in \mathcal{U}} \mathcal{J}(u). \tag{4.7}$$

4.3 Dynkin's Formula

We have from system (4.1) with $u \in \mathcal{U}^T$ that $\{(x(t), \theta(t)); t \in [0, T]\}$ is a Markov process evolving in $\mathbb{R}^n \times \mathcal{S}$ with sample paths that are continuous from the right. Define $\mathcal{X} = [0, T] \times \mathbb{R}^n \times \mathcal{S}$ and $\mathcal{C}^1(\mathcal{X})$ the set of all functions $g \in \mathbb{B}(\mathcal{X}, \mathbb{R})$ that, for each $i \in \mathcal{S}$ fixed, satisfy:

(i) For $x \in \mathbb{R}^n$ fixed, $g(t, x, i)$ is continuous for $t \in [0, T]$, and there exists a set D_g in $[0, T]$ that contains at most a finite number of points such that the partial derivative of $g(t, x, i)$ with respect to t, represented by $\frac{\partial}{\partial t} g(t, x, i)$, exists and is continuous for every $t \in \mathbb{R}^+ \setminus D_g$. For $\tau \in D_g$, the left-hand and right-hand derivatives, denoted by $\frac{\partial}{\partial t-} g(\tau, x, i)$ and $\frac{\partial}{\partial t+} g(\tau, x, i)$, respectively, exist and are finite numbers in $\mathbb{R}$. We extend the definition of $\frac{\partial}{\partial t} g(t, x, i)$ for $t = \tau \in D_g$ and assume that it has some arbitrary value $\frac{\partial}{\partial t} g(\tau, x, i) \in \mathbb{R}$.
(ii) For $t \in [0, T]$ fixed, $g(t, x, i)$ is continuously differentiable on $x \in \mathbb{R}^n$, and we denote its gradient by $\nabla_x g(t, x, i)$.

Define the operator $\mathfrak{L}^u$ (the superscript u is just to highlight the dependence on the control u) applied to a function $g \in \mathcal{C}^1(\mathcal{X})$ as follows:

$$\mathfrak{L}^u g(t, x, i) = \frac{\partial}{\partial t} g(t, x, i) + \nabla_x g(t, x, i)^* \big(A_i(t)x + B_i(t)u(t)\big) + \sum_{j \in \mathcal{S}} \lambda_{ij} g(t, x, j). \tag{4.8}$$

Set for $g \in \mathcal{C}^1(\mathcal{X})$, the stochastic process $\{C^g(t); t \in [0, T]\}$ as

$$C^g(t) := g(t, \vartheta_t) - g(0, \vartheta_0) - \int_0^t \mathfrak{L}^u g(s, \vartheta_s)\, ds. \tag{4.9}$$

We have the following result.

Proposition 4.4 *$\{C^g(t); t \in [0, T]\}$ is a martingale with respect to the filtration $\{\mathcal{F}_t; t \in \mathbb{R}^+\}$.*

Proof We have that $C^g(t)$ is $\mathcal{F}_t$-measurable, and, from the fact that $g \in \mathcal{C}^1(\mathcal{X})$, we have that $C^g(t)$ is integrable. Let us define for $i \in \mathcal{S}$, $x \in \mathbb{R}^n$, and $t \in \mathbb{R}^+$, the vector-valued function $g(t, x)$ as

$$g(t, x) := \big[g(t, x, 1) \quad \ldots \quad g(t, x, N)\big] \tag{4.10}$$

and the gradient matrix

$$\nabla_x g(t, x) := \big[\nabla_x g(t, x, 1) \quad \ldots \quad \nabla_x g(t, x, N)\big].$$

Recall from Sect. 2.5 that $\mathcal{S}_v = \{e_1, \ldots, e_N\}$ with $e_i \in \mathbb{R}^N$ formed by 1 in the ith component and zero elsewhere. For $\chi \in \mathcal{S}_v$, set $g(t, x, \chi) = g(t, x)\chi$. Recall also the definition of $\chi(t)$ and $M(t)$ in (2.23) and (2.24), respectively. From the representation result (2.25) in Lemma 2.14 we have that

$$\begin{aligned}
g\big(t, x(t), \chi(t)\big) = g\big(0, \chi(0)\big) &+ \int_0^t \frac{\partial}{\partial s} g\big(s, x(s)\big)\chi(s)\,ds \\
&+ \int_0^t \big(\nabla_x g\big(s, x(s)\big)\chi(s)\big)' dx(s) \\
&+ \int_0^t g\big(s, x(s)\big)\Pi'\chi(s-)\,ds \\
&+ \int_0^t g\big(s, x(s)\big)' dM(s).
\end{aligned} \tag{4.11}$$

Notice now that

$$\frac{\partial}{\partial s} g\big(s, x(s)\big)\chi(s) = \frac{\partial}{\partial s} g\big(s, x(s), \theta(s)\big) \tag{4.12}$$

$$\begin{aligned}
\big(\nabla_x g\big(s, x(s)\big)\chi(s)\big)' dx(s) = \nabla_x g\big(s, x(s), \theta(s)\big)' \big(&A_{\theta(s)} x(s) \\
&+ B_{\theta(s)} u(s)\big)\,ds
\end{aligned} \tag{4.13}$$

$$g\big(s, x(s)\big)\Pi'\chi(s-) = \sum_{j \in \mathcal{S}} \lambda_{\theta(s-)j}\, g\big(s, x(s), j\big). \tag{4.14}$$

Since for each $\omega \in \Omega$, $\lambda_{\theta(s-)(\omega)j} = \lambda_{\theta(s)(\omega)j}$ Lebesgue a.s. for $s \in [0, t]$, we have from (4.11), (4.12), (4.13), and (4.14) that

$$C^g(t) = \int_0^t g\big(s, x(s)\big)' dM(s). \tag{4.15}$$

Noticing now that $g(s, x(s))$ is continuous for $s \in [0, t]$ (since $g(\cdot,\cdot)$ and $x(s)$ are continuous) and that $M = \{M(s); s \in \mathbb{R}^+\}$ is a martingale with respect to the filtration $\{\mathcal{F}_s; s \in \mathbb{R}^+\}$, we get from Theorem 11.4.5 in [259] (see also (2.26) in Lemma 2.14) that $\{C^g(t); t \in \mathbb{R}^+\}$ is a martingale with respect to the filtration $\{\mathcal{F}_t; t \in \mathbb{R}^+\}$, completing the proof. □

As a consequence of Proposition 4.4, we have that $E_{\vartheta_0}(C^g(t)) = C^g(0) = 0$, which implies, from the Markov property, that Dynkin's formula holds and reads as

$$E_{\vartheta_s}\big(g(t, \vartheta_t)\big) - g(s, \vartheta_s) = E_{\vartheta_s}\left(\int_s^t \mathfrak{L}^u g(r, \vartheta_r)\,dr\right) \tag{4.16}$$

for $g \in \mathcal{C}^1(\mathcal{X})$.

Remark 4.5 Following the definition of the infinitesimal generator as in Sect. 2.3, equation (2.4), it is not difficult to show that (4.8) is the infinitesimal generator of $(x(t),\theta(t))$ with $x(t)$ satisfying (4.1). For that, we define the semigroup $T_h g(t,x,i) = E[g(t+h, x(t+h), \theta(t+h)) \mid x(t)=x, \theta(t)=i]$ and

$$\mathfrak{L}^u g(t,x,i) = \lim_{h\downarrow 0} \frac{T_h g(t,x(t),\theta(t)) - T_0 g(t,x(t),\theta(t))}{h}.$$

See, for instance, [159] for more details.

4.4 The Finite-Horizon Optimal Control Problem

The purpose of this section is to derive the solution of the quadratic optimal control problem for the finite-horizon case posed in Sect. 4.2. This solution will be associated to the solution of a set of coupled Riccati differential equations (CDRE), to be presented in the next theorem. For notational simplicity, we omit in this section the dependence of the solution on the final time T. In what follows, we denote by $\mathbb{D} \subset \mathbb{R}^+$ the union of the discontinuity points of A_i, B_i, C_i, D_i, $[D_i^* D_i]^{-1}$, $i \in \mathcal{S}$. As pointed out in [53], p. 7, for any $T \in \mathbb{R}^+$, $\mathbb{D} \cap [0,T]$ contains at most a finite number of points.

Theorem 4.6 *There exists a unique set of N positive semi-definite and continuous $n \times n$ matrices $\mathbf{X}(t) = (X_1(t), \ldots, X_N(t)) \in \mathbb{H}^{n+}$, $0 \le t \le T$, satisfying the following CDRE*:

$$\begin{aligned} &\dot{X}_i(t) + A_i^*(t)X_i(t) + X_i(t)A_i(t) - X_i(t)B_i(t)\big[D_i^*(t)D_i(t)\big]^{-1} B_i^*(t)X_i(t) \\ &\quad + \sum_{j\in\mathcal{S}} \lambda_{ij} X_j(t) + C_i^*(t)C_i(t) = 0, \quad i \in \mathcal{S},\ t \in [0,T] \setminus \mathbb{D}, \end{aligned} \tag{4.17}$$

with boundary condition $\mathbf{X}(T) = \mathbf{L}$.

Proof See Theorem A.1 in Appendix A. □

The next proposition derives an expression for the Dynkin formula applied to a function defined in terms of the solution of the CDRE.

Proposition 4.7 *Let $g \in \mathcal{C}^1(\mathcal{X})$ be such that*

$$g(t,x,i) = x^* X_i(t) x, \tag{4.18}$$

where $t \mapsto \mathbf{X}(t) = (X_1(t), \ldots, X_N(t)) \in \mathbb{H}^{n+}$ satisfies the CDRE given by (4.17) with terminal condition $\mathbf{X}(T) = \mathbf{L} \in \mathbb{H}^{n+}$. Then, for system (4.1) with $u \in \mathcal{U}^T$, the operator $\mathfrak{L}^u$ is given by

$$\begin{aligned}\big(\mathfrak{L}^u g\big)(t,\vartheta_t) &= x(t)^*\big(-C^*_{\theta(t)}(t)C_{\theta(t)}(t)\\ &\quad + X_{\theta(t)}(t)B_{\theta(t)}(t)\big(D^*_{\theta(t)}(t)D_{\theta(t)}(t)\big)^{-1}B^*_{\theta(t)}(t)X_{\theta(t)}(t)\big)x(t)\\ &\quad + u(t)^*B^*_{\theta(t)}(t)X_{\theta(t)}(t)x(t) + x(t)^*X_{\theta(t)}(t)B_{\theta(t)}(t)u(t) \end{aligned}\tag{4.19}$$

for any $(t,\vartheta_t)\in\mathcal{X}$. *Furthermore, Dynkin's formula* (4.16) *can be written as*

$$\begin{aligned}&x(s)^*X_{\theta(s)}(s)x(s) - E_{\vartheta_s}\big(x(t)^*X_{\theta(t)}(t)x(t)\big)\\ &\quad = E_{\vartheta_s}\bigg(\int_s^t \big(x(r)^*C^*_{\theta(r)}(r)C_{\theta(r)}(r)x(r)\\ &\qquad - x(r)^*X_{\theta(r)}(r)B_{\theta(r)}(r)\big(D^*_{\theta(r)}(r)D_{\theta(r)}(r)\big)^{-1}B^*_{\theta(r)}(r)X_{\theta(r)}(r)x(r)\\ &\qquad - u(r)^*B_{\theta(r)}(r)^*X_{\theta(r)}(r)x(r) - x(r)^*X_{\theta(r)}(r)B_{\theta(r)}(r)u(r)\big)\,dr\bigg). \end{aligned}\tag{4.20}$$

Proof From (4.8) and (4.17) we get (4.19). Applying Dynkin's formula (4.16) and using (4.19), we get (4.20). □

We now derive the cost expression for an arbitrary $u\in\mathcal{U}^T$ and the optimal solution for the finite-horizon case. In what follows, we set

$$\Gamma_i(t) = \big(D^*_i(t)D_i(t)\big)^{-1},\quad i\in\mathcal{S}. \tag{4.21}$$

Proposition 4.8 *For arbitrary* $u\in\mathcal{U}^T$, *the cost defined in* (4.4) *is given by*

$$\begin{aligned}\mathcal{J}_{[s,T],\mathbf{L}}(\vartheta_s,u) &= E_{\vartheta_s}\bigg(x(s)^*X_{\theta(s)}(s)x(s) + \int_s^T \big\|B^*_{\theta(r)}(r)X_{\theta(r)}(r)x(r)\\ &\quad + \big(D^*_{\theta(r)}(r)D_{\theta(r)}(r)\big)u(r)\big\|^2_{\Gamma_{\theta(r)}(r)}\,dr\bigg) \end{aligned}\tag{4.22}$$

with $\mathbf{X}(t)\in\mathbb{H}^{n+}$ *satisfying* (4.17).

Proof From (4.4) we have that

$$\begin{aligned}\mathcal{J}_{[s,T],\mathbf{L}}(\vartheta_s,u) &= E_{\vartheta_s}\bigg(\int_s^T \big[\big(x(r)^*C^*_{\theta(r)}(r)C_{\theta(r)}(r)x(r)\\ &\quad + u(r)^*\big(D^*_{\theta(r)}(r)D_{\theta(r)}(r)\big)u(r)\big]\,dr + x(T)^*L_{\theta(T)}x(T)\bigg). \end{aligned}\tag{4.23}$$

Now, from Proposition 4.7, setting $t=T$ in Dynkin's formula (4.20), we get that

$$\begin{aligned}&\mathcal{J}_{[s,T],\mathbf{L}}(\vartheta_s,u)\\ &\quad = x(s)^*X_{\theta(s)}(s)x(s) + E_{\vartheta_s}\bigg(\int_s^T \big[u(r)^*\big(D^*_{\theta(r)}(r)D_{\theta(r)}(r)\big)u(r) \end{aligned}$$

$$+ x(r)^* X_{\theta(r)}(r) B_{\theta(r)}(r) \big(D^*_{\theta(r)}(r) D_{\theta(r)}(r)\big)^{-1} B^*_{\theta(r)}(r) X_{\theta(r)}(r) x(r)$$

$$+ u(r)^* B^*_{\theta(r)}(r) X_{\theta(r)}(r) x(r) + x(r)^* X_{\theta(r)}(r) B_{\theta(r)}(r) u(r)\big] \, dr\bigg). \quad (4.24)$$

Using (4.21), the expression under integration can be rewritten as

$$\begin{aligned} & u(r)^* \big(D^*_{\theta(r)}(r) D_{\theta(r)}(r)\big) \Gamma_{\theta(r)}(r) \big(D^*_{\theta(r)}(r) D_{\theta(r)}(r)\big) u(r) \\ & \quad + x(r)^* X_{\theta(r)}(r) B_{\theta(r)}(r) \Gamma_{\theta(r)}(r) B^*_{\theta(r)}(r) X_{\theta(r)}(r) x(r) \\ & \quad + \big(D^*_{\theta(r)}(r) D_{\theta(r)}(r) u(r)\big)^* \Gamma_{\theta(r)}(r) B^*_{\theta(r)}(r) X_{\theta(r)}(r) x(r) \\ & \quad + x(r)^* X_{\theta(r)}(r) B_{\theta(r)}(r) \Gamma_{\theta(r)}(r) \big(D^*_{\theta(r)}(r) D_{\theta(r)}(r)\big) u(r), \end{aligned}$$

and, setting $y = B^*_{\theta(r)}(r) X_{\theta(r)}(r) x(r)$ and $w = (D^*_{\theta(r)}(r) D_{\theta(r)}(r)) u(r)$, it becomes

$$\begin{aligned} & w^* \Gamma_{\theta(r)}(r) w + y^* \Gamma_{\theta(r)}(r) y + w^* \Gamma_{\theta(r)}(r) y + y^* \Gamma_{\theta(r)}(r) w \\ & \quad = (y + w)^* \Gamma_{\theta(r)}(r)(y + w) \\ & \quad = \|y + w\|^2_{\Gamma_{\theta(r)}(r)} \\ & \quad = \big\| B^*_{\theta(r)}(r) X_{\theta(r)}(r) x(r) + \big(D^*_{\theta(r)}(r) D_{\theta(r)}(r)\big) u(r) \big\|^2_{\Gamma_{\theta(r)}(r)}. \end{aligned}$$

Substituting this into (4.24) yields

$$\begin{aligned} \mathcal{J}_{[s,T],\mathbf{L}}(\vartheta_s, u) &= x(s)^* X_{\theta(s)}(s) x(s) + E_{\vartheta_s}\bigg(\int_s^T \big[\big\| B^*_{\theta(r)}(r) X_{\theta(r)}(r) x(r) \\ & \quad + \big(D^*_{\theta(r)}(r) D_{\theta(r)}(r)\big) u(r) \big\|^2_{\Gamma_{\theta(r)}(r)}\big] \, dr\bigg), \end{aligned}$$

which completes the proof. □

The main result of the section reads as follows.

Theorem 4.9 *The optimal control in the admissible class $\mathcal{U}^T$ is given by*

$$\hat{u}(t) = -K_{\theta(t)}(t) x(t), \quad (4.25)$$

where $\mathbf{K}(t) = (K_1(t), \ldots, K_N(t)) \in \mathbb{H}^{n,m}$ is given by

$$K_i(t) = \big(D_i^*(t) D_i(t)\big)^{-1} B_i^*(t) X_i(t)$$

with $\mathbf{X}(t) \in \mathbb{H}^{n+}$ satisfying (4.17). *Furthermore the minimum cost is given by*

$$\mathcal{J}_{[s,T],\mathbf{L}}(\vartheta_s, \hat{u}) = \min_{u \in \mathcal{U}^T} \mathcal{J}_{[s,T],\mathbf{L}}(\vartheta_s, u) = x(s)^* X_{\theta(s)}(s) x(s). \quad (4.26)$$

Proof First, we notice that $\bar{u}(t, x, i) = K_i(t) x$ is an admissible control policy (that is, satisfies (4.2) and (4.3)). The rest of the proof is immediate from (4.22). □

4.5 The Infinite-Horizon Optimal Control Problem

Let us now turn our attention to the infinite-horizon optimal control problem. We recall that in this case we assume that the matrices A_i, B_i, C_i, and D_i are time-invariant. In this section it is convenient to highlight the dependence on the final time T of the solution of (4.17) and write it as $\mathbf{X}^T(t) = (X_1^T(t), \dots, X_N^T(t))$. It is also convenient to define the nonlinear operator $\mathcal{R} : \mathbb{H}^n \to \mathbb{H}^n$ in the following way: for any $\mathbf{S} = (S_1, \dots, S_N) \in \mathbb{H}^n$,

$$\mathcal{R}_i(\mathbf{S}) = A_i^* S_i + S_i A_i - S_i B_i \left(D_i^* D_i\right)^{-1} B_i^* S_i + \sum_{j=1}^{N} \lambda_{ij} S_j + C_i^* C_i. \tag{4.27}$$

Since in this case there is no fixed "time horizon," we expect, paralleling with the classical LQ problem, that the minimum cost (see (4.26)) should not depend on the starting time s whenever we preserve the same initial condition ϑ_s. This suggests that we should find a constant function $\mathbf{X}^T(t) = \mathbf{X}$ for all $t \in [0, T]$ satisfying (4.17) with terminal condition $\mathbf{X}^T(T) = \mathbf{X}$, a "matched" solution for the finite-horizon problem in the sense that the controller is indifferent between paying the terminal cost $x(T)^* X_{\theta(T)} x(T)$ or continuing optimally indefinitely. We have that $\mathbf{X}^T$ satisfies (4.17) with terminal condition $\mathbf{X}^T(T) = \mathbf{X}$ if and only if $\mathbf{X}$ satisfies the CARE

$$\mathcal{R}(\mathbf{X}) = 0. \tag{4.28}$$

From this we should expect that the solution of our infinite-time problem would be related to the solution of the CARE (4.28). The following definitions are related to the solutions of the CARE (4.28).

Definition 4.10 We say that $\mathbf{X} = (X_1, \dots, X_N) \in \mathbb{H}^{n+}$ is a maximal solution of (4.28) if $\mathcal{R}(\mathbf{X}) = 0$ and, for any $\mathbf{V} = (V_1, \dots, V_N) \in \mathbb{H}^{n*}$ such that $\mathcal{R}(\mathbf{V}) \geq 0$, we have that $\mathbf{X} \geq \mathbf{V}$.

In the sequel we recall the definitions of the set $\mathbb{K}$, the system $(\mathbf{A}, \mathbf{B}, \Pi)$ being mean-square stabilizable (SS) and the system $(\mathbf{C}, \mathbf{A}, \Pi)$ being mean-square detectable (SD) from Definitions 3.34 and 3.35.

Definition 4.11 We say that $\mathbf{X} = (X_1, \dots, X_N) \in \mathbb{H}^n$ is a positive semi-definite solution to the CARE if $\mathbf{X} \in \mathbb{H}^{n+}$ and $\mathcal{R}(\mathbf{X}) = 0$. Furthermore, $\mathbf{X}$ is the mean-square stabilizing solution of (4.28) if $\mathcal{R}(\mathbf{X}) = 0$, and for $\mathbf{K} = (K_1, \dots, K_N) \in \mathbb{H}^{n,m}$ defined as $K_i = (D_i^* D_i)^{-1} B_i^* X_i$, we have that $\mathbf{K} \in \mathbb{K}$.

The next theorem provides conditions for the existence and uniqueness in $\mathbb{H}^{n+}$ of a solution of (4.28) and, moreover, conditions for the existence of the mean-square stabilizing solution.

Theorem 4.12 *Suppose that* $(\mathbf{A}, \mathbf{B}, \Pi)$ *is SS. Then there exists the maximal solution* $\mathbf{X} = (X_1, \ldots, X_N) \in \mathbb{H}^{n+}$ *of* (4.28). *Suppose, in addition, that* $(\mathbf{C}, \mathbf{A}, \Pi)$ *is SD. Then* $\mathbf{X}$ *is the mean-square stabilizing solution and the unique positive semi-definite solution to* (4.28). *Moreover, for arbitrary terminal condition* $\mathbf{L} \in \mathbb{H}^{n+}$, *we have for the unique solution* $\mathbf{X}^T(t) \in \mathbb{H}^{n+}$, $t \in [0, T]$, *to* (4.17) *that* $\mathbf{X}^T(0) \to \mathbf{X}$ *as* $T \to \infty$.

Proof The existence of the maximal solution follows from Theorem A.5 in Appendix A, Sect. A.6. That this solution is mean-square stabilizing and the unique positive semi-definite solution to (4.28) follows from Corollary A.9 in Appendix A, Sect. A.6. Finally the convergence of $\mathbf{X}^T(0) \to \mathbf{X}$ as $T \to \infty$ follows from Corollary A.18 in Appendix A, Sect. A.6. □

The next theorem derives the optimal solution for the infinite-horizon problem posed in Sect. 4.2 and is the main result of this section. It says that the optimal control policy $\hat{u}$ is given by $\hat{u}(t) = -K_{\theta(t)}x(t)$ with $\mathbf{K} = (K_1, \ldots, K_N) \in \mathbb{H}^{n,m}$, $K_i = (D_i^* D_i)^{-1} B_i^* X_i$, and yields the cost $\mathcal{J}(\hat{u}) = \inf_{u \in \mathcal{U}} \mathcal{J}(u) = E[x(0)^* X_{\theta(0)} x(0)]$.

Theorem 4.13 *Suppose that* $(\mathbf{A}, \mathbf{B}, \Pi)$ *is SS and that* $(\mathbf{C}, \mathbf{A}, \Pi)$ *is SD. Let* $\mathbf{X} = (X_1, \ldots, X_N)$ *be the mean-square stabilizing solution to the CARE* (4.28). *Then*

$$\begin{aligned}\inf_{u \in \mathcal{U}} \mathcal{J}(u) &= \mathcal{J}(\hat{u}) \\ &= E\left(\int_0^{\infty} \left(\left\|C_{\theta(t)}x(t)\right\|^2 + \left\|D_{\theta(t)}\hat{u}(t)\right\|^2\right) dt\right) \\ &= E\left(x(0)^* X_{\theta(0)} x(0)\right),\end{aligned} \tag{4.29}$$

where $\hat{u}(t) = -K_{\theta(t)}x(t)$ *with* $\mathbf{K} = (K_1, \ldots, K_N) \in \mathbb{H}^{n,m}$ *such that* $K_i = (D_i^* D_i)^{-1} \times B_i^* X_i$, *and* $x(t)$ *is given by* (4.1) *with the control strategy* $\hat{u}$.

Proof First notice that $\hat{u}$, as defined above, stabilizes $(\mathbf{A}, \mathbf{B}, \Pi)$. Thus, $\hat{u} \in \mathcal{U}$, so that $\mathcal{U}$ is nonempty. Let us pick an arbitrary control strategy $u \in \mathcal{U}$ and consider the finite-time horizon case with an arbitrary T, the matched cost termination $\mathbf{X}^T(T) = \mathbf{L} = \mathbf{X}$, and the control policy u^T such that for $t \in [0, T]$, $u^T(t) = u(t) \in \mathbb{R}^m$. Clearly, $x^T(t) = x(t)$ for $t \in [0, T]$, where $x^T(t)$ satisfies system (4.1) under the control law u^T, $x(t)$ satisfies (4.1) under the control law u, and the same initial data ϑ_0 stands for both cases. Now, as a consequence of having $\mathbf{X}^T(T) = \mathbf{L} = \mathbf{X}$, the solution to the Riccati equation (4.17) is $\mathbf{X}^T(t) = \mathbf{X}$, $t \in [0, T]$, and, consequently, $X^T_{\theta(0)}(0) = X_{\theta(0)}$. Thus, from the cost defined in (4.4) and Proposition 4.8 we have that

$$\begin{aligned}\mathcal{J}_{T,\mathbf{X}}\left(u^T\right) &= E\left(\int_0^{T} \left(\left\|C_{\theta(t)}x(t)\right\|^2 + \left\|D_{\theta(t)}u(t)\right\|^2\right) dt\right) + E\left(x(T)^* X_{\theta(T)} x(T)\right) \\ &= E\left(x(0)^* X_{\theta(0)} x(0)\right)\end{aligned}$$

$$+E\left(\int_0^T \left\| B^*_{\theta(r)} X_{\theta(r)} x(r) + \left(D^*_{\theta(r)} D_{\theta(r)}\right) u(r) \right\|^2_{\Gamma_{\theta(r)}} dr\right). \tag{4.30}$$

Now, since $u \in \mathcal{U}$, we have from condition (C3) in Sect. 4.2 that

$$0 \le E(x(T)^* X_{\theta(T)} x(T)) \le \|\mathbf{X}\|_{\max} E\left(\|x(T)\|^2\right) \to 0$$

as $T \to \infty$. Thus, taking the limit as $T \to \infty$ in (4.30), by the definition in (4.6) we get that

$$\begin{aligned}
\mathcal{J}(u) &= \lim_{T\to\infty} \mathcal{J}_{T,\mathbf{X}}\left(u^T\right) \\
&= E\left(\int_0^\infty \left(\|C_{\theta(t)} x(t)\|^2 + \|D_{\theta(t)} u(t)\|^2\right) dt\right) \\
&= E\left(x(0)^* X_{\theta(0)} x(0)\right) \\
&\quad + E\left(\int_0^\infty \left\| B^*_{\theta(r)} X_{\theta(r)} x(r) + \left(D^*_{\theta(r)} D_{\theta(r)}\right) u(r) \right\|^2_{\Gamma_{\theta(r)}} dr\right)
\end{aligned} \tag{4.31}$$

for arbitrary $u \in \mathcal{U}$. Hence, bearing in mind that $\mathbf{X}$ is the mean-square stabilizing solution to the CARE (4.28), it follows that $\widehat{u} \in \mathcal{U}$, and thus the minimum of (4.31) over $u \in \mathcal{U}$ is achieved with $\hat{u}$. For this choice, the second term on the right-hand side of (4.31) is zero (recall that $\hat{u}(t) = -(D^*_{\theta(t)} D_{\theta(t)})^{-1} B^*_{\theta(t)} X_{\theta(t)} x(t))$. Thus, from this (4.29) follows. □

Remark 4.14 It has been shown in Proposition A.7 that there exists at most one mean-square stabilizing solution to the CARE (4.28). It is interesting to notice that the proof of Theorem 4.13 provides another way of showing this, from the optimal control point of view. Indeed, let us suppose that there exists a stabilizing solution $\mathbf{V} \neq \mathbf{X}$ to the CARE. As in the case of the stabilizing solution $\mathbf{X}$, we arrive at the conclusion that $E[x(0)^* V_{\theta(0)} x(0)]$ is the minimum of $\mathcal{J}(u)$ over $u \in \mathcal{U}$. But the minimum clearly does not depend on $\mathbf{X}$ and $\mathbf{V}$, so it follows that $E[x(0)^* X_{\theta(0)} x(0)] = E[x(0)^* V_{\theta(0)} x(0)]$ for any initial condition ϑ_0. In particular, for $x(0) = x$ and $\theta(0) = i$, x and i deterministic and arbitrary in $\mathbb{R}^n$ and $\mathcal{S}$, respectively, the above equation reduces to $x^* X_i x = x^* V_i x$. Since X_i and V_i are Hermitian for every $i \in \mathcal{S}$, we must have that $\mathbf{X} = \mathbf{V}$.

4.6 Historical Remarks

From a mathematical point of view, control theory initiates with the rigorous analysis carried out by J.C. Maxwell on the Watt's centrifugal governor. Since then, control engineering has played a fundamental role in the development of modern society. Modern optimal control theory has its roots on works of researchers such as R.E. Bellman, L.S. Pontryagin, and R.E. Kalman. The importance of taking disturbance into account has given rise to a great variety of topics, including stochastic control. In this scenario, we mention [23, 138, 142, 203, 303] as seminal papers in the building of the stochastic theory.

Regarding quadratic optimal control for MJLS, we mention [270] and [303] as fundamental papers that stimulated much research in this scenario. In the first, the jump linear quadratic (JLQ) control problem is considered only in the finite-horizon setting, and a stochastic maximum principle approach is used (see also [269]). In the other one, dynamic programming is used, and the infinite-horizon case is also treated. Although the objective had been carried out successfully, it seemed clear, prima facie, that the stability criteria used in [303] were not fully adequate. The inchoate idea in [303] was to consider the class above as a "natural" extension of the linear class and use as stability criteria the stability for each operation mode of the system plus a certain restrictive assumption which allows one to use fixed-point-type arguments to treat the coupled Riccati equations. In fact, the Riccati equation results used in [303] come from the seminal paper [302]. The restrictive assumption in [302] is removed in [156], which, to the best of the authors' knowledge, seems to be the first work to introduce the idea of the solution of the Riccati equation in the *mean-square sense*. Without any intention of being exhaustive, we mention, for instance, [6, 33, 59, 60, 63, 74, 94, 144, 176, 188, 191, 223, 229, 270, 303] as a representative sample of some works dealing with optimal control of MJLS (see also [93]).

The case with multiplicative noise is studied, for instance, in [121] and [88]. Some issues regarding the associated *coupled Riccati equation*, including numerical algorithms, can be found in [5, 66, 67, 72, 86, 103, 109, 110, 156, 246] (see also [121]). Quadratic optimal control problems of MJLS subject to *constraints on the state and control variables* are considered in [69]. An iterative *Monte Carlo technique* for deriving the optimal control of the infinite-horizon linear regulator problem of MJLS for the case in which the transition probability matrix of the Markov chain is not known is analyzed in [70]. The case in which the weighting matrices of the state and control for the quadratic part are allowed to be *indefinite* is considered, for instance, in [72]. The case in which the state space of the Markov chain is *infinite* countable was treated, for instance, in [78, 146, 292] (see also [16, 17, 149] for some issues regarding the associated Riccati equation). Although the avenues of research to a risk sensitivity approach for the optimal control of MJLS seem to be fascinating, this is a topic which has defied the researchers up to now.

For those interested in a fairly comprehensive introduction to the linear theory, we refer, for instance, to [97] and [100] (see also the classical [13] and [140]). For recent advances in the control theory, we refer to [295, 311] and references therein.

This chapter is, essentially, based on [146] (see also [159]).

Chapter 5
H_2 Optimal Control with Complete Observations

5.1 Outline of the Chapter

The purpose of this chapter is to revisit the infinite-horizon quadratic optimal control for continuous-time MJLS, studied in Chap. 4, but now from another point of view, usually known in the literature of linear systems as H_2 control. We assume here that the state variable $x(t)$ and the jump parameter $\theta(t)$ are available to the controller, so that is why we call this case "with complete observations." The case in which only the output $y(t)$ and jump parameter $\theta(t)$ are known will be considered in Chap. 6 and referred to as the incomplete observations case. The advantage of the H_2 approach is that it allows us to consider parametric uncertainties and solve the problem using LMIs optimization tools (see Chap. 2). In this chapter we suppose that the transition rate matrix Π may be subject to polytopic uncertainties. The organization of the chapter is as follows. In Sect. 5.2, we present the definitions of robust and quadratic mean-square stabilizability, bearing in mind the uncertainties on the transition rate matrix Π. The controllability, observability Gramians, and the H_2-norm for MJLS are introduced in Sect. 5.3. In this section, we also obtain a result showing how the H_2-norm can be derived from the controllability and observability Gramians. In Sect. 5.4, the H_2 control problem for both cases, with and without uncertainties on Π, is solved using LMIs optimization tools. For the case in which there are no uncertainties, in Sect. 5.5 we shall see that the H_2 formulation and the infinite-horizon quadratic cost formulation, analyzed in Chap. 4, coincide. A numerical example illustrating the technique presented here can be found in Sect. 10.2.

5.2 Robust and Quadratic Mean-Square Stabilizability

Consider $\mathbf{A} = (A_1, \dots, A_N) \in \mathbb{H}^n$, $\mathbf{B} = (B_1, \dots, B_N) \in \mathbb{H}^{m,n}$, and the following MJLS:

$$\mathcal{G} = \begin{cases} \dot{x}(t) = A_{\theta(t)}x(t) + B_{\theta(t)}u(t), \\ \vartheta_0 = (x_0, \theta_0), \end{cases} \tag{5.1}$$

O.L.V. Costa et al., *Continuous-Time Markov Jump Linear Systems*,
Probability and Its Applications, DOI 10.1007/978-3-642-34100-7_5,

where, as before, $\{\theta(t); t \geq 0\}$ is a continuous-time Markov chain with finite state space $\mathcal{S}$ and transition rate matrix $\Pi = [\lambda_{ij}]$. In this chapter, we will consider that the transition matrix Π will be either exactly known or belong to a convex set $\mathbb{V}$, where

$$\mathbb{V} = \left\{ \Pi; \Pi = \sum_{\kappa=1}^{\ell} \rho^{\kappa} \Pi^{\kappa}; \rho^{\kappa} \geq 0, \sum_{\kappa=1}^{\ell} \rho^{\kappa} = 1 \right\}, \tag{5.2}$$

and $\Pi^{\kappa} = [\lambda_{ij}^{\kappa}]$, $\kappa = 1, \ldots, \ell$, is a set of known transition rate matrices.

For $\mathbf{K} = (K_1, \ldots, K_N) \in \mathbb{H}^{n,m}$, consider in (5.1) $u(t) = -K_{\theta(t)}x(t)$ and $\widetilde{A}_i = A_i - B_i K_i$, $i \in \mathcal{S}$, so that (5.1) can be rewritten as $\dot{x}(t) = \widetilde{A}_{\theta(t)}x(t)$. As in Chap. 3, define $\mathbf{Q}(t) = (Q_1(t), \ldots, Q_N(t))$, where

$$Q_j(t) := E\big(x(t)x(t)^* 1_{\{\theta(t)=j\}}\big) \geq 0, \tag{5.3}$$

and consider the operators $\mathcal{L}$ and $\mathcal{T} = \mathcal{L}^*$ as in (3.21), with A_i replaced by $\widetilde{A}_i$. We recall that $\mathcal{L}$ and $\mathcal{T}$ possess the same eigenvalues and, by Proposition 3.7, that for every $t \in \mathbb{R}^+$,

$$\dot{\mathbf{Q}}(t) = \mathcal{L}\big(\mathbf{Q}(t)\big). \tag{5.4}$$

We recall from Definition 3.34 that

$$\mathbb{K} = \big\{\mathbf{K} \in \mathbb{H}^{n,m};\ \mathbf{K} \text{ stabilizes } (\mathbf{A}, \mathbf{B}, \Pi) \text{ in the mean-square sense}\big\}$$

and, for the case in which Π is not exactly known, introduce the following definitions.

Definition 5.1 We say that $(\mathbf{A}, \mathbf{B}, \Pi)$ is robustly mean-square stabilizable if there exists $\mathbf{K} = (K_1, \ldots, K_N) \in \mathbb{H}^{n,m}$ such that model (5.1) is MSS with $\widetilde{A}_i = A_i - B_i K_i$ for every $\Pi \in \mathbb{V}$. In this case we say that $\mathbf{K}$ stabilizes robustly $(\mathbf{A}, \mathbf{B}, \Pi)$ in the mean-square sense. We set $\mathbb{K}_{\mathrm{r}} = \{\mathbf{K} \in \mathbb{H}^{n,m}; \mathbf{K}$ stabilizes robustly $(\mathbf{A}, \mathbf{B}, \Pi)$ in the mean-square sense$\}$.

Definition 5.2 We say that $(\mathbf{A}, \mathbf{B}, \Pi)$ is quadratically mean-square stabilizable if there exist $\mathbf{K} = (K_1, \ldots, K_N) \in \mathbb{H}^{n,m}$ and $\mathbf{P} = (P_1, \ldots, P_N) > 0$, $\mathbf{P} \in \mathbb{H}^{n+}$, such that

$$(A_j - B_j K_j)P_j + P_j(A_j - B_j K_j)^* + \sum_{i \in \mathcal{S}} \lambda_{ij} P_i < 0, \quad j \in \mathcal{S}, \tag{5.5}$$

for every $\Pi = [\lambda_{ij}] \in \mathbb{V}$. In this case we say that $\mathbf{K}$ stabilizes quadratically $(\mathbf{A}, \mathbf{B}, \Pi)$ in the mean-square sense. We set $\mathbb{K}_q = \{\mathbf{K} \in \mathbb{H}^{n,m}; \mathbf{K}$ stabilizes quadratically $(\mathbf{A}, \mathbf{B}, \Pi)$ in the mean-square sense$\}$.

From Theorem 3.25 it is clear that $\mathbb{K}_q \subset \mathbb{K}_{\mathrm{r}}$. Note also that from Definition 5.2 it follows that (5.5) is satisfied for every $\Pi = [\lambda_{ij}] \in \mathbb{V}$ if and only if for each

$\kappa = 1, \dots, \ell$, we have that

$$(A_j - B_j K_j)P_j + P_j(A_j - B_j K_j)^* + \sum_{i\in\mathcal{S}} \lambda_{ij}^{\kappa} P_i < 0, \quad j \in \mathcal{S}. \tag{5.6}$$

5.3 Controllability, Observability Gramians, and the H_2-Norm

Consider the following system:

$$\mathcal{G} = \begin{cases} \dot{x}(t) = \widetilde{A}_{\theta(t)} x(t) + J_{\theta(t)} w(t), \\ z(t) = \widetilde{C}_{\theta(t)} x(t), \\ x(0) = 0, \qquad \theta(0) = \theta_0, \end{cases} \tag{5.7}$$

where $\widetilde{\mathbf{A}} = (\widetilde{A}_1, \dots, \widetilde{A}_N) \in \mathbb{H}^n$, $\widetilde{\mathbf{C}} = (\widetilde{C}_1, \dots, \widetilde{C}_N) \in \mathbb{H}^{n,p}$, and $\mathbf{J} = (J_1, \dots, J_N) \in \mathbb{H}^{r,n}$ with $J_i J_i^* > 0$ for each $i \in \mathcal{S}$. The initial distribution for θ_0 is given by $\nu = \{\nu_i; i \in \mathcal{S}\}$, satisfying $\nu_i > 0$, $i \in \mathcal{S}$, $\sum_{i\in\mathcal{S}} \nu_i = 1$.

If $\mathrm{Re}\{\lambda(\mathcal{L})\} < 0$ (that is, model (5.1) is MSS) and $w = \{w(t); t \geq 0\}$ is an impulse input (that is, $w(t) = v\delta(t)$ where v is an r-dimensional vector, and $\delta(t)$ the unitary impulse), then, as a consequence of Theorem 3.15(v), there exist $b > 0$ and $a > 0$ such that for each $t \in \mathbb{R}^+$,

$$E\big(\|z(t)\|^2\big) \leq a e^{-bt} E\big(\|x_0\|^2\big), \tag{5.8}$$

and clearly we have that $z = \{z(t); t \geq 0\} \in L_2^p(\Omega, \mathcal{F}, P)$. The next definition is a generalization of the H_2-norm from continuous-time deterministic systems to the stochastic Markovian jump case.

Definition 5.3 We define the H_2-norm of the system $\mathcal{G}$ as

$$\|\mathcal{G}\|_2^2 = \sum_{s=1}^{r} \sum_{j\in\mathcal{S}} \nu_j \|z_{s,j}\|_2^2,$$

where $z_{s,j}$ represents the output $\{z(t);\ t \geq 0\}$ given by (5.7) when

(a) the input is given by $w = \{w(t); t \geq 0\}$, $w(t) = e_s\delta(t)$, $\delta(t)$ the unitary impulse, and e_s the r-dimensional unitary vector formed by 1 at the sth position and zero elsewhere, and
(b) $\theta_0 = j$.

For the deterministic case ($N = 1$), the above definition reduces to the usual H_2-norm. At the end of this section we also present a connection between the above definition and the expected value of the stationary variance of the output when the input is the unit-variance white noise.

We will next show that the H_2-norm as defined above can be calculated via the solution of the continuous-time coupled observability and controllability Gramians, a result that mirrors its deterministic counterpart. For this, define $\mathbf{M} = (\widetilde{C}_1^*\widetilde{C}_1, \ldots, \widetilde{C}_N^*\widetilde{C}_N) \in \mathbb{H}^{n+}$, $\nu\mathbf{N} = (\nu_1 J_1 J_1^*, \ldots, \nu_N J_N J_N^*) \in \mathbb{H}^{n+}$, $\mathbf{S} = (S_1, \ldots, S_N) \in \mathbb{H}^{n+}$, and $\mathbf{P} = (P_1, \ldots, P_N) \in \mathbb{H}^{n+}$ as the unique solution of the equations (see Theorem 3.25)

$$\mathcal{T}(\mathbf{S}) + \mathbf{M} = 0 \quad \text{(observability Gramian)},$$

$$\mathcal{L}(\mathbf{P}) + \nu\mathbf{N} = 0 \quad \text{(controllability Gramian)}.$$

From Theorem 3.25 and $\nu\mathbf{N} > 0$ it follows that $\mathbf{P} > 0$. We have the following result, establishing a connection between the H_2-norm with the observability and controllability Gramians.

Theorem 5.4 $\|\mathcal{G}\|_2^2 = \sum_{j\in\mathcal{S}} \nu_j \operatorname{tr}(J_j^* S_j J_j) = \sum_{j\in\mathcal{S}} \operatorname{tr}(\widetilde{C}_j P_j \widetilde{C}_j^*)$.

Proof Let us show the first equality. For $i \in \mathcal{S}$,

$$\mathcal{T}_i(\mathbf{S}) + \widetilde{C}_i^*\widetilde{C}_i = \widetilde{A}_i^* S_i + S_i \widetilde{A}_i + \sum_{j\in\mathcal{S}} \lambda_{ij} S_j + \widetilde{C}_i^*\widetilde{C}_i = 0.$$

Consider $z = \{z(t); t \geq 0\}$, an impulse response of (5.7). Then, from (5.3) and Proposition 5.4, recalling that $\mathcal{T}^* = \mathcal{L}$, we get

$$\begin{aligned}
E\big(z(t)^* z(t)\big) &= E\big(x(t)^* \widetilde{C}_{\theta(t)}^* \widetilde{C}_{\theta(t)} x(t)\big) = -E\big(x(t)^* \mathcal{T}_{\theta(t)}(\mathbf{S}) x(t)\big) \\
&= -\sum_{i\in\mathcal{S}} E\big(x(t)^* \mathcal{T}_i(\mathbf{S}) x(t) 1_{\{\theta(t)=i\}}\big) \\
&= -\sum_{i\in\mathcal{S}} \operatorname{tr}\big(E\big(x(t)x(t)^* 1_{\{\theta(t)=i\}}\big) \mathcal{T}_i(\mathbf{S})\big) \\
&= -\sum_{i\in\mathcal{S}} \operatorname{tr}\big(Q_i(t) \mathcal{T}_i(\mathbf{S})\big) \\
&= -\langle \mathbf{Q}(t); \mathcal{T}(\mathbf{S})\rangle \\
&= -\langle \mathcal{L}(\mathbf{Q}(t)); \mathbf{S}\rangle \\
&= -\langle \dot{\mathbf{Q}}(t); \mathbf{S}\rangle.
\end{aligned}$$

Taking the integral over t from 0 to ∞ and recalling that $\mathbf{Q}(t) \to 0$ as $t \to \infty$ (since the system is MSS) and that $\theta(0) = j$ and $x(0) = J_j e_s$, we get that

$$\begin{aligned}
\|z_{s,j}\|_2^2 &= \int_0^\infty E\big(\|z(t)\|^2\big)\, dt \\
&= -\int_0^\infty \langle \dot{\mathbf{Q}}(t); \mathbf{S}\rangle\, dt
\end{aligned}$$

$$
\begin{aligned}
&= -\langle \mathbf{Q}(t); \mathbf{S}\rangle\big|_0^{\infty} \\
&= \langle \mathbf{Q}(0); \mathbf{S}\rangle \\
&= \sum_{i\in\mathcal{S}} \operatorname{tr}\bigl(Q_i(0)S_i\bigr) \\
&= e_s^* J_j^* S_j J_j e_s.
\end{aligned}
$$

Therefore,

$$
\|\mathcal{G}\|_2^2 = \sum_{s=1}^{r}\sum_{j\in\mathcal{S}} \nu_j \|z_{s,j}\|_2^2 = \sum_{s=1}^{r}\sum_{j\in\mathcal{S}} \nu_j e_s^* J_j^* S_j J_j e_s = \sum_{j\in\mathcal{S}} \nu_j \operatorname{tr}\bigl(J_j^* S_j J_j\bigr),
$$

proving the first equality. For the second equality, we have that

$$
\begin{aligned}
\|\mathcal{G}\|_2^2 &= \sum_{j=1}^{N} \nu_j \operatorname{tr}\bigl(J_j^* S_j J_j\bigr) \\
&= \sum_{j=1}^{N} \operatorname{tr}\bigl(\nu_j J_j^* S_j J_j + P_j\bigl(\widetilde{C}_j^* \widetilde{C}_j + \mathcal{T}_j(\mathbf{S})\bigr)\bigr) \\
&= \sum_{j=1}^{N} \operatorname{tr}\bigl(\nu_j J_j J_j^* S_j + P_j \widetilde{C}_j^* \widetilde{C}_j\bigr) + \bigl\langle \mathbf{P}; \mathcal{T}(\mathbf{S})\bigr\rangle \\
&= \langle \mathbf{P}; \mathbf{M}\rangle + \langle \nu\mathbf{N}; \mathbf{S}\rangle + \bigl\langle \mathcal{L}(\mathbf{P}); \mathbf{S}\bigr\rangle \\
&= \langle \mathbf{P}; \mathbf{M}\rangle + \bigl\langle \bigl(\mathcal{L}(\mathbf{P}) + \nu\mathbf{N}\bigr); \mathbf{P}\bigr\rangle \\
&= \langle \mathbf{P}; \mathbf{M}\rangle = \sum_{j=1}^{N} \operatorname{tr}\bigl(\widetilde{C}_j P_j \widetilde{C}_j^*\bigr),
\end{aligned}
$$

completing the proof of the theorem. □

If the Markov process $\{\theta(t); t \geq 0\}$ is ergodic, then (see (3.4)) $P(\theta(t) = i) \to \pi_i$ as $t \to \infty$ exponentially fast for some $\pi_i > 0$, $i \in \mathcal{S}$, $\sum_{i\in\mathcal{S}} \pi_i = 1$. In this case, if $\nu_i = \pi_i$ above, we get that, like in the case with no jumps, the definition of the H_2-norm coincides with the expected value of the stationary variance of the output when the input is the unit-variance white noise. Indeed, consider the stochastic differential equation

$$
dx(t) = \widetilde{A}_{\theta(t)} x(t)\,dt + J_{\theta(t)}\,dw(t),
$$

where $W = \{(w(t), \mathcal{F}_t), t \in \mathbb{R}^+\}$ is a Wiener process with incremental covariance equal to the identity. We recall from Proposition 3.28 that for every $t \in \mathbb{R}^+$,

$$
\dot{\mathbf{Q}}(t) = \mathcal{L}\bigl(\mathbf{Q}(t)\bigr) + \mathbf{R}(t),
$$

where $\mathbf{R}(t) = (R_1(t), \ldots, R_N(t))$, with $R_i(t) = J_i J_i^* p_i(t)$, and $p_i(t) = P(\theta(t) = i)$. Moreover (see Proposition 3.29), $\mathbf{Q}(t) \to \mathbf{P}$ as $t \to \infty$, and

$$\lim_{t\to\infty} E\left(\|z(t)\|^2\right) = \lim_{t\to\infty} \sum_{i\in\mathcal{S}} \operatorname{tr}\left(\widetilde{C}_i Q_i(t) \widetilde{C}_i^*\right) = \sum_{i\in\mathcal{S}} \operatorname{tr}\left(\widetilde{C}_i P_i \widetilde{C}_i^*\right) = \|\mathcal{G}\|_2^2.$$

However, as it will be seen in the next sections, for the H_2-optimal control problem with Π exactly known, the optimal controller will not depend on $\mathbf{J} = (J_1, \ldots, J_N)$. In particular, we have that the choice of $\nu = \{\nu_i; i \in \mathcal{S}\}$ in the definition of the H_2-norm will not affect the optimal solution of the problem, as long as $\nu_i > 0$ for all $i \in \mathcal{S}$ (see Remark 5.12 below).

5.4 H_2 Control via Convex Analysis

5.4.1 Preliminaries

Consider the jump controlled system

$$\mathcal{G} = \begin{cases} \dot{x}(t) = A_{\theta(t)}x(t) + B_{\theta(t)}u(t) + J_{\theta(t)}w(t), \\ z(t) = C_{\theta(t)}x(t) + D_{\theta(t)}u(t), \\ x(0) = 0, \qquad \theta(0) = \theta_0, \end{cases} \tag{5.9}$$

where $\mathbf{A} = (A_1, \ldots, A_N) \in \mathbb{H}^n$, $\mathbf{B} = (B_1, \ldots, B_N) \in \mathbb{H}^{m,n}$, $\mathbf{J} = (J_1, \ldots, J_N) \in \mathbb{H}^{r,n}$, $\mathbf{C} = (C_1, \ldots, C_N) \in \mathbb{H}^{n,p}$, $\mathbf{D} = (D_1, \ldots, D_N) \in \mathbb{H}^{m,p}$, with $D_i^* D_i > 0$ and $C_i^* D_i = 0$ for each $i \in \mathcal{S}$.

For $\mathbf{K} = (K_1, \ldots, K_N) \in \mathbb{K}$, set $\mathcal{G}_{\mathbf{K}}$ as system (5.1) with $u(t) = -K_{\theta(t)}x(t)$. The H_2-optimal control problem for Π exactly known is defined as

$$\min\left\{\|\mathcal{G}_{\mathbf{K}}\|_2;\ \mathbf{K} \in \mathbb{K}\right\}, \tag{5.10}$$

whereas the H_2-guaranteed cost control for continuous-time Markovian jump linear systems with uncertain transition rate matrix Π is defined as follows.

Definition 5.5 Find $\mathbf{K} \in \mathbb{K}_\mathrm{r}$ and $\beta < \infty$ such that

$$\|\mathcal{G}_{\mathbf{K}}^{\Pi}\|_2^2 \le \beta \quad \text{for every } \Pi \in \mathbb{V}, \tag{5.11}$$

where $\|\mathcal{G}_{\mathbf{K}}^{\Pi}\|_2$ represents the H_2-norm of system (5.9) with feedback control law defined by $\mathbf{K} \in \mathbb{K}_\mathrm{r}$ and transition probability matrix given by $\Pi \in \mathbb{V}$ (see (5.2) for the definition of $\mathbb{V}$).

5.4.2 Π Exactly Known

Define $\mathbf{U} = (U_1, \ldots, U_N) \in \mathbb{H}^{(n+m)+}$ and $\mathbf{W} = (W_1, \ldots, W_N) \in \mathbb{H}^{(n+m)+}$ as

$$U_i = \begin{bmatrix} C_i^* C_i & 0 \\ 0 & D_i^* D_i \end{bmatrix}, \quad W_i = \begin{bmatrix} W_{i1} & W_{i2} \\ W_{i2}^* & W_{i3} \end{bmatrix}, \quad i \in \mathcal{S},$$

and set

$$\mathcal{C} = \{\mathbf{W} \geq 0;\ W_{j1} > 0, \Theta_j(\mathbf{W}) \leq 0, j \in \mathcal{S}\}, \tag{5.12}$$

where, for $j \in \mathcal{S}$,

$$\Theta_j(\mathbf{W}) = A_j W_{j1} + W_{j1} A_j^* - B_j W_{j2}^* - W_{j2} B_j^* + \sum_{i \in \mathcal{S}} \lambda_{ij} W_{i1} + \nu_j J_j J_j^*.$$

Clearly, $\mathcal{C}$ is a convex set. We have the following result.

Proposition 5.6 $\mathbb{K} \neq \emptyset$ *iff* $\mathcal{C} \neq \emptyset$. *Moreover,*

$$\mathbb{K} = \{(W_{12}^* W_{11}^{-1}, \ldots, W_{N2}^* W_{N1}^{-1});\ \mathbf{W} = (W_1, \ldots, W_N) \in \mathcal{C}\}.$$

Proof If $\mathbb{K} \neq \emptyset$, then we can find $\mathbf{K} = (K_1, \ldots, K_N) \in \mathbb{K}$ such that system (3.77) is MSS with $\widetilde{A}_i = A_i - B_i K_i$. By Theorem 3.21 there exists $\mathbf{P} = (P_1, \ldots, P_N) > 0$ in $\mathbb{H}^{n+}$ such that

$$(A_i - B_i K_i) P_i + P_i (A_i - B_i K_i)^* + \sum_{j \in \mathcal{S}} \lambda_{ji} P_j + \nu_i J_i J_i^* = 0. \tag{5.13}$$

Then by choosing $W_{i1} = P_i$, $W_{i2} = P_i K_i^*$, and $W_{i3} = K_i P_i^{-1} K_i^*$ we get from Lemma 2.26 that $\mathbf{W} \geq 0$, $W_{i1} > 0$, and from (5.13) that $\Theta_i(\mathbf{W}) \leq 0$, $i \in \mathcal{S}$, showing that $\mathcal{C} \neq \emptyset$ and that $K_i = W_{i2}^* W_{i1}^{-1}$. On the other hand, if $\mathcal{C} \neq \emptyset$, then, recalling that $\nu_j J_j J_j^* > 0$, we get from $\Theta_i(\mathbf{W}) \leq 0$ that

$$A_i W_{i1} + W_{i1} A_i^* - B_i W_{i2}^* - W_{i2} B_i^* + \sum_{j \in \mathcal{S}} \lambda_{ji} W_{j1} < 0,$$

and from Lemma 3.37(SS3) we get that $\mathbb{K} \neq \emptyset$. Moreover, with $K_i = W_{i2}^* W_{i1}^{-1}$ and $\mathbf{K} = (K_1, \ldots, K_N) \in \mathbb{H}^{n,m}$, we have that $\mathbf{K} \in \mathbb{K}$. □

Define the following convex optimization problem.

$$\mu := \min \left\{ \sum_{i \in \mathcal{S}} \operatorname{tr}(U_i W_i);\ \mathbf{W} = (W_1, \ldots, W_N) \in \mathcal{C} \right\}. \tag{5.14}$$

We have the following theorem, which parallels the results presented in [75].

Theorem 5.7 *The following assertions are equivalent*:

(a) *There exists* $\widehat{\mathbf{W}} = (\widehat{W}_1, \dots, \widehat{W}_N)$ *such that*

$$\widehat{\mathbf{W}} = \arg\min \left\{ \sum_{i \in \mathcal{S}} \mathrm{tr}(U_i W_i);\ \mathbf{W} = (W_1, \dots, W_N) \in \boldsymbol{\mathcal{C}} \right\}. \tag{5.15}$$

(b) *There exists* $\widehat{\mathbf{K}} = (\widehat{K}_1, \dots, \widehat{K}_N) \in \mathbb{K}$ *such that*

$$\|\mathcal{G}_{\widehat{\mathbf{K}}}\|_2^2 = \min\big\{ \|\mathcal{G}_{\mathbf{K}}\|_2^2; \mathbf{K} \in \mathbb{K} \big\}. \tag{5.16}$$

Moreover, if (a) *holds, then* (5.16) *holds with*

$$\widehat{K}_i = \widehat{W}_{i2}^* \widehat{W}_{i1}^{-1}, \quad i \in \mathcal{S}, \tag{5.17}$$

and

$$\mu = \sum_{i \in \mathcal{S}} \mathrm{tr}(U_i \widehat{W}_i) = \min\big\{ \|\mathcal{G}_{\mathbf{K}}\|_2^2;\ \mathbf{K} \in \mathbb{K} \big\}.$$

On the other hand, if (b) *holds for* $\widehat{\mathbf{K}} = (\widehat{K}_1, \dots, \widehat{K}_N) \in \mathbb{K}$, *then* (5.15) *is satisfied with*

$$\widehat{W}_i = \begin{bmatrix} \widehat{P}_i & \widehat{P}_i \widehat{K}_i^* \\ \widehat{K}_i \widehat{P}_i & \widehat{K}_i \widehat{P}_i \widehat{K}_i^* \end{bmatrix}, \quad i \in \mathcal{S}, \tag{5.18}$$

and $\widehat{\mathbf{W}} = (\widehat{W}_1, \dots, \widehat{W}_N)$, *where* $\widehat{\mathbf{P}} = (\widehat{P}_1, \dots, \widehat{P}_N) \in \mathbb{H}^{n+}$ *satisfies*

$$(A_j - B_j \widehat{K}_j)\widehat{P}_j + \widehat{P}_j (A_j - B_j \widehat{K}_j)^* + \sum_{i \in \mathcal{S}} \lambda_{ij} \widehat{P}_i + \nu_j J_j J_j^* = 0.$$

Proof For any $\mathbf{W} \in \boldsymbol{\mathcal{C}}$, $\mathrm{tr}(U_i W_i) = \mathrm{tr}(C_i W_{i1} C_i^*) + \mathrm{tr}(D_i W_{i3} D_i^*)$ and, from (5.12), $\mathcal{L}(\mathbf{Q}) + \nu \mathbf{N} \le 0$, where $\widetilde{A}_i = A_i - B_i K_i$, $K_i = W_{i2}^* W_{i1}^{-1}$, $\mathbf{Q} = (Q_1, \dots, Q_N)$, $Q_i = W_{i1} > 0$. Therefore, by Theorem 3.25, $W_{i1} \ge P_i$ for $i \in \mathcal{S}$, where $\mathbf{P} = (P_1, \dots, P_N) \in \mathbb{H}^{n+}$ satisfies $\mathcal{L}(\mathbf{P}) + \nu \mathbf{N} = 0$. By Theorem 5.4,

$$\begin{aligned} \|\mathcal{G}_{\mathbf{K}}\|_2^2 &= \sum_{j \in \mathcal{S}} \mathrm{tr}\big((C_j - D_j K_j) P_j (C_j - D_j K_j)^*\big) \\ &\le \sum_{j \in \mathcal{S}} \mathrm{tr}\big((C_j - D_j K_j) W_{j1} (C_j - D_j K_j)^*\big) \\ &\le \sum_{j \in \mathcal{S}} \mathrm{tr}\big(C_j W_{j1} C_j^* + D_j W_{j3} D_j^*\big) = \sum_{j \in \mathcal{S}} \mathrm{tr}(R_j W_j), \end{aligned}$$

showing that $\mu \ge \min\{\|\mathcal{G}_{\mathbf{K}}\|_2^2; \mathbf{K} \in \mathbb{K}\}$. On the other hand, for any $\mathbf{K} = (K_1, \dots, K_N) \in \mathbb{K}$, we have from Theorem 5.4 that

$$\|\mathcal{G}_{\mathbf{K}}\|_2^2 = \sum_{j \in \mathcal{S}} \mathrm{tr}\big((C_j - D_j K_j) P_j (C_j - D_j K_j)^*\big),$$

where $\mathbf{P} = (P_1, \dots, P_N) > 0$ satisfies $\mathcal{L}(\mathbf{P}) + \nu\mathbf{N} = 0$ with $\widetilde{A}_i = A_i - B_i K_i$. Defining $\mathbf{W}$ as $W_{i1} = P_i$, $W_{i2} = P_i K_i^*$, and $W_{i3} = W_{i2}^* W_{i1}^{-1} W_{i2}$, we get that $\mathbf{W} \in \mathcal{C}$ and

$$\begin{aligned}\sum_{j\in\mathcal{S}} \operatorname{tr}(R_j W_j) &= \sum_{j\in\mathcal{S}} \operatorname{tr}\bigl(C_j W_{j1} C_j^*\bigr) + \operatorname{tr}\bigl(D_j W_{j3} D_j^*\bigr) \\ &= \sum_{j\in\mathcal{S}} \operatorname{tr}\bigl((C_j - D_j K_j) P_{j1} (C_j - D_j K_j)^*\bigr) = \|\mathcal{G}_{\mathbf{K}}\|_2^2,\end{aligned}$$

that is, $\mu \le \min\{\|\mathcal{G}_{\mathbf{K}}\|_2^2; \mathbf{K} \in \mathbb{K}\}$. Thus, $\mu = \min\{\|\mathcal{G}_{\mathbf{K}}\|_2^2; \mathbf{K} \in \mathbb{K}\}$, and one solution can be recovered from the other according to (5.17) and (5.18). □

5.4.3 *Π Not Exactly Known*

We consider now the case in which Π is not exactly known but belongs to $\mathbb{V}$ (see (5.2)). Set for $\kappa = 1, \dots, \ell$,

$$\mathcal{C}^\kappa = \bigl\{\mathbf{W} \ge 0;\ W_{j1} > 0,\ \Theta_j^\kappa(\mathbf{W}) \le 0,\ j \in \mathcal{S}\bigr\},$$

where

$$\Theta_j^\kappa(\mathbf{W}) = A_j W_{j1} + W_{j1} A_j^* - B_j W_{j2}^* - W_{j2} B_j^* + \sum_{i\in\mathcal{S}} \lambda_{ij}^\kappa W_{i1} + \nu_j J_j J_j^*$$

for $j \in \mathcal{S}$, with $\Pi^\kappa = [\lambda_{ij}^\kappa]$. It is clear that $\mathcal{C}^\kappa$ is a convex set, and thus

$$\mathcal{C}_q := \bigcap_{\kappa=1}^{\ell} \mathcal{C}^\kappa$$

is also a convex set. The following result follows the same steps as the proof of Proposition 5.6 (see also [170]).

Proposition 5.8 $\mathbb{K}_q \ne \emptyset$ *iff* $\mathcal{C}_q \ne \emptyset$. *Moreover,*

$$\mathbb{K}_q = \bigl\{\bigl(W_{12}^* W_{11}^{-1}, \dots, W_{N2}^* W_{N1}^{-1}\bigr);\ \mathbf{W} = (W_1, \dots, W_N) \in \mathcal{C}_q\bigr\}.$$

Proof If $\mathbb{K}_q \ne \emptyset$, then from Definition 5.2 we can find $\mathbf{K} = (K_1, \dots, K_N) \in \mathbb{K}$ such that for some $\mathbf{P} = (P_1, \dots, P_N) > 0$, we have that for each $\kappa = 1, \dots, \ell$, (5.6) is satisfied. From (5.6) we can find a constant $c > 0$ such that, with $\bar{\mathbf{P}} = c\mathbf{P}$,

$$(A_j - B_j K_j)\bar{P}_j + \bar{P}_j (A_j - B_j K_j)^* + \sum_{i\in\mathcal{S}} \lambda_{ij}^\kappa \bar{P}_i + \nu_j J_j J_j^* < 0, \quad j \in \mathcal{S}. \quad (5.19)$$

Taking $W_{i1} = \bar{P}_i$, $W_{i2} = (K_i \bar{P}_i)^*$ and $W_{i3} = K_i \bar{P}_i^{-1} K_i^*$, we get that $\mathbf{W} \in \mathcal{C}_q$ with $K_i = W_{i2}^* W_{i1}^{-1}$. Similarly, if $\mathbf{W} \in \mathcal{C}_q$ then by choosing $\bar{P}_i = W_{i1}$ and $K_i = W_{i2}^* W_{i1}^{-1}$ we get that (5.19) is satisfied, and thus $\mathbf{K} = (K_1, \dots, K_N) \in \mathbb{K}$. □

We have the following corollary concerning problem (5.11).

Corollary 5.9 *Let* $\widehat{\mathbf{W}} = (\widehat{W}_1, \dots, \widehat{W}_N) = \arg\min\{\sum_{i\in\mathcal{S}} \operatorname{tr}(U_i W_i); \mathbf{W} = (W_1, \dots, W_N) \in \mathcal{C}_q\}$. *Then* $\widehat{K}_i = \widehat{W}_{i2}^* \widehat{W}_{i1}^{-1}$, $i \in \mathcal{S}$, *solves the* H_2*-guaranteed cost control problem with* $\beta = \sum_{i\in\mathcal{S}} \operatorname{tr}(U_i \widehat{W}_i)$.

Proof For any $\mathbf{W} \in \mathcal{C}_q$ and every $\Pi = [\lambda_{ij}] \in \mathbb{V}$ (see Theorem 5.7 and Proposition 5.8), $\operatorname{tr}(U_i W_i) = \operatorname{tr}(C_i W_{i1} C_i^*) + \operatorname{tr}(D_i W_{i3} D_i^*)$, and

$$\left(A_j - B_j W_{j2}^* W_{j1}^{-1}\right) W_{j1} + W_{j1}\left(A_j - B_j W_{j2}^* W_{j1}^{-1}\right)^* + \sum_{i\in\mathcal{S}} \lambda_{ij} W_{i1} + \nu_j J_j J_j^* \le 0.$$

Therefore, by Theorem 3.25, $W_{i1} \ge P_i$ for $i \in \mathcal{S}$, where $\mathbf{P} = (P_1, \dots, P_N) \in \mathbb{H}^{n+}$ satisfies

$$\left(A_j - B_j W_{j2}^* W_{j1}^{-1}\right) P_j + P_j\left(A_j - B_j W_{j2}^* W_{j1}^{-1}\right)^* + \sum_{i\in\mathcal{S}} \lambda_{ij} P_i + \nu_j J_j J_j^* = 0$$

for arbitrary $\Pi \in \mathbb{V}$. By Theorem 5.7, $\|\mathcal{G}_{\mathbf{K}}^{\Pi}\|_2^2 \le \sum_{j\in\mathcal{S}} \operatorname{tr}(R_j W_j)$, and since Π is arbitrary in $\mathbb{V}$, $\max_{\Pi\in\mathbb{V}} \|\mathcal{G}_{\mathbf{K}}^{\Pi}\|_2^2 \le \sum_{j\in\mathcal{S}} \operatorname{tr}(R_j W_j)$, showing that every $\mathbf{W} \in \mathcal{C}_q$ generates an upper bound to $\|\mathcal{G}_{\mathbf{K}}^{\Pi}\|_2^2$. Clearly, the smallest one is given by the optimal solution of the problem posed above, so that $\|\mathcal{G}_{\mathbf{K}}^{\Pi}\|_2^2 \le \beta$ for all $\Pi \in \mathbb{V}$. □

5.5 The Convex Approach and the CARE

This section establishes a link between the convex approach, based on the solution of problem (5.14) that corresponds to the H_2 control problem of an MJLS with Π exactly known and on the CARE (4.28). As seen in Chap. 4, this equation arises when one solves the infinite-horizon quadratic optimization problem of an MJLS

$$\dot{x}(t) = A_{\theta(t)} x(t) + B_{\theta(t)} u(t)$$

with minimization cost given by $\mathcal{J}(u)$ defined in (4.6). In what follows, we recall the definitions of maximal (Definition 4.10) and (mean-square) stabilizing (Definition 4.11) solutions for the CARE (4.28). We have the following result (see also Theorem 3.1 in [246]).

Proposition 5.10 *Suppose that* $(\mathbf{A}, \mathbf{B}, \Pi)$ *is SS. Then there exists a maximal solution* $\mathbf{V} = (V_1, \dots, V_N)$ *to* (4.28). *Moreover, if* $\mathbf{V}$ *is the mean-square stabilizing* *solution of* (4.28), *then* (4.28) *has a unique mean-square stabilizing solution. Furthermore, defining the following sequence of CARE,*

$$A_i^* X_i + X_i A_i - X_i B_i (D_i^* D_i)^{-1} B_i^* X_i + \sum_{j \in \mathcal{S}} \lambda_{ij} X_j + C_i^* C_i + \frac{I}{k} = 0, \tag{5.20}$$

$i \in \mathcal{S}$, *we have that there exists a mean-square stabilizing solution* $\mathbf{V}^k = (V_1^k, \ldots, V_N^k)$ *to* (5.20), $V_i^k \geq V_i^{k+1}$, *and* $\mathbf{V}^k \to \mathbf{V}$ *as* $k \to \infty$.

Proof The first part follows from Theorem A.5 and Remark A.10. From the hypothesis that $(\mathbf{A}, \mathbf{B}, \Pi)$ is SS and Theorem A.5 it is clear that (5.20) has a maximal solution, which we write as $\mathbf{V}^k = (V_1^k, \ldots, V_N^k)$. From the identity (A.12) and (5.20) we have that $\mathbf{V}^k$ satisfies the following set of Lyapunov equations

$$\begin{aligned} &\big(A_i - B_i \mathcal{K}_i(\mathbf{V}^k)\big)^* V_i^k + V_i^k \big(A_i - B_i \mathcal{K}_i(\mathbf{V}^k)\big) + \sum_{j \in \mathcal{S}} \lambda_{ij} V_j^k \\ &\quad + \mathcal{K}_i^*(\mathbf{V}^k) D_i^* D_i \mathcal{K}_i(\mathbf{V}^k) + C_i^* C_i + \frac{I}{k} = 0, \quad i \in \mathcal{S}, \end{aligned} \tag{5.21}$$

and by Theorem 3.21, $\mathrm{Re}\{\lambda(\mathcal{L}^k)\} < 0$, where for $\mathbf{P} \in \mathbb{H}^n$, $\mathcal{L}^k(\mathbf{P}) = (\mathcal{L}_1^k(\mathbf{P}), \ldots, \mathcal{L}_N^k(\mathbf{P}))$ with $\mathcal{L}_j^k(\mathbf{P}) = (A_j - B_j \mathcal{K}_j(\mathbf{V}^k)) P_j + P_j (A_j - B_j \mathcal{K}_j(\mathbf{V}^k))^* + \sum_{i \in \mathcal{S}} \lambda_{ij} P_i$. From identity (A.12) and (5.20) we have that

$$\begin{aligned} &\big(A_i - B_i \mathcal{K}_i(\mathbf{V}^k)\big)^* V_i^{k+1} + V_i^{k+1} \big(A_i - B_i \mathcal{K}_i(\mathbf{V}^k)\big) \\ &\quad + \mathcal{K}_i^*(\mathbf{V}^k) D_i^* D_i \mathcal{K}_i(\mathbf{V}^k) + \sum_{j \in \mathcal{S}} \lambda_{ij} V_j^{k+1} + C_i^* C_i + \frac{I}{k+1} \\ &\quad - \big(\mathcal{K}_i(\mathbf{V}^k) - \mathcal{K}_i(\mathbf{V}^{k+1})\big)^* D_i^* D_i \big(\mathcal{K}_i(\mathbf{V}^k) - \mathcal{K}_i(\mathbf{V}^{k+1})\big) = 0. \end{aligned} \tag{5.22}$$

Writing $\mathbf{Q} = \mathbf{V}^k - \mathbf{V}^{k+1}$, we get from (5.21) and (5.22) that

$$\begin{aligned} &\big(A_i - B_i \mathcal{K}_i(\mathbf{V}^k)\big)^* Q_i + Q_i \big(A_i - B_i \mathcal{K}_i(\mathbf{V}^k)\big) + \sum_{j \in \mathcal{S}} \lambda_{ij} Q_j + \frac{I}{k(k+1)} \\ &\quad + \big(\mathcal{K}_i(\mathbf{V}^k) - \mathcal{K}_i(\mathbf{V}^{k+1})\big)^* D_i^* D_i \big(\mathcal{K}_i(\mathbf{V}^k) - \mathcal{K}_i(\mathbf{V}^{k+1})\big) = 0. \end{aligned} \tag{5.23}$$

By Proposition 3.20, there exists a unique solution $\mathbf{Q}$ for (5.23), and moreover $\mathbf{Q} \in \mathbb{H}^{n+}$, which shows that $\mathbf{V}^k \geq \mathbf{V}^{k+1}$. Similarly, we can show that $\mathbf{V}^k \geq \mathbf{V}$. Taking the limit as $k \to \infty$, from Lemma 2.17 we have that $\mathbf{V}^k \to \hat{\mathbf{V}} \in \mathbb{H}^{n+}$, with $\mathcal{R}(\hat{\mathbf{V}}) = 0$ and $\hat{\mathbf{V}} \geq \mathbf{V}$. From Theorem A.5 we get that $\mathbf{V} \geq \hat{\mathbf{V}}$ and thus $\hat{\mathbf{V}} = \mathbf{V}$, completing the proof. □

We can now show the main result of this section, which relates the CARE and the convex problem (5.14).

Theorem 5.11 *The following assertions are equivalent*:

(a) *There exists a mean-square stabilizing solution* $\widehat{\mathbf{V}} = (\widehat{V}_1, \dots, \widehat{V}_N) \in \mathbb{H}^{n+}$ *to* (4.28).

(b) *There exists* $\widehat{\mathbf{W}} = (\widehat{W}_1, \dots, \widehat{W}_N)$ *such that*

$$\widehat{\mathbf{W}} = \arg\min\left\{\sum_{i\in\mathcal{S}} \mathrm{tr}(U_i W_i);\ \mathbf{W} = (W_1, \dots, W_N) \in \mathcal{C}\right\}.$$

Moreover, if $\widehat{\mathbf{V}} = (\widehat{V}_1, \dots, \widehat{V}_N) \in \mathbb{H}^{n+}$ *is the mean-square stabilizing solution to* (4.28), *then setting* $\widehat{K}_i = (D_i^* D_i)^{-1} B_i^* \widehat{V}_i, i \in \mathcal{S}$, $\widehat{\mathbf{K}} = (\widehat{K}_1, \dots, \widehat{K}_N)$, $\widehat{\mathbf{P}} = (\widehat{P}_1, \dots, \widehat{P}_N) > 0$ *satisfying*

$$(A_j - B_j \widehat{K}_j)\widehat{P}_j + \widehat{P}_j (A_j - B_j \widehat{K}_j)^* + \sum_{i=1}^{N} \lambda_{ij} \widehat{P}_i + \nu_j J_j J_j^* = 0, \quad j \in \mathcal{S},$$

and $\widehat{\mathbf{W}} = (\widehat{W}_1, \dots, \widehat{W}_N)$ *as*

$$\widehat{W}_i = \begin{bmatrix} \widehat{P}_i & \widehat{P}_i \widehat{K}_i^* \\ \widehat{K}_i \widehat{P}_i & \widehat{K}_i \widehat{P}_i \widehat{K}_i^* \end{bmatrix}, \quad i \in \mathcal{S},$$

we have that $\widehat{\mathbf{W}}$ *satisfies* (b). *On the other hand, if* $\widehat{\mathbf{W}}$ *satisfies* (b), *then setting* $\widehat{K}_i = \widehat{W}_{i2}^* \widehat{W}_{i1}^{-1}$ *and* $\widehat{\mathbf{V}} = (\widehat{V}_1, \dots, \widehat{V}_N) \in \mathbb{H}^{n+}$ *satisfying*

$$\begin{aligned}(A_i - B_i \widehat{K}_i)^* \widehat{V}_i + \widehat{V}_i (A_i - B_i \widehat{K}_i) + \sum_{j\in\mathcal{S}} \lambda_{ij} \widehat{V}_j \\ + (C_i - D_i \widehat{K}_i)^* (C_i - D_i \widehat{K}_i) = 0, \quad i \in \mathcal{S}, \qquad (5.24)\end{aligned}$$

we have that $\widehat{\mathbf{V}}$ *is the mean-square stabilizing solution of* (4.28).

Proof We prove that (a) is equivalent to (b). The second part of the theorem follows as a byproduct of this proof. Let us first prove that (b) implies (a). Indeed, let $\widehat{\mathbf{W}} = (\widehat{W}_1, \dots, \widehat{W}_N) = \arg\min\{\sum_{i\in\mathcal{S}} \mathrm{tr}(U_i W_i);\ \mathbf{W} = (W_1, \dots, W_N) \in \mathcal{C}\}$, and write $\widehat{K}_i = \widehat{W}_{i2}^* \widehat{W}_{i1}^{-1}, i \in \mathcal{S}, \widehat{\mathbf{K}} = (\widehat{K}_1, \dots, \widehat{K}_N)$. As seen in Proposition 5.6, $\widehat{\mathbf{K}} \in \mathbb{K}$, and by Theorem 3.25, there exists a unique solution $\widehat{\mathbf{V}} = (\widehat{V}_1, \dots, \widehat{V}_N) \in \mathbb{H}^{n+}$ satisfying (5.24). From the fact that $(\mathbf{A}, \mathbf{B}, \Pi)$ is mean-square stabilizable (indeed, $\widehat{\mathbf{K}} \in \mathbb{K} \neq \emptyset$) we know, from Proposition 5.10, that there exists a maximal solution $\mathbf{V} = (V_1, \dots, V_N)$ to (4.28). Moreover, from (4.28) we have that

$$\begin{aligned}(A_i - B_i \widehat{K}_i)^* V_i + V_i (A_i - B_i \widehat{K}_i) + \sum_{j\in\mathcal{S}} \lambda_{ij} V_j \\ + (C_i - D_i \widehat{K}_i)^* (C_i - D_i \widehat{K}_i) - (K_i - \widehat{K}_i)^* \big(D_i^* D_i\big)(K_i - \widehat{K}_i) = 0, \qquad (5.25)\end{aligned}$$

where $K_i = (D_i^* D_i)^{-1} B_i^* V_i$, $i \in \mathcal{S}$, and $\mathbf{K} = (K_1, \ldots, K_N)$. From $\widehat{\mathbf{K}} \in \mathbb{K}$, (5.24), (5.25), and Theorem 3.25 we get that $\widehat{\mathbf{V}} \geq \mathbf{V}$. As stated in Proposition 5.10, $\mathbf{V} = \lim_{k\to\infty} \mathbf{V}^k$, where $\mathbf{V}^k$ is the mean-square stabilizing solution of (5.20) with $K_i^k = (D_i^* D_i)^{-1} B_i^* V_i^k$, $\mathbf{K}^k = (K_1^k, \ldots, K_N^k) \in \mathbb{K}$. For each $k = 1, 2, \ldots$, let $\widehat{\mathbf{V}}^k = (\widehat{V}_1^k, \ldots, \widehat{V}_N^k) \in \mathbb{H}^{n+}$ be the unique solution satisfying (see Theorem 3.25)

$$\begin{aligned}(A_i - B_i K_i^k)^* \widehat{V}_i^k + \widehat{V}_i^k (A_i - B_i K_i^k) + \sum_{j\in\mathcal{S}} \lambda_{ij} \widehat{V}_j^k \\ + (C_i - D_i K_i^k)^* (C_i - D_i K_i^k) = 0, \quad i \in \mathcal{S}.\end{aligned} \tag{5.26}$$

From $\mathbf{K}^k \in \mathbb{K}$, (5.26), (5.20), and Theorem 3.25 it follows that $\mathbf{V}^k \geq \widehat{\mathbf{V}}^k$. By the optimality of $\widehat{\mathbf{W}}$, Theorem 5.4, Theorem 5.7, and the fact that $\mathbf{K}^k \in \mathbb{K}$ for each $k = 1, 2, \ldots$, we get that

$$\|\mathcal{G}_{\widehat{\mathbf{K}}}\|_2^2 = \sum_{i\in\mathcal{S}} \nu_i \operatorname{tr}(J_i^* \widehat{V}_i J_i) \leq \|\mathcal{G}_{\mathbf{K}^k}\|_2^2 = \sum_{i\in\mathcal{S}} \nu_i \operatorname{tr}(J_i^* \widehat{V}_i^k J_i) \leq \sum_{i\in\mathcal{S}} \nu_i \operatorname{tr}(J_i^* V_i^k J_i),$$

and taking the limit as $k \to \infty$ and recalling that $\mathbf{V}^k \to \mathbf{V}$ as $k \to \infty$, we obtain that

$$\sum_{i\in\mathcal{S}} \nu_i \operatorname{tr}(J_i^* \widehat{V}_i J_i) \leq \sum_{i\in\mathcal{S}} \nu_i \operatorname{tr}(J_i^* V_i J_i).$$

Thus, $\widehat{\mathbf{V}} \geq \mathbf{V}$ and $\sum_{i\in\mathcal{S}} \nu_i \operatorname{tr}(J_i^* (\widehat{V}_i - V_i) J_i) \leq 0$, which yields $\operatorname{tr}(J_i^* (\widehat{V}_i - V_i) J_i) = 0$ for each $i \in \mathcal{S}$. Since $J_i J_i^* > 0$ by assumption, it follows that $\operatorname{tr}(\widehat{V}_i - V_i) = 0$ for $i \in \mathcal{S}$, with $\widehat{\mathbf{V}} \geq \mathbf{V}$, which can only hold if $\widehat{\mathbf{V}} = \mathbf{V}$. Subtracting (5.24) and (5.25) yields for $i \in \mathcal{S}$,

$$(K_i - \widehat{K}_i)^* (D_i^* D_i)(K_i - \widehat{K}_i) = 0,$$

and since $D_i^* D_i > 0$ by assumption, it follows that $K_i = \widehat{K}_i$, showing that (b) implies (a). Let us now prove that (a) implies (b). Suppose that $\widehat{\mathbf{V}} = (\widehat{V}_1, \ldots, \widehat{V}_N) \in \mathbb{H}^{n+}$ is a mean-square stabilizing solution of (4.28). Then $\widehat{\mathbf{K}} = (\widehat{K}_1, \ldots, \widehat{K}_N) \in \mathbb{K}$, where for $i \in \mathcal{S}$, $\widehat{K}_i = (D_i^* D_i)^{-1} B_i^* \widehat{V}_i$. For any $\mathbf{G} = (G_1, \ldots, G_N) \in \mathbb{K}$, define $\mathbf{S} = (S_1, \ldots, S_N) \in \mathbb{H}^{n+}$ as the unique solution of

$$\begin{aligned}(A_i - B_i G_i)^* S_i + S_i (A_i - B_i G_i) + \sum_{j\in\mathcal{S}} \lambda_{ij} S_j \\ + (C_i - D_i G_i)^* (C_i - D_i G_i) = 0, \quad i \in \mathcal{S}.\end{aligned} \tag{5.27}$$

Equation (4.28) with $\widehat{\mathbf{V}}$ can be written as

$$\begin{aligned}(A_i - B_i G_i)^* \widehat{V}_i + \widehat{V}_i (A_i - B_i G_i) + \sum_{j\in\mathcal{S}} \lambda_{ij} \widehat{V}_j + (C_i - D_i G_i)^* (C_i - D_i G_i) \\ - (\widehat{K}_i - G_i)^* (D_i^* D_i)(\widehat{K}_i - G_i) = 0, \quad i \in \mathcal{S},\end{aligned} \tag{5.28}$$

and from (5.27), (5.28), and Theorem 3.25 we have that $\mathbf{S} \geq \widehat{\mathbf{V}}$. Moreover, since

$$(A_i - B_i\widehat{K}_i)^* \widehat{V}_i + \widehat{V}_i(A_i - B_i\widehat{K}_i) + \sum_{j \in \mathcal{S}} \lambda_{ij}\widehat{V}_j + (C_i - D_i\widehat{K}_i)^*(C_i - D_i\widehat{K}_i) = 0,$$

$i \in \mathcal{S}$, we have, from Theorem 5.4, that

$$\|\mathcal{G}_{\widehat{\mathbf{K}}}\|_2^2 = \sum_{i \in \mathcal{S}} \nu_i \operatorname{tr}\bigl(J_i^* \widehat{V}_i J_i\bigr) \leq \sum_{i \in \mathcal{S}} \nu_i \operatorname{tr}\bigl(J_i^* S_i J_i\bigr) = \|\mathcal{G}_{\mathbf{G}}\|_2^2,$$

and since $\mathbf{G}$ is arbitrary in $\mathbb{K}$, it follows that

$$\|\mathcal{G}_{\widehat{\mathbf{K}}}\|_2^2 = \min\bigl\{\|\mathcal{G}_{\mathbf{G}}\|_2^2; \mathbf{G} \in \mathbb{K}\bigr\}.$$

□

Remark 5.12 Since the solution of the CARE (4.28) does not depend on $J_j J_j^*$, we can see that the optimal controller does not depend on $\nu\mathbf{N}$. In particular, the choice of ν in Definition 5.3 will not affect the optimal solution for the H_2-optimal control problem.

5.6 Historical Remarks

One of the earliest and most powerful design control techniques for multi-variable systems has been the so-called LQG technique. It soon became extremely popular in the control community, as it provided powerful tools to design dynamic controllers in a coherent and systematical manner. The LQG theory is based on an stochastic setup for the system. More recently, a deterministic interpretation for the LQG control problem has been provided, which yielded to what is now known as the H_2 optimal control problem. The subject of H_2 optimal control is by now vast and immense, and the reader is referred to [251] for a comprehensive treatment on this subject (see also [117]). One of the advantages of the deterministic setup adopted in the H_2 optimal control problem is that it allows one to use LMIs optimization tools for dealing with parameter uncertainties of the system (see, for instance, [168]).

It seems that the H_2 control problem for MJLS was first introduced in the literature in [75] for the discrete-time case and in [76] for the continuous-time case. Since then, several other results related to H_2 for MJLS have appeared as, for instance, dealing with parametric uncertainties on the systems matrices and/or transition rate matrix [76, 113, 255, 305], when a nonobserved part of the Markov state is grouped in a number of clusters of observations [111], the optimal time-weighted H_2 model reduction problem [267], mixed H_2/H_∞ control [85, 185, 216], mode-independent output feedback control [3], and the separation principle [80].

The results of this chapter are based mainly on those presented in [76].

Chapter 6
Quadratic and H_2 Optimal Control with Partial Observations

6.1 Outline of the Chapter

This chapter deals with the finite-horizon quadratic optimal control problem and the H_2 control problem for continuous-time MJLS when the state variable $x(t)$ is not directly accessible to the controller. It is assumed that only an output $y(t)$ and jump process $\theta(t)$ are available. The main goal is to derive the so-called *separation principle* for this problem. We consider the admissible controllers as those in the class of Markov jump output dynamic observer-based control systems. Tracing a parallel with the classical LQG theory, it will be shown that the optimal control is obtained from two sets of coupled differential (for the finite-horizon case) and algebraic (for the H_2 case) Riccati equations. One set is associated with the optimal control problem when the state variable is available, as analyzed in Chaps. 4 and 5, and the other set is associated with the optimal filtering problem. An outline of the content of this chapter is as follows. Section 6.2 is devoted to the finite-horizon case, while Sect. 6.3 deals with the H_2 case. For the finite-horizon case, the description and the problem statement are presented in Sect. 6.2.1. In Sect. 6.2.2, we study the filtering problem. The main result for the finite-horizon case, which is the separation principle, is derived in Sect. 6.2.3. The H_2 case follows similar steps, with Sect. 6.3.1 presenting the problem formulation, Sect. 6.3.2 considering the H_2 filtering problem, and Sect. 6.3.3 presenting the separation principle for the H_2-control problem with partial observations. In Sect. 6.3.4, we present an LMIs approach for the problem.

O.L.V. Costa et al., *Continuous-Time Markov Jump Linear Systems*, Probability and Its Applications, DOI 10.1007/978-3-642-34100-7_6,

6.2 Finite-Horizon Quadratic Optimal Control with Partial Observations

6.2.1 Problem Statement

Let us consider, for $t \in \mathbb{R}^+$, the class of MJLS given by

$$\mathcal{G} = \begin{cases} dx(t) = A_{\theta(t)}(t)x(t)\,dt + B_{\theta(t)}(t)u(t)\,dt + J_{\theta(t)}(t)\,dw(t), \\ dy(t) = H_{\theta(t)}(t)x(t)\,dt + G_{\theta(t)}(t)\,dw(t), \\ \quad z(t) = C_{\theta(t)}(t)x(t) + D_{\theta(t)}(t)u(t), \end{cases} \tag{6.1}$$

with $x(0) = x_0$, where, as before, $\{x(t)\}$ denotes the *state process* in $\mathbb{R}^n$, $\{y(t)\}$ is the *observation process* in $\mathbb{R}^q$, $\{u(t)\}$ is the *control process* in $\mathbb{R}^m$, and $\{z(t)\}$ is the *output process* in $\mathbb{R}^p$. Moreover, we assume that $\mathbf{A}(t) = (A_1(t), \dots, A_N(t)) \in \mathbb{H}^n$, $\mathbf{B}(t) = (B_1(t), \dots, B_N(t)) \in \mathbb{H}^{m,n}$, $\mathbf{C}(t) = (C_1(t), \dots, C_N(t)) \in \mathbb{H}^{n,p}$, $\mathbf{D}(t) = (D_1(t), \dots, D_N(t)) \in \mathbb{H}^{m,p}$, $\mathbf{J}(t) = (J_1(t), \dots, J_N(t)) \in \mathbb{H}^{r,n}$, $\mathbf{H}(t) = (H_1(t), \dots, H_N(t)) \in \mathbb{H}^{n,q}$, $\mathbf{G}(t) = (G_1(t), \dots, G_N(t)) \in \mathbb{H}^{r,q}$, and all matrices are of class **PC**. We also assume that, for each $i \in \mathcal{S}$ and $t \in \mathbb{R}^+$,

$$D_i^*(t)D_i(t) > 0, \qquad G_i(t)G_i^*(t) > 0, \qquad J_i^*(t)G_i(t) = 0, \qquad C_i^*(t)D_i(t) = 0,$$

and that the matrices $(D_i^*(t)D_i(t))^{-1}$ and $(G_i(t)G_i^*(t))^{-1}$ are of class **PC**. We write $\mathbb{D} \subset \mathbb{R}^+$ as the union of the discontinuity points of A_i, B_i, J_i, H_i, G_i, C_i, D_i, $(D_i^* D_i)^{-1}$, and $(G_i G_i^*)^{-1}$, $i \in \mathcal{S}$. As pointed out in [53], p. 7, for any $T \in \mathbb{R}^+$, $\mathbb{D} \cap [0, T]$ contains at most a finite number of points. The r-dimensional Wiener process $W = \{(w(t), \mathcal{F}_t), t \in \mathbb{R}^+\}$ and the homogeneous Markov process $\theta = \{(\theta(t), \mathcal{F}_t), t \in \mathbb{R}^+\}$ are as defined in Chap. 3. We recall that $\{\theta(t)\}$ and $\{w(t)\}$ are statistically mutually independent. Finally, we suppose that $x(0)$ and $\theta(0)$ are independent random variables taking values in respectively $\mathbb{R}^n$ and $\mathcal{S}$ with

$$E\big[x(0)\big] = \mu, \qquad E\big[x(0)x(0)^*\big] = S, \qquad \nu_i = P\big(\theta(0) = i\big) > 0 \quad \text{for all } i \in \mathcal{S}.$$

Remark 6.1 From the hypothesis that $P(\theta(0) = i) > 0$ for $i \in \mathcal{S}$ we have that $p_i(t) := P(\theta(t) = i) > 0$ for all $i \in \mathcal{S}$ and $t \in \mathbb{R}^+$. Indeed, first notice that $\dot{p}_i(t) = \sum_{j \in \mathcal{S}} \lambda_{ji} p_j(t)$, and therefore,

$$p_i(t) = e^{\lambda_{ii} t} p_i(0) + \int_0^t e^{\lambda_{ii}(t-s)} \Bigg(\sum_{j \in \mathcal{S};\, j \neq i} \lambda_{ji} p_j(s) \Bigg)\, ds. \tag{6.2}$$

Thus $p_i(t) > 0$, bearing in mind that $p_i(0) > 0$ and the second part of (6.2) is non-negative.

In this chapter it is assumed that we do not have access to the state $x(t)$, but only to the measurement and jump processes, $\{y(t)\}$ and $\{\theta(t)\}$, for all t. We define

$\mathcal{F}_t := \sigma\{(\theta(s), y(s)); 0 \le s \le t\}$. The idea here is, as in the LQG problem with partial observations, to use an optimal estimate, $\widehat{x}(t)$. In addition, it is important that the filter does not introduce too much nonlinearity in the control problem. For this reason, we shall consider the class $\mathcal{G}_K$ of Markovian jump controllers given by

$$\mathcal{G}_K = \begin{cases} d\widehat{x}(t) = \widehat{A}_{\theta(t)}(t)\widehat{x}(t)\,dt + \widehat{B}_{\theta(t)}(t)\,dy(t), \\ u(t) = \widehat{C}_{\theta(t)}(t)\widehat{x}(t). \end{cases} \tag{6.3}$$

The quadratic cost associated to the closed-loop system $\mathcal{G}_{\mathrm{cl}}$ with control law $\{u(t), 0 \le t \le T\}$ given by (6.3) is

$$\mathcal{J}(u) := E\left\{\int_0^T \|z(t)\|^2\,dt + x(T)^* L_{\theta(T)} x(T)\right\}, \tag{6.4}$$

where $\mathbf{L} = (L_1, \dots, L_N) \in \mathbb{H}^{n+}$. The finite-horizon optimal quadratic control (OC) problem we study in this chapter can be stated as follows.

Problem 6.2 Find $\widehat{\mathbf{A}}(t) = (\widehat{A}_1(t), \dots, \widehat{A}_N(t))$, $\widehat{\mathbf{B}}(t) = (\widehat{B}_1(t), \dots, \widehat{B}_N(t))$, and $\widehat{\mathbf{C}}(t) = (\widehat{C}_1(t), \dots, \widehat{C}_N(t))$ within the class **PC** such that the control law $\{u(t), 0 \le t \le T\}$ induced by (6.3) minimizes the cost function $\mathcal{J}(u)$. This *minimal (optimal) cost* will be denoted by $\mathcal{J}^{\mathrm{op}} := \min_{u \in \mathcal{G}_K} \mathcal{J}(u)$, that is,

$$\mathcal{J}^{\mathrm{op}} = \min_{u \in \mathcal{G}_K}\left\{\int_0^T E\left[\left\|C_{\theta(t)}(t)x(t)\right\|^2 + \left\|D_{\theta(t)}(t)u(t)\right\|^2\right]dt + E\left[x(T)^* L_{\theta(T)} x(T)\right]\right\}. \tag{6.5}$$

6.2.2 Filtering Problem

Consider equation (6.1) with observable $\{y(t)\}$ and $\{\theta(t)\}$, and the class of admissible control policies $\mathcal{G}_K$ as in (6.3). The *optimal filtering* (OF) *problem* here consists of finding $\widehat{\mathbf{A}}(t) = (\widehat{A}_1(t), \dots, \widehat{A}_N(t))$, $\widehat{\mathbf{B}}(t) = (\widehat{B}_1(t), \dots, \widehat{B}_N(t))$, and $\widehat{\mathbf{C}}(t) = (\widehat{C}_1(t), \dots, \widehat{C}_N(t))$ in (6.3) with *deterministic* $\widehat{x}(0)$, such that $E(\|v(t)\|^2)$ is minimized for each $t \in [0, T]$, where

$$v(t) = x(t) - \widehat{x}(t). \tag{6.6}$$

With $u(t) = \widehat{C}_{\theta(t)}(t)\widehat{x}(t)$ given as in (6.3), define

$$\begin{cases} d\widehat{x}_{\mathrm{op}}(t) = A_{\theta(t)}(t)\widehat{x}_{\mathrm{op}}(t)\,dt + B_{\theta(t)}(t)u(t)\,dt + K^f_{\theta(t)}(t)\,d\nu(t), \\ \widehat{x}_{\mathrm{op}}(0) = E\left[x(0)\right] = \mu, \end{cases} \tag{6.7}$$

where

$$\widetilde{x}_{\mathrm{op}}(t) := x(t) - \widehat{x}_{\mathrm{op}}(t), \tag{6.8}$$

and recalling that $p_i(t) > 0$ (see Remark 6.1), we set

$$K_i^f(t) := Y_i(t) H_i^*(t) \big(G_i(t) G_i^*(t) p_i(t)\big)^{-1}, \tag{6.9}$$

where

$$Y_i(t) := E\big[\widetilde{x}_{\mathrm{op}}(t) \widetilde{x}_{\mathrm{op}}(t)^* 1_{\{\theta(t)=i\}}\big]. \tag{6.10}$$

The innovation process is given by

$$\begin{aligned} dv(t) &:= dy(t) - H_{\theta(t)}(t) \widehat{x}_{\mathrm{op}}(t)\, dt \\ &= H_{\theta(t)}(t) \widetilde{x}_{\mathrm{op}}(t)\, dt + G_{\theta(t)}(t)\, dw(t). \end{aligned} \tag{6.11}$$

From (6.1), (6.7), (6.11), and (6.8) we have that

$$\begin{cases} d\widetilde{x}_{\mathrm{op}}(t) = \bar{A}_{\theta(t)}(t) \widetilde{x}_{\mathrm{op}}(t)\, dt + \big(J_{\theta(t)}(t) - K_{\theta(t)}^f(t) G_{\theta(t)}(t)\big)\, dw(t), \\ \widetilde{x}_{\mathrm{op}}(0) = x_0 - \mu, \end{cases} \tag{6.12}$$

where

$$\bar{A}_{\theta(t)}(t) := A_{\theta(t)}(t) - K_{\theta(t)}^f(t) H_{\theta(t)}(t). \tag{6.13}$$

Lemma 6.3 *We have that* $\mathbf{Y}(t) = (Y_1(t), \ldots, Y_N(t)) \in \mathbb{H}^{n+}$ *is a unique positive semi-definite set of* $n \times n$ *matrices satisfying the following interconnected Riccati equations*:

$$\begin{aligned} \dot{Y}_i(t) &= A_i(t) Y_i(t) + Y_i(t) A_i^*(t) + \sum_{j \in \mathcal{S}} \lambda_{ji} Y_j(t) + J_i(t) J_i^*(t) p_i(t) \\ &\quad - Y_i(t) H_i^*(t) \big(G_i(t) G_i^*(t) p_i(t)\big)^{-1} H_i(t) Y_i(t), \quad t \in \mathbb{R}^+ \setminus \mathbb{D}, \end{aligned} \tag{6.14}$$

with $Y_i(0) = (S - \mu\mu^*) p_i(0)$.

Proof The existence, positive-definiteness, continuity, and uniqueness follow from Theorem A.11. By Itô's rule we get

$$\begin{aligned} dY_i(t) = E\Big(&\big(d\widetilde{x}_{\mathrm{op}}(t)\big) \widetilde{x}_{\mathrm{op}}(t)^* 1_{\{\theta(t)=i\}} + \widetilde{x}_{\mathrm{op}}(t) \big(d\widetilde{x}_{\mathrm{op}}(t)^*\big) 1_{\{\theta(t)=i\}} \\ &+ \widetilde{x}_{\mathrm{op}}(t) \widetilde{x}_{\mathrm{op}}(t)^*\, d(1_{\{\theta(t)=i\}}) + 1_{\{\theta(t)=i\}}\, d\widetilde{x}_{\mathrm{op}}(t)\, d\widetilde{x}_{\mathrm{op}}(t)^*\Big). \end{aligned} \tag{6.15}$$

Recalling that $J_i(t) G_i^*(t) = 0$, from (6.12) we get that

$$E\big(\widetilde{x}_{\mathrm{op}}(t) \widetilde{x}_{\mathrm{op}}(t)^*\, d(1_{\{\theta(t)=i\}})\big) = \sum_{j \in \mathcal{S}} \lambda_{ji} Y_j(t)\, dt$$

and

$$\begin{aligned}
&E\big(1_{\{\theta(t)=i\}}\,d\widetilde{x}_{\mathrm{op}}(t)\,d\widetilde{x}_{\mathrm{op}}(t)^*\big)\\
&\quad= E\big(1_{\{\theta(t)=i\}}\big(J_{\theta(t)}(t)J_{\theta(t)}^*(t)-K_{\theta(t)}^f(t)G_{\theta(t)}(t)G_{\theta(t)}^*(t)\big(K_{\theta(t)}^f(t)\big)^*\big)\big)\,dt\\
&\quad= \big(J_iJ_i^*-K_i^f(t)G_i(t)G_i^*(t)\big(K_i^f(t)\big)^*\big)p_i(t)\,dt,
\end{aligned} \tag{6.16}$$

as well as

$$\begin{aligned}
&E\big(d\widetilde{x}_{\mathrm{op}}(t)\widetilde{x}_{\mathrm{op}}(t)^*1_{\{\theta(t)=i\}}\big)\\
&\quad= E\big(\big(\bar{A}_{\theta(t)}(t)\widetilde{x}_{\mathrm{op}}(t)\,dt+\big(J_{\theta(t)}(t)-K_{\theta(t)}^f(t)G_{\theta(t)}(t)\big)\,dw(t)\big)\widetilde{x}_{\mathrm{op}}(t)^*1_{\{\theta(t)=i\}}\big)\\
&\quad= \bar{A}_i(t)Y_i(t)\,dt.
\end{aligned} \tag{6.17}$$

From (6.9) and (6.15)–(6.17) we get (6.14). □

The following orthogonality results will be very important in the sequel.

Lemma 6.4 *For* $\widehat{x}(t)$, $\widehat{x}_{\mathrm{op}}(t)$, *and* $\widetilde{x}_{\mathrm{op}}(t)$ *given by* (6.3), (6.7), *and* (6.8), *respectively, and for* $i\in\mathcal{S}$ *and* $t\geq 0$, *we have that*

$$E\big(\widetilde{x}_{\mathrm{op}}(t)\widehat{x}_{\mathrm{op}}(t)^*1_{\{\theta(t)=i\}}\big)=0, \tag{6.18}$$

$$E\big(\widetilde{x}_{\mathrm{op}}(t)\widehat{x}(t)^*1_{\{\theta(t)=i\}}\big)=0. \tag{6.19}$$

Proof Denote by $\widetilde{\mathbb{D}}$ the union of $\mathbb{D}$ with the discontinuity points of the matrices $\widehat{A}_i(t)$, $\widehat{B}_i(t)$, $\widehat{C}_i(t)$, $i\in\mathcal{S}$. First notice that $E[\widetilde{x}_{\mathrm{op}}(0)\widehat{x}_{\mathrm{op}}^*(0)1_{\{\theta_0=i\}}]=0$. This follows from the fact that $\widehat{x}(0)$ is deterministic and from the independence of $x(0)$ and θ_0, bearing in mind that $E[\widetilde{x}_{\mathrm{op}}(0)]=0$. Similarly, we can prove that $E(\widetilde{x}_{\mathrm{op}}(0)\widehat{x}^*(0)1_{\{\theta_0=i\}})=0$. In what follows, we will use that, by (6.9),

$$K_i^f(t)G_i(t)G_i^*(t)p_i(t)=Y_i(t)H_i^*(t). \tag{6.20}$$

Let us show (6.18) and (6.19). Define

$$Q_i(t)=E\big(\widetilde{x}_{\mathrm{op}}(t)\widehat{x}_{\mathrm{op}}(t)^*1_{\{\theta(t)=i\}}\big), \tag{6.21}$$

$$\bar{Q}_i(t)=E\big(\widetilde{x}_{\mathrm{op}}(t)\widehat{x}(t)^*1_{\{\theta(t)=i\}}\big), \tag{6.22}$$

and consider $t\in\mathbb{R}^+\setminus\widetilde{\mathbb{D}}$. By Itô's rule we get

$$\begin{aligned}
dQ_i(t) &= E\big(d\widetilde{x}_{\mathrm{op}}(t)\widehat{x}_{\mathrm{op}}(t)^*1_{\{\theta(t)=i\}}+\widetilde{x}_{\mathrm{op}}(t)\,d\widehat{x}_{\mathrm{op}}(t)^*1_{\{\theta(t)=i\}}\\
&\qquad+\widetilde{x}_{\mathrm{op}}(t)\widehat{x}_{\mathrm{op}}(t)^*\,d(1_{\{\theta(t)=i\}})+1_{\{\theta(t)=i\}}\,d\widetilde{x}_{\mathrm{op}}(t)\,d\widehat{x}_{\mathrm{op}}(t)^*\big)\\
&= E\big(\big(\bar{A}_{\theta(t)}(t)\widetilde{x}_{\mathrm{op}}(t)\,dt\\
&\qquad+\big(J_{\theta(t)}(t)-K_{\theta(t)}^f(t)G_{\theta(t)}(t)\big)\,dw(t)\big)\widehat{x}_{\mathrm{op}}(t)^*1_{\{\theta(t)=i\}}
\end{aligned}$$

$$+\widetilde{x}_{\mathrm{op}}(t)\big(A_{\theta(t)}(t)\widehat{x}_{\mathrm{op}}\,dt+B_{\theta(t)}(t)u(t)\,dt+K^f_{\theta(t)}(t)\,dv(t)\big)^*1_{\{\theta(t)=i\}}$$
$$+\widetilde{x}_{\mathrm{op}}(t)\widehat{x}_{\mathrm{op}}(t)^*\,d(1_{\{\theta(t)=i\}})+1_{\{\theta(t)=i\}}\,d\widetilde{x}_{\mathrm{op}}(t)\,d\widehat{x}_{\mathrm{op}}(t)^*\big).$$

Now, we have that

$$E\big(\widetilde{x}_{\mathrm{op}}(t)\widehat{x}_{\mathrm{op}}(t)^*\,d(1_{\{\theta(t)=i\}})\big)=\sum_{j\in\mathcal{S}}\lambda_{ji}Q_j(t)\,dt$$

and, recalling that $J_i(t)G_i^*(t)=0$,

$$\begin{aligned}E\big(1_{\{\theta(t)=i\}}\,d\widetilde{x}_{\mathrm{op}}(t)\,d\widehat{x}_{\mathrm{op}}(t)^*\big)&=-E\big(1_{\{\theta(t)=i\}}K^f_{\theta(t)}(t)G_{\theta(t)}(t)G^*_{\theta(t)}(t)\big(K^f_{\theta(t)}(t)\big)^*\big)\,dt\\&=-K_i^f(t)G_i(t)G_i^*(t)\big(K_i^f(t)\big)^*p_i(t)\,dt.\end{aligned}$$

In addition, we have that

$$\begin{aligned}&E\big(\big(\bar{A}_{\theta(t)}(t)\widetilde{x}_{\mathrm{op}}(t)\,dt+\big(J_{\theta(t)}(t)-K^f_{\theta(t)}(t)G_{\theta(t)}(t)\big)\,dw(t)\big)\widehat{x}_{\mathrm{op}}(t)^*1_{\{\theta(t)=i\}}\big)\\&\quad=\bar{A}_i(t)Q_i(t)\,dt\end{aligned}$$

and, from (6.20),

$$\begin{aligned}&E\big(\widetilde{x}_{\mathrm{op}}(t)\big(A_{\theta(t)}(t)\widehat{x}_{\mathrm{op}}\,dt+B_{\theta(t)}(t)u(t)\,dt+K^f_{\theta(t)}(t)\,dv(t)\big)^*1_{\{\theta(t)=i\}}\big)\\&\quad=E\big(\widetilde{x}_{\mathrm{op}}(t)\big(A_{\theta(t)}(t)\widehat{x}_{\mathrm{op}}\,dt+B_{\theta(t)}(t)u(t)\,dt\\&\qquad+K^f_{\theta(t)}(t)\big(H_{\theta(t)}(t)\widetilde{x}_{\mathrm{op}}(t)\,dt+G_{\theta(t)}(t)\,dw(t)\big)\big)^*1_{\{\theta(t)=i\}}\big)\\&\quad=Q_i(t)A_i^*(t)\,dt+E\big(\widetilde{x}_{\mathrm{op}}(t)u(t)^*1_{\{\theta(t)=i\}}\big)B_i^*(t)\,dt+Y_i(t)H_i^*(t)\big(K_i^f(t)\big)^*\,dt\\&\quad=Q_i(t)A_i^*(t)\,dt+\bar{Q}_i(t)\widehat{C}_i^*(t)B_i^*(t)\,dt+K_i^f(t)G_i(t)G_i^*(t)\big(K_i^f(t)\big)^*p_i(t)\,dt.\end{aligned}$$

Therefore, for $t\in\mathbb{R}^+\setminus\widetilde{\mathbb{D}}$,

$$\dot{Q}_i(t)=\bar{A}_i(t)Q_i(t)+Q_i(t)A_i^*(t)+\bar{Q}_i(t)\widehat{C}_i^*(t)B_i^*(t)+\sum_{j\in\mathcal{S}}\lambda_{ji}Q_j(t).\qquad(6.23)$$

Notice from (6.1), (6.3), and (6.8) that

$$\begin{aligned}d\widehat{x}(t)&=\widehat{A}_{\theta(t)}(t)\widehat{x}(t)\,dt\\&\quad+\widehat{B}_{\theta(t)}(t)\big(H_{\theta(t)}(t)\big(\widetilde{x}_{\mathrm{op}}(t)+\widehat{x}_{\mathrm{op}}(t)\big)\,dt+G_{\theta(t)}(t)\,dw(t)\big).\end{aligned}\qquad(6.24)$$

Similarly, we have, by Itô's rule, that

$$\begin{aligned}d\bar{Q}_i(t)=E\big(&d\widetilde{x}_{\mathrm{op}}(t)\widehat{x}(t)^*1_{\{\theta(t)=i\}}+\widetilde{x}_{\mathrm{op}}(t)\,d\widehat{x}(t)^*1_{\{\theta(t)=i\}}\\&+\widetilde{x}_{\mathrm{op}}(t)\widehat{x}(t)^*\,d(1_{\{\theta(t)=i\}})+1_{\{\theta(t)=i\}}\,d\widetilde{x}_{\mathrm{op}}(t)\,d\widehat{x}(t)^*\big)\end{aligned}$$

$$
\begin{aligned}
&= E\big(\big(\bar{A}_{\theta(t)}(t)\widetilde{x}_{\text{op}}(t)\,dt + \big(J_{\theta(t)}(t) - K^{f}_{\theta(t)}(t)G_{\theta(t)}(t)\big)\,dw(t)\big)\widehat{x}_{\text{op}}(t)^* \\
&\quad \times 1_{\{\theta(t)=i\}} \\
&\quad + \widetilde{x}_{\text{op}}(t)\big(\widehat{A}_{\theta(t)}(t)\widehat{x}(t)\,dt + \widehat{B}_{\theta(t)}(t)\big(H_{\theta(t)}(t)\big(\widetilde{x}_{\text{op}}(t) + \widehat{x}_{\text{op}}(t)\big)\,dt \\
&\quad + G_{\theta(t)}(t)\,dw(t)\big)\big)^* 1_{\{\theta(t)=i\}} \\
&\quad + \widetilde{x}_{\text{op}}(t)\widehat{x}_{\text{op}}(t)^*\,d(1_{\{\theta(t)=i\}}) + 1_{\{\theta(t)=i\}}\,d\widetilde{x}_{\text{op}}(t)\,d\widehat{x}_{\text{op}}(t)^*\big).
\end{aligned}
$$

We have, as in Lemma 3.6, that

$$
E\big(\widetilde{x}_{\text{op}}(t)\widehat{x}(t)^*\,d(1_{\{\theta(t)=i\}})\big) = \sum_{j\in\mathcal{S}} \lambda_{ji}\bar{Q}_j(t)\,dt,
$$

and, recalling that $J_i(t)G_i^*(t) = 0$, we get from (6.24) that

$$
\begin{aligned}
E\big(1_{\{\theta(t)=i\}}\,d\widetilde{x}_{\text{op}}(t)\,d\widehat{x}(t)^*\big) &= -E\big(1_{\{\theta(t)=i\}}K^{f}_{\theta(t)}(t)G_{\theta(t)}(t)G^*_{\theta(t)}(t)\widehat{B}^*_{\theta(t)}(t)\big)\,dt \\
&= -K^{f}_{i}(t)G_i(t)G_i^*(t)\widehat{B}_i^*(t)p_i(t)\,dt.
\end{aligned}
$$

In addition, we have that

$$
\begin{aligned}
&E\big(\big(\bar{A}_{\theta(t)}(t)\widetilde{x}_{\text{op}}(t)\,dt + \big(J_{\theta(t)}(t) - K^{f}_{\theta(t)}(t)G_{\theta(t)}(t)\big)\,dw(t)\big)\widehat{x}(t)^* 1_{\{\theta(t)=i\}}\big) \\
&\quad = \bar{A}_i(t)\bar{Q}_i(t)\,dt
\end{aligned}
$$

and, from (6.20),

$$
\begin{aligned}
&E\big(\widetilde{x}_{\text{op}}(t)\big(\widehat{A}_{\theta(t)}(t)\widehat{x}(t)\,dt + \widehat{B}_{\theta(t)}(t)\big(H_{\theta(t)}(t)\big(\widetilde{x}_{\text{op}}(t) + \widehat{x}_{\text{op}}(t)\big)\,dt \\
&\qquad + G_{\theta(t)}(t)\,dw(t)\big)\big)^* 1_{\{\theta(t)=i\}}\big) \\
&\quad = \big(\bar{Q}_i(t)\widehat{A}_i^*(t) + Y_i(t)H_i^*(t)\widehat{B}_i^*(t) + Q_i(t)H_i^*(t)\widehat{B}_i^*(t)\big)\,dt \\
&\quad = \big(\bar{Q}_i(t)\widehat{A}_i^*(t) + p_i(t)K_i^{f}(t)G_i(t)G_i^*(t)\widehat{B}_i^*(t) + Q_i(t)H_i^*(t)\widehat{B}_i^*(t)\big)\,dt.
\end{aligned}
$$

Therefore, for $t \in \mathbb{R}^+ \setminus \widetilde{\mathbb{D}}$,

$$
\dot{\bar{Q}}_i(t) = \bar{A}_i(t)\bar{Q}_i(t) + \bar{Q}_i(t)\widehat{A}_i^*(t) + Q_i(t)H_i^*(t)\widehat{B}_i^*(t) + \sum_{j\in\mathcal{S}} \lambda_{ji}\bar{Q}_j(t). \tag{6.25}
$$

Since $Q_i(0) = 0$ and $\bar{Q}_i(0) = 0$, $i \in \mathcal{S}$, it follows from the uniqueness of the solution of (6.23) and (6.25) (see Theorem 2.4 or Theorem B1.2-6, p. 470, in [53]) that $Q_i(t) = 0$ and $\bar{Q}_i(t) = 0$, for all $i \in \mathcal{S}$ and $t \in \mathbb{R}^+$. This completes the proof of (6.18) and (6.19). □

Lemma 6.5 *Let* $v(t)$ *and* $\mathbf{Y}(t) = (Y_1(t), \ldots, Y_N(t))$ *be as in* (6.6) *and* (6.10), *respectively. Then for every* $t \in \mathbb{R}^+$,

$$
E\big(\|v(t)\|^2\big) \geq \sum_{i\in\mathcal{S}} \operatorname{tr}\{Y_i(t)\}.
$$

Proof Bearing in mind that $v(t) := x(t) - \widehat{x}(t)$, with $x(t)$ solution of (6.1) and $\widehat{x}(t)$ solution of (6.3), we have:

$$\begin{aligned}
E\big(\|v(t)\|^2\big) &= E\big(\|x(t) - \widehat{x}(t)\|^2\big) \\
&= E\big(\|\widetilde{x}_{\text{op}}(t) + \widehat{x}_{\text{op}}(t) - \widehat{x}(t)\|^2\big) \\
&= \operatorname{tr}\big\{E\big[\big(\widetilde{x}_{\text{op}}(t) + \big(\widehat{x}_{\text{op}}(t) - \widehat{x}(t)\big)\big)\big(\widetilde{x}_{\text{op}}(t) + \big(\widehat{x}_{\text{op}}(t) - \widehat{x}(t)\big)\big)^*\big]\big\} \\
&= \operatorname{tr}\big\{E\big(\widetilde{x}_{\text{op}}(t)\widetilde{x}_{\text{op}}(t)^*\big) + E\big(\widetilde{x}_{\text{op}}(t)\widehat{x}_{\text{op}}(t)^*\big) - E\big(\widetilde{x}_{\text{op}}(t)\widehat{x}(t)^*\big) \\
&\quad + E\big(\widehat{x}_{\text{op}}(t)\widetilde{x}_{\text{op}}(t)^*\big) - E\big(\widehat{x}(t)\widetilde{x}_{\text{op}}(t)^*\big) \\
&\quad + E\big(\big(\widehat{x}_{\text{op}}(t) - \widehat{x}(t)\big)\big(\widehat{x}_{\text{op}}(t) - \widehat{x}(t)\big)^*\big)\big\}.
\end{aligned}$$

Thus, from Lemma 6.4 it follows that

$$\begin{aligned}
E\big(\|v(t)\|^2\big) &= \operatorname{tr}\big\{E\big(\widetilde{x}_{\text{op}}(t)\widetilde{x}_{\text{op}}(t)^*\big) + E\big(\big(\widehat{x}_{\text{op}}(t) - \widehat{x}(t)\big)\big(\widehat{x}_{\text{op}}(t) - \widehat{x}(t)\big)^*\big)\big\} \\
&= \operatorname{tr}\bigg\{\sum_{i\in\mathcal{S}} E\big(\widetilde{x}_{\text{op}}(t)\widetilde{x}_{\text{op}}(t)^* 1_{\{\theta(t)=i\}}\big) \\
&\quad + E\big(\big(\widehat{x}_{\text{op}}(t) - \widehat{x}(t)\big)\big(\widehat{x}_{\text{op}}(t) - \widehat{x}(t)\big)^*\big)\bigg\} \\
&= \operatorname{tr}\bigg\{\sum_{i\in\mathcal{S}} Y_i(t) + E\big(\big(\widehat{x}_{\text{op}}(t) - \widehat{x}(t)\big)\big(\widehat{x}_{\text{op}}(t) - \widehat{x}(t)\big)^*\big)\bigg\} \\
&= \sum_{i\in\mathcal{S}} \operatorname{tr}\big\{Y_i(t)\big\} + E\big(\|\widehat{x}_{\text{op}}(t) - \widehat{x}(t)\|^2\big) \\
&\geq \sum_{i\in\mathcal{S}} \operatorname{tr}\big\{Y_i(t)\big\},
\end{aligned} \tag{6.26}$$

and the result follows. □

In view of the previous result, we have the following theorem.

Theorem 6.6 *An optimal solution for the OF problem described before is*: $\widehat{x}(0) = \mu$, $\widehat{C}_i(t)$ *arbitrary in the class* **PC**, *and*

$$\begin{aligned}
\widehat{A}_i(t) &= A_i(t) - K_i^f(t)H_i(t) + B_i(t)\widehat{C}_i(t), \\
\widehat{B}_i(t) &= K_i^f(t).
\end{aligned}$$

Proof From (6.26), $E(\|v(t)\|^2)$ is minimal when $E(\|\widehat{x}_{\text{op}}(t) - \widehat{x}(t)\|^2) = 0$, that is, when $\widehat{x}(t) = \widehat{x}_{\text{op}}(t)$ almost surely, and the minimum is given by $\sum_{i\in\mathcal{S}} \operatorname{tr}\{Y_i(t)\}$ for each t. □

6.2.3 A Separation Principle for MJLS

In order to show the separation principle, we go back to the OC problem posed in Sect. 6.2.1. We have the following lemma.

Lemma 6.7 *The problem of minimizing the cost*

$$\mathcal{J}(u) = \int_0^T E(\|z(t)\|^2)\,dt + E(x(T)^* L_{\theta(T)} x(T))$$

with $\{x(t)\}$ *and* $\{z(t)\}$ *as in* (6.1) *and* $\{u(t)\}$ *as in* (6.3) *is equivalent to minimizing*

$$\mathcal{J}_{\rm op}(u) = \int_0^T E(\|\widehat{z}_{\rm op}(t)\|^2)\,dt + E(\widehat{x}^*_{\rm op}(T) L_{\theta(T)} \widehat{x}_{\rm op}(T))$$

subject to

$$\begin{cases} d\widehat{x}_{\rm op}(t) = A_{\theta(t)}(t)\widehat{x}_{\rm op}(t)\,dt + B_{\theta(t)}(t)u(t)\,dt + K^f_{\theta(t)}(t)\,d\nu(t), \\ \widehat{x}_{\rm op}(0) = E(x(0)) = \mu, \\ \widehat{z}_{\rm op}(t) = C_{\theta(t)}(t)\widehat{x}_{\rm op}(t) + D_{\theta(t)}(t)u(t), \end{cases} \tag{6.27}$$

with $d\nu(t)$ *defined as in* (6.11) *and* $\{u(t)\}$ *as in* (6.3).

Proof First, notice that

$$\begin{aligned} E(\|z(t)\|^2) &= E(\|C_{\theta(t)}(t)x(t)\|^2) + E(\|D_{\theta(t)}(t)u(t)\|^2) \\ &= \operatorname{tr}(E(C_{\theta(t)}(t)x(t)x(t)^* C^*_{\theta(t)}(t))) + E(\|D_{\theta(t)}(t)u(t)\|^2) \\ &= \sum_{i\in\mathcal{S}} \operatorname{tr}(C_i^*(t)C_i(t)E(x(t)x(t)^* 1_{\{\theta(t)=i\}})) \\ &\quad + E(\|D_{\theta(t)}(t)u(t)\|^2). \end{aligned} \tag{6.28}$$

Now, from Lemma 6.4 and due to $\widetilde{x}_{\rm op}(t) = x(t) - \widehat{x}_{\rm op}(t)$, we get that

$$\begin{aligned} E(x(t)x(t)^* 1_{\{\theta(t)=i\}}) &= E\{(\widetilde{x}_{\rm op}(t) + \widehat{x}_{\rm op}(t))(\widetilde{x}_{\rm op}(t)^* + \widehat{x}_{\rm op}(t)^*)1_{\{\theta(t)=i\}}\} \\ &= E(\widetilde{x}_{\rm op}(t)\widetilde{x}_{\rm op}(t)^* 1_{\{\theta(t)=i\}}) + E(\widetilde{x}_{\rm op}(t)\widehat{x}_{\rm op}(t)^* 1_{\{\theta(t)=i\}}) \\ &\quad + E(\widehat{x}_{\rm op}(t)\widetilde{x}_{\rm op}(t)^* 1_{\{\theta(t)=i\}}) + E(\widehat{x}_{\rm op}(t)\widehat{x}_{\rm op}(t)^* 1_{\{\theta(t)=i\}}) \\ &= Y_i(t) + E(\widehat{x}_{\rm op}(t)\widehat{x}_{\rm op}(t)^* 1_{\{\theta(t)=i\}}). \end{aligned}$$

Then, from (6.28) we have that

$$E(\|z(t)\|^2) = \sum_{i\in\mathcal{S}} \operatorname{tr}\{C_i^*(t)C_i(t)(Y_i(t) + E(\widehat{x}_{\rm op}(t)\widehat{x}_{\rm op}(t)^* 1_{\{\theta(t)=i\}}))\}$$

$$
\begin{aligned}
&+ E\big(\big\|D_{\theta(t)}(t)u(t)\big\|^2\big)\\
&= \sum_{i\in\mathcal{S}} \operatorname{tr}\big\{C_i(t)Y_i(t)C_i^*(t) + C_i(t)E\big(\widehat{x}_{\mathrm{op}}(t)\widehat{x}_{\mathrm{op}}(t)^* 1_{\{\theta(t)=i\}}\big)C_i^*(t)\big\}\\
&\quad + E\big(\big\|D_{\theta(t)}(t)u(t)\big\|^2\big)\\
&= \sum_{i\in\mathcal{S}} \operatorname{tr}\big(C_i(t)Y_i(t)C_i^*(t)\big)\\
&\quad + \operatorname{tr}\big\{E\big(C_{\theta(t)}(t)\widehat{x}_{\mathrm{op}}(t)\widehat{x}_{\mathrm{op}}(t)^* C_{\theta(t)}^*(t)\big)\big\} + E\big(\big\|D_{\theta(t)}(t)u(t)\big\|^2\big)\\
&= \sum_{i\in\mathcal{S}} \operatorname{tr}\big(C_i(t)Y_i(t)C_i^*(t)\big)\\
&\quad + E\big(\big\|C_{\theta(t)}(t)\widehat{x}_{\mathrm{op}}(t)\big\|^2\big) + E\big(\big\|D_{\theta(t)}(t)u(t)\big\|^2\big)\\
&= \sum_{i\in\mathcal{S}} \operatorname{tr}\big(C_i(t)Y_i(t)C_i^*(t)\big) + E\big(\big\|C_{\theta(t)}(t)\widehat{x}_{\mathrm{op}}(t) + D_{\theta(t)}(t)u(t)\big\|^2\big),
\end{aligned}
$$

recalling that $C_i^*(t)D_i(t) = 0$ for $i \in \mathcal{S}$ and all $t \geq 0$. Now, from (6.27) we conclude that

$$
E\big(\big\|z(t)\big\|^2\big) = E\big(\big\|\widehat{z}_{\mathrm{op}}(t)\big\|^2\big) + \sum_{i\in\mathcal{S}} \operatorname{tr}\big(C_i(t)Y_i(t)C_i^*(t)\big).
$$

Similarly, we have that

$$
\begin{aligned}
E\big(x(T)^* L_{\theta(T)} x(T)\big) &= E\big(\big(\widetilde{x}_{\mathrm{op}}(T)^* + \widehat{x}_{\mathrm{op}}(T)^*\big)L_{\theta(T)}\big(\widetilde{x}_{\mathrm{op}}(T) + \widehat{x}_{\mathrm{op}}(T)\big)\big)\\
&= E\big(\widetilde{x}_{\mathrm{op}}(T)^* L_{\theta(T)}\widetilde{x}_{\mathrm{op}}(T)\big) + E\big(\widetilde{x}_{\mathrm{op}}(T)^* L_{\theta(T)}\widehat{x}_{\mathrm{op}}(T)\big)\\
&\quad + E\big(\widehat{x}_{\mathrm{op}}(T)^* L_{\theta(T)}\widetilde{x}_{\mathrm{op}}(T)\big) + E\big(\widehat{x}_{\mathrm{op}}(T)^* L_{\theta(T)}\widehat{x}_{\mathrm{op}}(T)\big)\\
&= E\big(\widetilde{x}_{\mathrm{op}}(T)^* L_{\theta(T)}\widetilde{x}_{\mathrm{op}}(T)\big) + E\big(\widehat{x}_{\mathrm{op}}(T)^* L_{\theta(T)}\widehat{x}_{\mathrm{op}}(T)\big)
\end{aligned}
$$

since, by Lemma 6.4, $E(\widetilde{x}_{\mathrm{op}}(T)^* L_{\theta(T)}\widehat{x}_{\mathrm{op}}(T)) = E(\widehat{x}_{\mathrm{op}}(T)^* L_{\theta(T)}\widetilde{x}_{\mathrm{op}}(T)) = 0$. But

$$
\begin{aligned}
E\big(\widetilde{x}_{\mathrm{op}}(T)^* L_{\theta(T)}\widetilde{x}_{\mathrm{op}}(T)\big) &= \sum_{i\in\mathcal{S}} \operatorname{tr}\big\{E\big(L_{\theta(T)}\widetilde{x}_{\mathrm{op}}(T)\widetilde{x}_{\mathrm{op}}(T)^* 1_{\{\theta(T)=i\}}\big)\big\}\\
&= \sum_{i\in\mathcal{S}} \operatorname{tr}\big\{L_i E\big(\widetilde{x}_{\mathrm{op}}(T)\widetilde{x}_{\mathrm{op}}(T)^* 1_{\{\theta(T)=i\}}\big)\big\}\\
&= \sum_{i\in\mathcal{S}} \operatorname{tr}\big(L_i Y_i(T)\big).
\end{aligned}
$$

Therefore,

$$
E\big(x(T)^* L_{\theta(T)} x(T)\big) = \sum_{i\in\mathcal{S}} \operatorname{tr}\big(L_i Y_i(T)\big) + E\big(\widehat{x}_{\mathrm{op}}(T)^* L_{\theta(T)}\widehat{x}_{\mathrm{op}}(T)\big).
$$

Finally, the cost function is given by

$$\begin{aligned}\mathcal{J}(u) &= \int_0^T E\big(\|z(t)\|^2\big)\,dt + E\big(x(T)^* L_{\theta(T)} x(T)\big) \\ &= \int_0^T \Big\{E\big(\|\widehat{z}_{\mathrm{op}}(t)\|^2\big) + \sum_{i\in\mathcal{S}} \operatorname{tr}\big(C_i(t)Y_i(t)C_i^*(t)\big)\Big\}\,dt + \sum_{i\in\mathcal{S}} \operatorname{tr}\big(L_i Y_i(T)\big) \\ &\quad + E\big(\widehat{x}_{\mathrm{op}}(T)^* L_{\theta(T)} \widehat{x}_{\mathrm{op}}(T)\big) \\ &= \int_0^T E\big(\|\widehat{z}_{\mathrm{op}}(t)\|^2\big)\,dt + E\big(\widehat{x}_{\mathrm{op}}(T)^* L_{\theta(T)} \widehat{x}_{\mathrm{op}}(T)\big) \\ &\quad + \sum_{i\in\mathcal{S}} \operatorname{tr}\Big(\int_0^T C_i(t)Y_i(t)C_i^*(t)dt + L_i Y_i(T)\Big). \end{aligned} \tag{6.29}$$

Notice now that $\sum_{i\in\mathcal{S}} \operatorname{tr}(\int_0^T C_i(t)Y_i(t)C_i^*(t)\,dt + L_i Y_i(T))$ does not depend on the control u. Therefore, minimizing

$$\mathcal{J}(u) = \int_0^T E\big(\|z(t)\|^2\big)\,dt + E\big(x(T)^* L_{\theta(T)} x(T)\big)$$

is equivalent to minimizing

$$\mathcal{J}_{\mathrm{op}}(u) = \int_0^T E\big(\|\widehat{z}_{\mathrm{op}}(t)\|^2\big)\,dt + E\big(\widehat{x}_{\mathrm{op}}(T)^* L_{\theta(T)} \widehat{x}_{\mathrm{op}}(T)\big)$$

subject to (6.27), i.e., we recast the problem as one with complete observations via $\widehat{x}_{\mathrm{op}}(t)$, the optimal filter. □

Remark 6.8 Notice that feedback controls of the form $u(t) = \widehat{C}_{\theta(t)}\widehat{x}_{\mathrm{op}}(t)$ can be written as in (6.3), since, in this case, we would have from (6.27) that

$$d\widehat{x}_{\mathrm{op}}(t) = \big(A_{\theta(t)}(t) + B_{\theta(t)}(t)\widehat{C}_{\theta(t)} - K^f_{\theta(t)}(t)H_{\theta(t)}(t)\big)\widehat{x}_{\mathrm{op}}(t)\,dt + K^f_{\theta(t)}(t)\,dy(t),$$

which is as in (6.3).

We proceed now to study the complete observations problem posed in Lemma 6.7, following an approach as in [159]. In this case we have an *observable state variable* system whose evolution in time is described by (6.27) with $\{u(t)\}$ as in (6.3) and

$$dv(t) = H_{\theta(t)}(t)\widetilde{x}_{\mathrm{op}}(t)\,dt + G_{\theta(t)}\,dw(t), \tag{6.30}$$

where $\widetilde{x}_{\mathrm{op}}(t) := x(t) - \widehat{x}_{\mathrm{op}}(t)$ with $x(t)$ in (6.1). Recall from Lemma 6.4 that for $i \in \mathcal{S}$,

$$E\big[\widetilde{x}_{\mathrm{op}}(t)\widehat{x}_{\mathrm{op}}(t)^* 1_{\{\theta(t)=i\}}\big] = 0. \tag{6.31}$$

The cost functional is given by

$$\mathcal{J}_{\mathrm{op}}(u) = E\left\{\int_0^T \left\|\widehat{z}_{\mathrm{op}}(t)\right\|^2 dt + \widehat{x}_{\mathrm{op}}(T)^* L_{\theta(T)} \widehat{x}_{\mathrm{op}}(T)\right\}, \tag{6.32}$$

and we set the *optimal cost* as

$$\mathcal{J}_{\mathrm{op}}^{\mathrm{op}} = \inf_{u \in \mathcal{U}} \mathcal{J}_{\mathrm{op}}(u). \tag{6.33}$$

We recall from Theorem 4.6 that there exists a unique set of N positive semi-definite and continuous $n \times n$ matrices $\mathbf{X}(t) = (X_1(t), \dots, X_N(t)) \in \mathbb{H}^{n+}$, $0 \le t \le T$, satisfying the CDRE (4.17), repeated here for convenience:

$$\begin{cases} \dot{X}_i(t) + A_i^*(t)X_i(t) + X_i(t)A_i(t) + \sum_{j \in \mathcal{S}} \lambda_{ij} X_j(t) + C_i^*(t)C_i(t) \\ \quad - X_i(t)B_i(t)\left[D_i^*(t)D_i(t)\right]^{-1} B_i^*(t)X_i(t) = 0, \quad t \in [0,T] \setminus \mathbb{D}, \\ X_i(T) = L_i, \quad i \in \mathcal{S}. \end{cases} \tag{6.34}$$

Let us define

$$\widehat{C}_i^{\mathrm{op}}(t) = -\left[D_i^*(t)D_i(t)\right]^{-1} B_i^*(t)X_i(t). \tag{6.35}$$

We will show in the sequel the following result.

Theorem 6.9 *Consider the stochastic optimal control problem defined via* (6.27) *and the cost functional* (6.32). *Then the optimal control policy in the class determined by* (6.3) *is given by* (*see Remark* 6.8)

$$u^0(t) = \widehat{C}_{\theta(t)}^{\mathrm{op}}(t)\widehat{x}_{\mathrm{op}}(t), \quad 0 \le t \le T. \tag{6.36}$$

In addition, from (6.33) *we have that the optimal cost is given by*

$$\mathcal{J}_{\mathrm{op}}^{\mathrm{op}} = \sum_{i \in \mathcal{S}} \left\{ p_i(0)\mu^* X_i(0)\mu + \int_0^T p_i(t)\,\mathrm{tr}\left(K_i^f(t)G_i(t)G_i^*(t)\left(K_i^f(t)\right)^* X_i(t)\right) dt \right\}, \tag{6.37}$$

where $\mathbf{X}(t) = (X_1(t), \dots, X_N(t))$ *are the unique positive semi-definite and continuous $n \times n$ matrices satisfying* (6.34).

The following auxiliary results aim to prove Theorem 6.9. In what follows, set $\mathcal{F}_t := \sigma\{(\theta(s), \widehat{x}_{\mathrm{op}}(s)); 0 \le s \le t\}$.

Lemma 6.10 *Let* $\mathbf{X}(t)=(X_1(t),\dots,X_N(t))\in\mathbb{H}^{n+}$ *be as in Theorem* 6.9. *Then*

$$\begin{aligned}E\big(\widehat{x}_{\mathrm{op}}(t)^* \, dX_{\theta(t)}(t)\widehat{x}_{\mathrm{op}}(t)\big)\\&=E\left(\widehat{x}_{\mathrm{op}}(t)^*\left[\sum_{j\in\mathcal{S}}\lambda_{\theta(t)j}X_j(t)\,dt+\dot{X}_{\theta(t)}(t)\,dt\right]\widehat{x}_{\mathrm{op}}(t)\right).\end{aligned}$$

Proof We have from (2.13) that

$$\begin{aligned}&E\big(X_{\theta(t+h)}(t+h)-X_{\theta(t)}(t)\big|\mathcal{F}_t\big)\\&\quad=E\big(X_{\theta(t+h)}(t+h)-X_{\theta(t)}(t+h)+X_{\theta(t)}(t+h)-X_{\theta(t)}(t)\big|\mathcal{F}_t\big)\\&\quad=E\big(X_{\theta(t+h)}(t+h)-X_{\theta(t)}(t+h)\big|\mathcal{F}_t\big)+X_{\theta(t)}(t+h)-X_{\theta(t)}(t)\\&\quad=\sum_{j\in\mathcal{S}}\lambda_{\theta(t)j}X_j(t+h)h+\dot{X}_{\theta(t)}(t)h+o(h),\end{aligned}$$

since $X_{\theta(t)}(t+h)-X_{\theta(t)}(t)=\dot{X}_{\theta(t)}(t)h+o(h)$. This implies also that

$$\begin{aligned}&E\big(X_{\theta(t+h)}(t+h)-X_{\theta(t)}(t)\big|\mathcal{F}_t\big)\\&\quad=\sum_{j\in\mathcal{S}}\lambda_{\theta(t)j}\big(X_j(t)+\dot{X}_j(t)h+o(h)\big)h+\dot{X}_{\theta(t)}(t)h+o(h)\\&\quad=\sum_{j\in\mathcal{S}}\lambda_{\theta(t)j}X_j(t)h+\dot{X}_{\theta(t)}(t)h+o(h),\end{aligned}$$

and the result follows (bearing in mind the notation in (3.22) and (3.23)). □

Lemma 6.11 *We have that*

$$E\big(\mathrm{tr}\big(K^f_{\theta(t)}(t)\,d\nu(t)\widehat{x}_{\mathrm{op}}(t)^*X_{\theta(t)}(t)\big)\big)=0 \tag{6.38}$$

and

$$\begin{aligned}&E\big(\mathrm{tr}\big(K^f_{\theta(t)}(t)\,d\nu(t)\,d\nu(t)^*\big(K^f_{\theta(t)}(t)\big)^*X_{\theta(t)}(t)\big)\big)\\&\quad=E\big(\mathrm{tr}\big(K^f_{\theta(t)}(t)G_{\theta(t)}(t)G^*_{\theta(t)}(t)\big(K^f_{\theta(t)}(t)\big)^*X_{\theta(t)}(t)\big)\big)\,dt.\end{aligned}$$

Proof The first result follows from (6.30) and (6.31), since

$$\begin{aligned}&E\big(\mathrm{tr}\big(K^f_{\theta(t)}(t)\,d\nu(t)\widehat{x}_{\mathrm{op}}(t)^*X_{\theta(t)}(t)\big)\big)\\&\quad=\sum_{i\in\mathcal{S}}\mathrm{tr}\big(K^f_i(t)H_iE\big(\widetilde{x}_{\mathrm{op}}(t)\widehat{x}_{\mathrm{op}}(t)^*1_{\{\theta(t)=i\}}\big)X_i(t)\big)\,dt=0,\end{aligned}$$

whereas the second result follows from (6.30), bearing in mind the assumptions made on W and θ. □

Lemma 6.12 *Consider $\widehat{x}_{\mathrm{op}}(t)$ defined as in* (6.27) *and*

$$V(t) := E\big(\widehat{x}_{\mathrm{op}}(t)^* X_{\theta(t)}(t)\widehat{x}_{\mathrm{op}}(t)\big) \tag{6.39}$$

with $\mathbf{X}(t) = (X_1(t), \dots, X_N(t)) \in \mathbb{H}^{n+}$ *as in Theorem* 6.9. *Then*

$$dV(t) = -E\big(\|\widehat{z}_{\mathrm{op}}(t)\|^2\big)\,dt + E\big(\|D_{\theta(t)}(t)\big(u(t) - \widehat{C}^{\mathrm{op}}_{\theta(t)}(t)\widehat{x}_{\mathrm{op}}(t)\big)\|^2\big)\,dt \\ + E\big(\mathrm{tr}\big(K^f_{\theta(t)}(t)G_{\theta(t)}(t)G^*_{\theta(t)}(t)\big(K^f_{\theta(t)}(t)\big)^* X_{\theta(t)}(t)\big)\big)\,dt.$$

Proof From Lemmas 6.10 and 6.11, in conjunction with (6.34), we have that

$$\begin{aligned} dV(t) &= E\big(d\widehat{x}_{\mathrm{op}}(t)^* X_{\theta(t)}(t)\widehat{x}_{\mathrm{op}}(t)\big) + E\big(\widehat{x}_{\mathrm{op}}(t)^*\, dX_{\theta(t)}(t)\widehat{x}_{\mathrm{op}}(t)\big) \\ &\quad + E\big(\widehat{x}_{\mathrm{op}}(t)^* X_{\theta(t)}(t)\, d\widehat{x}_{\mathrm{op}}(t)\big) + E\big(d\widehat{x}_{\mathrm{op}}(t)^* X_{\theta(t)}(t)\, d\widehat{x}_{\mathrm{op}}(t)\big) \\ &= -E\big(\|\widehat{z}_{\mathrm{op}}(t)\|^2\big)\,dt + E\big(\|D_{\theta(t)}(t)\big(u(t) - \widehat{C}^{\mathrm{op}}_{\theta(t)}(t)\widehat{x}_{\mathrm{op}}(t)\big)\|^2\big)\,dt \\ &\quad + 2E\big(\mathrm{tr}\big(K^f_{\theta(t)}(t)\,d\nu(t)\widehat{x}_{\mathrm{op}}(t)^* X_{\theta(t)}(t)\big)\big) \\ &\quad + E\big(\mathrm{tr}\big(K^f_{\theta(t)}(t)\,d\nu(t)\,d\nu(t)^*\big(K^f_{\theta(t)}(t)\big)^* X_{\theta(t)}(t)\big)\big) \\ &= -E\big(\|\widehat{z}_{\mathrm{op}}(t)\|^2\big)\,dt + E\big(\|D_{\theta(t)}(t)\big(u(t) - \widehat{C}^{\mathrm{op}}_{\theta(t)}(t)\widehat{x}_{\mathrm{op}}(t)\big)\|^2\big)\,dt \\ &\quad + E\big(\mathrm{tr}\big(K^f_{\theta(t)}(t)G_{\theta(t)}(t)G^*_{\theta(t)}(t)\big(K^f_{\theta(t)}(t)\big)^* X_{\theta(t)}(t)\big)\big)\,dt, \end{aligned}$$

and the result follows. □

We can now proceed to the proof of Theorem 6.9.

Proof of Theorem 6.9 From Lemma 6.12, it follows that

$$\begin{aligned} V(T) - V(0) = &-\int_0^T E\big(\|\widehat{z}_{\mathrm{op}}(t)\|^2\big)\,dt \\ &+ \int_0^T E\big(\|D_{\theta(t)}(t)\big(u(t) - \widehat{C}^{\mathrm{op}}_{\theta(t)}(t)\widehat{x}_{\mathrm{op}}(t)\big)\|^2\big)\,dt \\ &+ \int_0^T E\big(\mathrm{tr}\big(K^f_{\theta(t)}(t)G_{\theta(t)}(t)G^*_{\theta(t)}(t)\big(K^f_{\theta(t)}(t)\big)^* X_{\theta(t)}(t)\big)\big)\,dt \end{aligned}$$

or, equivalently,

$$\begin{aligned} \mathcal{J}_{\mathrm{op}}(u) = {}& E\big(\widehat{x}_{\mathrm{op}}(0)^* X_{\theta(0)}(0)\widehat{x}_{\mathrm{op}}(0)\big) \\ &+ \int_0^T E\big(\|D_{\theta(t)}(t)\big(u(t) - \widehat{C}^{\mathrm{op}}_{\theta(t)}\widehat{x}_{\mathrm{op}}(t)\big)\|^2\big)\,dt \\ &+ \int_0^T E\big(\mathrm{tr}\big(K^f_{\theta(t)}(t)G_{\theta(t)}(t)G^*_{\theta(t)}(t)\big(K^f_{\theta(t)}(t)\big)^* X_{\theta(t)}(t)\big)\,dt. \end{aligned} \tag{6.40}$$

From (6.40) it follows that minimizing the cost functional $\mathcal{J}_{\mathrm{op}}(u)$ within the class determined by (6.3) is equivalent to choosing (see Remark 6.8)

$$u(t) = \widehat{C}^{\mathrm{op}}_{\theta(t)} \widehat{x}_{\mathrm{op}}(t).$$

This, in turn, gives the optimal cost

$$\begin{aligned}
\mathcal{J}_{\mathrm{op}}(u) &= E\big(\widehat{x}_{\mathrm{op}}(0)^* X_{\theta(0)}(0)\widehat{x}_{\mathrm{op}}(0)\big) \\
&\quad + \int_0^T E\big(\mathrm{tr}\big(K^f_{\theta(t)}(t)G_{\theta(t)}(t)G^*_{\theta(t)}(t)\big(K^f_{\theta(t)}(t)\big)^* X_{\theta(t)}(t)\big)\big)\,dt \\
&= \sum_{i\in\mathcal{S}} \bigg\{ p_i(0)\mu^* X_i(0)\mu \\
&\quad + \int_0^T p_i(t)\,\mathrm{tr}\big(K^f_i(t)G_i(t)G^*_i(t)\big(K^f_i(t)\big)^* X_i(t)\big)\,dt \bigg\},
\end{aligned}$$

and the result follows. □

From Theorem 6.6, Lemma 6.7, and Theorem 6.9 we have the following separation principle result.

Theorem 6.13 (The Separation Principle) *An optimal solution for the OC problem posed in Sect. 6.2.1 is* $\widehat{\mathbf{A}}^{\mathrm{op}} = (\widehat{A}^{\mathrm{op}}_1(t), \ldots, \widehat{A}^{\mathrm{op}}_N(t))$, $\widehat{\mathbf{B}}^{\mathrm{op}} = (\widehat{B}^{\mathrm{op}}_1(t), \ldots, \widehat{B}^{\mathrm{op}}_N(t))$, $\widehat{\mathbf{C}}^{\mathrm{op}} = (\widehat{C}^{\mathrm{op}}_1(t), \ldots, \widehat{C}^{\mathrm{op}}_N(t))$ *given, for* $i \in \mathcal{S}$ *and* $t \in \mathbb{R}^+$, *by*

$$\begin{aligned}
\widehat{A}^{\mathrm{op}}_i(t) &= A_i(t) + B_i(t)\widehat{C}^{\mathrm{op}}_i(t) - \widehat{B}^{\mathrm{op}}_i(t)H_i(t), \\
\widehat{B}^{\mathrm{op}}_i(t) &= Y_i(t)H^*_i(t)\big(G_i(t)G^*_i(t)p_i(t)\big)^{-1}, \\
\widehat{C}^{\mathrm{op}}_i(t) &= -\big[D^*_i(t)D_i(t)\big]^{-1} B^*_i(t)X_i(t),
\end{aligned}$$

with $\widehat{x}(0) = \mu$, *where* $\mathbf{Y}(t) = (Y_1(t), \ldots, Y_N(t)) \in \mathbb{H}^{n+}$ *are the unique positive semi-definite and continuous* $n \times n$ *matrices satisfying* (6.14), *and* $\mathbf{X}(t) = (X_1(t), \ldots, X_N(t)) \in \mathbb{H}^{n+}$ *are the unique positive semi-definite and continuous* $n \times n$ *matrices satisfying* (6.34). *The optimal cost is given by* $\mathcal{J}^{\mathrm{op}}_{\mathrm{op}} = \sum_{i\in\mathcal{S}} \mathcal{J}^{\mathrm{op}}_{\mathrm{op}}(i)$ *with*

$$\begin{aligned}
\mathcal{J}^{\mathrm{op}}_{\mathrm{op}}(i) &= p_i(0)\mu^* X_i(0)\mu + \int_0^T p_i(t)\,\mathrm{tr}\big(K^f_i(t)G_i(t)G^*_i(t)\big(K^f_i(t)\big)^* X_i(t)\big)\,dt \\
&\quad + \int_0^T \mathrm{tr}\big(C_i(t)Y_i(t)C^*_i(t)\big)\,dt + \mathrm{tr}\big(L_iY_i(T)\big), \quad i \in \mathcal{S}.
\end{aligned}$$

6.3 The H_2 Control Problem with Partial Observations

6.3.1 Problem Statement

In the remainder of this chapter we will study the H_2 control problem with partial observations, which can be seen as an infinite-horizon time-invariant version of the problem studied in the previous section. We will consider a Markov jump linear system in which all the matrices are time-invariant,

$$\mathcal{G} = \begin{cases} \dot{x}(t) = A_{\theta(t)}x(t) + B_{\theta(t)}u(t) + J_{\theta(t)}w(t), \\ y(t) = H_{\theta(t)}x(t) + G_{\theta(t)}w(t), \\ z(t) = C_{\theta(t)}x(t) + D_{\theta(t)}u(t), \end{cases} \tag{6.41}$$

assuming that the output and "operation modes" ($y(t)$ and $\theta(t)$, respectively) are known at each time t. As before, we suppose that $D_i^* D_i > 0$, $C_i^* D_i = 0$, $G_i G_i^* > 0$, and $J_i G_i^* = 0$ for each $i \in \mathcal{S}$. In addition, we assume from now on that the initial distribution for the Markov chain coincides with the invariant distribution π_i, so that $p_i(t) = P(\theta(t) = i) = \pi_i$ for all t. We will consider the dynamic Markov jump controllers $\mathcal{G}_K$ for system (6.41) given by

$$\mathcal{G}_K = \begin{cases} \dot{\widehat{x}}(t) = \widehat{A}_{\theta(t)}\widehat{x}(t) + \widehat{B}_{\theta(t)}y(t), \\ u(t) = \widehat{C}_{\theta(t)}\widehat{x}(t), \end{cases} \tag{6.42}$$

with $\widehat{\mathbf{A}} = (\widehat{A}_1, \dots, \widehat{A}_N)$, $\widehat{\mathbf{B}} = (\widehat{B}_1, \dots, \widehat{B}_N)$, and $\widehat{\mathbf{C}} = (\widehat{C}_1, \dots, \widehat{C}_N)$.

Remark 6.14 The advantage of considering Markov jump controllers as in (6.42) is that they are not sample path dependent, which allows us to obtain the stochastic stability of the closed-loop system.

From (6.41) and (6.42) we have that the closed-loop system is

$$\begin{aligned} \begin{bmatrix} \dot{x}(t) \\ \dot{\widehat{x}}(t) \end{bmatrix} &= \begin{bmatrix} A_{\theta(t)} & B_{\theta(t)}\widehat{C}_{\theta(t)} \\ \widehat{B}_{\theta(t)}H_{\theta(t)} & \widehat{A}_{\theta(t)} \end{bmatrix} \begin{bmatrix} x(t) \\ \widehat{x}(t) \end{bmatrix} + \begin{bmatrix} J_{\theta(t)} \\ \widehat{B}_{\theta(t)}G_{\theta(t)} \end{bmatrix} w(t), \\ z(t) &= \begin{bmatrix} C_{\theta(t)} & D_{\theta(t)}\widehat{C}_{\theta(t)} \end{bmatrix} \begin{bmatrix} x(t) \\ \widehat{x}(t) \end{bmatrix}. \end{aligned} \tag{6.43}$$

Setting, for each $i \in \mathcal{S}$,

$$\begin{aligned} \Gamma_i &= \begin{bmatrix} A_i & B_i\widehat{C}_i \\ \widehat{B}_i H_i & \widehat{A}_i \end{bmatrix}, \qquad \Psi_i = \begin{bmatrix} J_i \\ \widehat{B}_i G_i \end{bmatrix}, \\ \Lambda_i &= \begin{bmatrix} C_i & D_i\widehat{C}_i \end{bmatrix}, \qquad \mathbf{v}(t) = \begin{bmatrix} x(t) \\ \widehat{x}(t) \end{bmatrix}, \end{aligned} \tag{6.44}$$

we have from (6.43) that the Markovian jump closed-loop system $\mathcal{G}_{\mathrm{cl}}$ is given by

$$\mathcal{G}_{\mathrm{cl}} = \begin{cases} \dot{\mathbf{v}}(t) = \Gamma_{\theta(t)}\mathbf{v}(t) + \Psi_{\theta(t)}w(t), \\ z(t) = \Lambda_{\theta(t)}\mathbf{v}(t). \end{cases} \tag{6.45}$$

Definition 6.15 We say that $(\widehat{\mathbf{A}}, \widehat{\mathbf{B}}, \widehat{\mathbf{C}})$ in the definition of the controller $\mathcal{G}_K$ given by (6.42) is admissible if the closed-loop MJLS $\mathcal{G}_{\text{cl}}$ (6.45) is MSS according to Definition 3.2.

For the class of admissible controllers, we recall from Definition 5.3 that the H_2-norm of the closed loop system $\mathcal{G}_{\text{cl}}$ (given by (6.45) with $\mathbf{v}(0) = 0$), denoted by $\|\mathcal{G}_{\text{cl}}\|_2$, is defined as

$$\|\mathcal{G}_{\text{cl}}\|_2^2 = \sum_{s=1}^{r} \sum_{i \in \mathcal{S}} \pi_i \|z_{s,i}\|_2^2 = \sum_{s=1}^{r} \|z_s\|_2^2, \tag{6.46}$$

where z_s represents the output $\{z(t); t \geq 0\}$ of (6.45) when:

(a) the input is given by $w(t) = \{e_s \delta(t); t \geq 0\}$, $\delta(t)$ the unitary impulse, and e_s the r-dimensional unitary vector formed by 1 at the sth position and zero elsewhere, and
(b) $\theta_0 = i$ with probability π_i for each $i \in \mathcal{S}$.

Since system (6.45) is MSS, we have that the norm $\|z_{s,j}\|_2^2$ in (6.46) is finite. As remarked in Chap. 5, for the case with no jumps, the definition above coincides with the usual H_2-norm. The H_2-norm can be computed in terms of the coupled *observability Gramian* $\mathbf{S} = (S_1, \ldots, S_N) \in \mathbb{H}^{n+}$:

$$\Gamma_i^* S_i + S_i \Gamma_i + \sum_{j \in \mathcal{S}} \lambda_{ij} S_j + \Lambda_i^* \Lambda_i = 0, \quad i \in \mathcal{S}, \tag{6.47}$$

or, alternatively, from the coupled *controllability Gramian* $\mathbf{P} = (P_1, \ldots, P_N) \in \mathbb{H}^{n+}$:

$$\Gamma_j P_j + P_j \Gamma_j^* + \sum_{i \in \mathcal{S}} \lambda_{ij} P_i + \pi_j \Psi_j \Psi_j^* = 0, \quad j \in \mathcal{S}, \tag{6.48}$$

as stated in the next theorem.

Theorem 6.16 *For the closed-loop MJLS* (6.45) *with an admissible controller, we have that*

$$\|\mathcal{G}_{\text{cl}}\|_2^2 = \sum_{j \in \mathcal{S}} \pi_j \operatorname{tr}(\Psi_j^* S_j \Psi_j) = \sum_{j \in \mathcal{S}} \operatorname{tr}(\Lambda_j P_j \Lambda_j^*). \tag{6.49}$$

Proof This is a consequence of Theorem 5.4. □

The optimal H_2 control (OC) problem with partial observations we want to study is defined as follows.

Definition 6.17 Find $(\widehat{\mathbf{A}}, \widehat{\mathbf{B}}, \widehat{\mathbf{C}})$ in (6.42) such that the closed-loop MJLS $\mathcal{G}_{\text{cl}}$ in (6.45) is MSS and minimizes $\|\mathcal{G}_{\text{cl}}\|_2^2$.

We next present an alternative definition for the H_2 control problem with partial observations, which shows that it is possible to rephrase the problem in the same manner as in the classical stochastic control way. Suppose that in model (6.41), $W = \{(w(t), \mathcal{F}_t), t \in \mathbb{R}^+\}$ is an r-dimensional Wiener process with incremental covariance operator $I dt$ and independent of the initial condition x_0 and the Markov process $\theta(t)$. Let $\mathcal{G}_K$ be an admissible controller given by (6.42), $\mathbf{v}(t)$ be as in (6.45), and

$$P_i(t) = E\big(\mathbf{v}(t)\mathbf{v}(t)^* 1_{\{\theta(t)=i\}}\big), \quad i \in \mathcal{S}. \tag{6.50}$$

By Proposition 3.28, $\mathbf{P}(t) = (P_1(t), \dots, P_N(t)) \in \mathbb{H}^{n+}$ satisfies

$$\dot{P}_j(t) = \Gamma_j P_j(t) + P_j(t)\Gamma_j^* + \sum_{i \in \mathcal{S}} \lambda_{ij} P_i(t) + \pi_j \Psi_j \Psi_j^*, \quad j \in \mathcal{S}.$$

Moreover, since the closed-loop system is MSS, we have that $\mathbf{P}(t) \overset{t\uparrow\infty}{\longrightarrow} \mathbf{P}$, where $\mathbf{P} = (P_1, \dots, P_N) \in \mathbb{H}^{n+}$ is a unique solution of the coupled Lyapunov equations (6.48). Notice that

$$\begin{aligned} E\big(\|z(t)\|^2\big) &= E\big(\operatorname{tr}\big(z(t)z(t)^*\big)\big) \\ &= \operatorname{tr}\big(E\big(\Lambda_{\theta(t)}\mathbf{v}(t)\mathbf{v}(t)^*\Lambda_{\theta(t)}^*\big)\big) \\ &= \sum_{i \in \mathcal{S}} \operatorname{tr}\big(E\big(\Lambda_i\big[\mathbf{v}(t)\mathbf{v}(t)^* 1_{\{\theta(t)=i\}}\big]\Lambda_i^*\big)\big) \\ &= \sum_{i \in \mathcal{S}} \operatorname{tr}\big(\Lambda_i P_i(t)\Lambda_i^*\big) \overset{t\uparrow\infty}{\longrightarrow} \sum_{i \in \mathcal{S}} \operatorname{tr}\big(\Lambda_i P_i \Lambda_i^*\big) = \|\mathcal{G}_{\mathrm{cl}}\|_2^2, \end{aligned} \tag{6.51}$$

and thus an alternative definition for the H_2 control problem is as follows.

Definition 6.18 Find $(\widehat{\mathbf{A}}, \widehat{\mathbf{B}}, \widehat{\mathbf{C}})$ in (6.42) such that the closed-loop MJLS $\mathcal{G}_{\mathrm{cl}}$ (6.45) is MSS and minimizes $\lim_{t\to\infty} E(\|z(t)\|^2)$.

Remark 6.19 Note that $\lim_{t\to\infty} E(\|z(t)\|^2)$ does not depend on the initial condition x_0 for system (6.41).

In order to solve the problem, we will have to assume the existence of stabilizing solutions associated to the control and filtering coupled algebraic Riccati equations (see Appendix A for a necessary and sufficient condition for the existence of this solution). For the control problem case, this was presented in Definition 4.11, which states that $\mathbf{X} = (X_1, \dots, X_N) \in \mathbb{H}^{n+}$ is the stabilizing solution of the control CARE if it satisfies, for each $i \in \mathcal{S}$,

$$A_i^* X_i + X_i A_i - X_i B_i\big(D_i^* D_i\big)^{-1} B_i^* X_i + \sum_{j \in \mathcal{S}} \lambda_{ij} X_j + C_i^* C_i = 0 \tag{6.52}$$

and $\operatorname{Re}\{\lambda(\mathcal{L})\} < 0$, where $\mathcal{L} = (\mathcal{L}_1, \dots, \mathcal{L}_N)$ is defined as in (3.21) replacing A_i by $A_i - B_i \mathcal{K}_i(\mathbf{X})$, with

$$\mathcal{K}_i(\mathbf{X}) = \left(D_i^* D_i\right)^{-1} B_i^* X_i \tag{6.53}$$

for each $i \in \mathcal{S}$. For the filtering case, the definition is as follows.

Definition 6.20 (Filtering case) We say that $\mathbf{Y} = (Y_1, \dots, Y_N) \in \mathbb{H}^{n+}$ is the stabilizing solution of the filtering CARE if it satisfies, for each $j \in \mathcal{S}$,

$$A_j Y_j + Y_j A_j^* - Y_j H_j^* \left(\pi_j G_j G_j^*\right)^{-1} H_j Y_j + \sum_{i \in \mathcal{S}} \lambda_{ij} Y_i + \pi_j J_j J_j^* = 0 \tag{6.54}$$

and $\operatorname{Re}\{\lambda(\mathcal{L})\} < 0$, where $\mathcal{L} = (\mathcal{L}_1, \dots, \mathcal{L}_N)$ is defined as in (3.21) replacing A_i by $A_i - \mathcal{K}_i^f(\mathbf{Y}) H_i$, with

$$\mathcal{K}_i^f(\mathbf{Y}) = Y_i H_i^* \left(G_i G_i^* \pi_i\right)^{-1} \tag{6.55}$$

for each $i \in \mathcal{S}$.

6.3.2 Filtering H_2 Problem

Since our problem is with partial observations, the following optimal filtering (OF) problem will be crucial for deriving the results. Consider the following MJLS:

$$\mathcal{G}_v = \begin{cases} dx(t) = \left(A_{\theta(t)} x(t) + B_{\theta(t)} u(t)\right) dt + J_{\theta(t)}\, dw(t), \\ dy(t) = H_{\theta(t)} x(t)\, dt + G_{\theta(t)}\, dw(t), \\ v(t) = R_{\theta(t)}^{1/2} \left(F_{\theta(t)} x(t) + u(t)\right), \end{cases} \tag{6.56}$$

where we assume, as before, that $W = \{(w(t), \mathcal{F}_t), t \in \mathbb{R}^+\}$ is an r-dimensional Wiener process with incremental covariance operator $I\,dt$, independent of the initial condition x_0 and the Markov process $\theta(t)$, and that the initial distribution for the Markov chain is given by π_i, so that $p_i(t) = \pi_i$ for all t. We also assume that $\mathbf{F} = (F_1, \dots, F_N)$ stabilizes $(\mathbf{A}, \mathbf{B}, \Pi)$ in the mean-square sense. It is desired to minimize $\lim_{t \to \infty} E(\|v(t)\|^2)$ by considering stochastically stabilizing Markov jump linear filters as in (6.42) with deterministic $\widehat{x}(0)$. Notice that minimizing $\|v(t)\|$ amounts to choosing $u(t)$ which best approximates (weighted by $R_{\theta(t)}$) the term $F_{\theta(t)} x(t)$.

The definition of this filtering problem traces a close parallel with the Output Estimation problem in the classical H_2 optimal control literature (see, for instance, [324]). In Sect. 6.3.3, we will present the separation of the cost function in two components, one of which will have to do with the filtering problem posed here.

Suppose that there exists $\mathbf{Y} = (Y_1, \dots, Y_N) \in \mathbb{H}^{n+}$, the stabilizing solution (see Definition 6.20) of the filtering CARE (6.54). Let $\mathbf{K}^f = (K_1^f, \dots, K_N^f) \in \mathbb{H}^{p,n}$ be as in (6.55) (for simplicity, we drop from now on the dependence on $\mathbf{Y}$).

As mentioned in Remark 6.19, $\lim_{t\to\infty} E(\|v(t)\|^2)$ does not depend on the initial conditions $\widehat{x}(0)$, x_0. Therefore, we make convenient choices for these initial conditions. We assume that

$$x_0 = \frac{1}{\pi_{\theta_0}^{1/2}} Y_{\theta_0}^{1/2} \eta, \qquad \widehat{x}(0) = 0, \tag{6.57}$$

where η is a zero-mean n-dimensional random vector independent of θ_0 and with identity covariance matrix. Notice that with this choice, $E(x_0) = 0$, $E(x_0 x_0^* 1_{\{\theta_0=i\}}) = Y_i$, and $E(\|x_0\|^2) = \operatorname{tr}(\sum_{i\in\mathcal{S}} E(x_0 x_0^* 1_{\{\theta_0=i\}})) \le n\|\mathbf{Y}\|_1$.

Consider now the MJLS $\mathcal{G}_v$ given by (6.56) and

$$\begin{cases} d\widehat{x}_e(t) = \big(A_{\theta(t)}\widehat{x}_e(t) + B_{\theta(t)}u(t)\big)\,dt + K^f_{\theta(t)}\big(dy(t) - H_{\theta(t)}\widehat{x}_e(t)\,dt\big), \\ \widehat{x}_e(0) = 0, \end{cases} \tag{6.58}$$

where $u(t)$ is given by (6.42). The associated error related with the estimator given in (6.58) is defined by $\widetilde{x}_e(t) = x(t) - \widehat{x}_e(t)$, and from (6.56) and (6.58) we have that it satisfies

$$\begin{cases} d\widetilde{x}_e(t) = \big(A_{\theta(t)} - K^f_{\theta(t)}H_{\theta(t)}\big)\widetilde{x}_e(t)\,dt + \big(J_{\theta(t)} - K^f_{\theta(t)}G_{\theta(t)}\big)\,dw(t), \\ \widetilde{x}_e(0) = x_0. \end{cases} \tag{6.59}$$

Set $\mathbf{Y}(t) = (Y_1(t), \ldots, Y_N(t)) \in \mathbb{H}^{n+}$ as $Y_i(t) = E(\widetilde{x}_e(t)\widetilde{x}_e(t)^* 1_{\{\theta(t)=i\}})$ for $t \ge 0$ and $i \in \mathcal{S}$. From Proposition 3.28 we have that $\mathbf{Y}(t)$ satisfies the equation

$$\dot{\mathbf{Y}}(t) = \mathcal{L}\big(\mathbf{Y}(t)\big) + \mathbf{V}, \tag{6.60}$$

where $\mathcal{L}$ is defined as in (3.21) replacing A_i by $A_i - K^f_i H_i$, and $\mathbf{V} = (V_1, \ldots, V_N) \in \mathbb{H}^{n+}$ is defined as $V_i = \pi_i(K^f_i G_i G_i^* K^{f^*}_i + J_i J_i^*)$. But notice that, by the CARE (6.54) and (6.55),

$$0 = \big(A_j - K^f_j H_j\big)Y_j + Y_j\big(A_j - K^f_j H_j\big)^* + \sum_{i\in\mathcal{S}} \lambda_{ij} Y_i + \pi_j\big(J_j J_j^* + K^f_j G_j G_j^* (K^f_j)^*\big)$$

for all $j \in \mathcal{S}$, that is,

$$0 = \mathcal{L}(\mathbf{Y}) + \mathbf{V}. \tag{6.61}$$

From the initial conditions (6.57) we get that $\mathbf{Y}(0) = \mathbf{Y}$ since

$$Y_i(0) = E\big(\widetilde{x}_e(0)\widetilde{x}_e(0)^* 1_{\{\theta_0=i\}}\big) = E\big(x_0 x_0^* 1_{\{\theta_0=i\}}\big) = Y_i. \tag{6.62}$$

From (6.60), (6.61), and (6.62) it follows that $\mathbf{Y}(t) = \mathbf{Y}$ for all $t \in \mathbb{R}^+$. We have the following propositions.

Proposition 6.21 *For $\widehat{x}(t)$, $\widehat{x}_e(t)$, and $\widetilde{x}_e(t)$ given by* (6.42),(6.58), *and* (6.59), *respectively, and for all $i \in \mathcal{S}$ and $t \in \mathbb{R}^+$, we have that*

$$E\big(\widetilde{x}_e(t)\widehat{x}_e(t)^* 1_{\{\theta(t)=i\}}\big) = 0,$$
$$E\big(\widetilde{x}_e(t)\widehat{x}(t)^* 1_{\{\theta(t)=i\}}\big) = 0.$$

Proof See Lemma 6.4. □

Proposition 6.22 *Let $v(t)$, $\mathbf{P}(t) = (P_1(t), \dots, P_N(t))$ and $\mathbf{Y}(t)=(Y_1(t), \dots, Y_N(t)) = \mathbf{Y}$ be as in* (6.56), (6.50), *and* (6.60), *respectively. Then for every $t \in \mathbb{R}^+$,*

$$E\big(\|v(t)\|^2\big) = \sum_{i\in\mathcal{S}} \operatorname{tr}\big(\Lambda_i P_i(t)\Lambda_i^*\big) \ge \sum_{i\in\mathcal{S}} \operatorname{tr}\big(R_i^{1/2} F_i Y_i F_i^* R_i^{1/2}\big), \tag{6.63}$$

where $\Lambda_i = R_i^{1/2}[F_i \; \widehat{C}_i]$.

Proof See Lemma 6.5. □

Set $\widehat{\mathbf{A}}^{\mathrm{op}} = (\widehat{A}_1^{\mathrm{op}}, \dots, \widehat{A}_N^{\mathrm{op}})$, $\widehat{\mathbf{B}}^{\mathrm{op}} = (\widehat{B}_1^{\mathrm{op}}, \dots, \widehat{B}_N^{\mathrm{op}})$, and $\widehat{\mathbf{C}}^{\mathrm{op}} = (\widehat{C}_1^{\mathrm{op}}, \dots, \widehat{C}_N^{\mathrm{op}})$ as follows:

$$\widehat{A}_i^{\mathrm{op}} = A_i - K_i^f H_i - B_i F_i, \quad \widehat{B}_i^{\mathrm{op}} = K_i^f, \quad \widehat{C}_i^{\mathrm{op}} = -F_i, \quad i \in \mathcal{S}. \tag{6.64}$$

We have the following proposition.

Proposition 6.23 *The controller $(\widehat{\mathbf{A}}^{\mathrm{op}}, \widehat{\mathbf{B}}^{\mathrm{op}}, \widehat{\mathbf{C}}^{\mathrm{op}})$ is admissible according to Definition* 6.15.

Proof Consider the MJLS

$$\begin{bmatrix} \dot{x}^{\mathrm{op}}(t) \\ \dot{e}^{\mathrm{op}}(t) \end{bmatrix} = \begin{bmatrix} A_{\theta(t)} - B_{\theta(t)}F_{\theta(t)} & B_{\theta(t)}F_{\theta(t)} \\ 0 & A_{\theta(t)} - K_{\theta(t)}^f H_{\theta(t)} \end{bmatrix} \begin{bmatrix} x^{\mathrm{op}}(t) \\ e^{\mathrm{op}}(t) \end{bmatrix}.$$

Since $\mathbf{Y}$ is the stabilizing solution of (6.54), we have that the subsystem

$$\dot{e}^{\mathrm{op}}(t) = \big(A_{\theta(t)} - K_{\theta(t)}^f H_{\theta(t)}\big)e^{\mathrm{op}}(t)$$

is MSS, and thus $e^{\mathrm{op}} = \{e^{\mathrm{op}}(t); t \in \mathbb{R}^+\} \in L_2^n(\Omega, \mathcal{F}, P)$. By hypothesis, $\mathbf{F}$ stabilizes $(\mathbf{A}, \mathbf{B}, \Pi)$, and thus from the fact that $e^{\mathrm{op}} \in L_2^n(\Omega, \mathcal{F}, P)$ and

$$\dot{x}^{\mathrm{op}}(t) = (A_{\theta(t)} - B_{\theta(t)}F_{\theta(t)})x^{\mathrm{op}}(t) + B_{\theta(t)}F_{\theta(t)}e^{\mathrm{op}}(t)$$

we have from Theorem 3.27 that $x^{\mathrm{op}} = \{x^{\mathrm{op}}(t); t \in \mathbb{R}^+\} \in L_2^n(\Omega, \mathcal{F}, P)$. Setting $\widehat{x}^{\mathrm{op}}(t) = x^{\mathrm{op}}(t) - e^{\mathrm{op}}(t)$ and $\widehat{x}^{\mathrm{op}} = \{\widehat{x}^{\mathrm{op}}(t); t \in \mathbb{R}^+\}$, it follows that $\widehat{x}^{\mathrm{op}} \in$

$L_2^n(\Omega, \mathcal{F}, P)$ and that

$$\begin{bmatrix} \dot{x}^{\mathrm{op}}(t) \\ \dot{\widehat{x}}^{\mathrm{op}}(t) \end{bmatrix} = \begin{bmatrix} A_{\theta(t)} & B_{\theta(t)}\widehat{C}^{\mathrm{op}}_{\theta(t)} \\ \widehat{B}^{\mathrm{op}}_{\theta(t)}H_{\theta(t)} & \widehat{A}^{\mathrm{op}}_{\theta(t)} \end{bmatrix} \begin{bmatrix} x^{\mathrm{op}}(t) \\ \widehat{x}^{\mathrm{op}}(t) \end{bmatrix},$$

proving the desired result. □

By combining the previous propositions we have the following result, which provides a solution for the H_2 filtering problem.

Theorem 6.24 *An optimal solution for the OF problem posed above is given by the admissible controller* $(\widehat{\mathbf{A}}^{\mathrm{op}}, \widehat{\mathbf{B}}^{\mathrm{op}}, \widehat{\mathbf{C}}^{\mathrm{op}})$. *The associated optimal cost is*

$$\min_{G_K} \|\mathcal{G}_v\|_2^2 = \|\mathcal{G}_v^{\mathrm{op}}\|_2^2 = \sum_{i\in\mathcal{S}} \mathrm{tr}\big(R_i^{1/2} F_i Y_i F_i^* R_i^{1/2}\big). \tag{6.65}$$

Proof Let us denote by $\widehat{x}^{\mathrm{op}}(t), u^{\mathrm{op}}(t)$ the process generated by (6.42) when $(\widehat{\mathbf{A}}^{\mathrm{op}}, \widehat{\mathbf{B}}^{\mathrm{op}}, \widehat{\mathbf{C}}^{\mathrm{op}})$ is as in (6.64), by $x^{\mathrm{op}}(t)$ the process generated by (6.56) when we apply the control $u^{\mathrm{op}}(t)$, and $e^{\mathrm{op}}(t) = x^{\mathrm{op}}(t) - \widehat{x}^{\mathrm{op}}(t)$. This leads to the following equations:

$$\begin{aligned} dx^{\mathrm{op}}(t) &= \big(A_{\theta(t)}x^{\mathrm{op}}(t) - B_{\theta(t)}F_{\theta(t)}\widehat{x}^{\mathrm{op}}(t)\big)\,dt + J_{\theta(t)}\,dw(t), \\ d\widehat{x}^{\mathrm{op}}(t) &= \big(A_{\theta(t)} - K^f_{\theta(t)}H_{\theta(t)}\big)\widehat{x}^{\mathrm{op}}(t)\,dt + K^f_{\theta(t)}\big(H_{\theta(t)}x^{\mathrm{op}}(t)\,dt + G_{\theta(t)}\,dw(t)\big) \\ &\quad - B_{\theta(t)}F_{\theta(t)}\widehat{x}^{\mathrm{op}}(t)\,dt, \\ &= A_{\theta(t)}\widehat{x}^{\mathrm{op}}(t)\,dt + K^f_{\theta(t)}H_{\theta(t)}e^{\mathrm{op}}(t)\,dt + K^f_{\theta(t)}G_{\theta(t)}\,dw(t) \\ &\quad - B_{\theta(t)}F_{\theta(t)}\widehat{x}^{\mathrm{op}}(t)\,dt, \end{aligned}$$

and thus,

$$\begin{aligned} dx^{\mathrm{op}}(t) &= (A_{\theta(t)} - B_{\theta(t)}F_{\theta(t)})x^{\mathrm{op}}(t)\,dt + B_{\theta(t)}F_{\theta(t)}e^{\mathrm{op}}(t)\,dt + J_{\theta(t)}\,dw(t), \\ de^{\mathrm{op}}(t) &= \big(A_{\theta(t)} - K^f_{\theta(t)}H_{\theta(t)}\big)e^{\mathrm{op}}(t)\,dt + \big(J_{\theta(t)} - K^f_{\theta(t)}G_{\theta(t)}\big)\,dw(t), \end{aligned}$$

that is,

$$\begin{aligned} \begin{bmatrix} dx^{\mathrm{op}}(t) \\ de^{\mathrm{op}}(t) \end{bmatrix} &= \begin{bmatrix} A_{\theta(t)} - B_{\theta(t)}F_{\theta(t)} & B_{\theta(t)}F_{\theta(t)} \\ 0 & A_{\theta(t)} - K^f_{\theta(t)}H_{\theta(t)} \end{bmatrix} \begin{bmatrix} x^{\mathrm{op}}(t) \\ e^{\mathrm{op}}(t) \end{bmatrix} dt \\ &\quad + \begin{bmatrix} J_{\theta(t)} \\ J_{\theta(t)} - K^f_{\theta(t)}G_{\theta(t)} \end{bmatrix} dw(t). \end{aligned}$$

By Proposition 6.23 the Markov jump closed-loop system obtained from $(\widehat{\mathbf{A}}^{\mathrm{op}}, \widehat{\mathbf{B}}^{\mathrm{op}}, \widehat{\mathbf{C}}^{\mathrm{op}})$ is MSS. We also have that

$$v^{\mathrm{op}}(t) := \bar{F}_{\theta(t)}\big(x^{\mathrm{op}}(t) - \widehat{x}^{\mathrm{op}}(t)\big) = \bar{F}_{\theta(t)}e^{\mathrm{op}}(t)$$

with

$$\bar{F}_i := R_i^{1/2} F_i, \quad i \in \mathcal{S}.$$

Writing $\mathbf{Y}^{\mathrm{op}}(t) = (Y_1^{\mathrm{op}}(t), \dots, Y_N^{\mathrm{op}}(t))$ with

$$Y_i^{\mathrm{op}}(t) = E\big(e^{\mathrm{op}}(t) e^{\mathrm{op}}(t)^* 1_{\{\theta(t)=i\}}\big), \quad i \in \mathcal{S},\ t \geq 0,$$

it follows from the same arguments as in (6.60), (6.61), and (6.62) that $\mathbf{Y}^{\mathrm{op}}(t) = \mathbf{Y}$ for all $t \in \mathbb{R}^+$. Thus,

$$\begin{aligned}
\|\mathcal{G}_v^{\mathrm{op}}\|_2^2 &= \lim_{t\to\infty} \operatorname{tr}\big(E\big(v^{\mathrm{op}}(t) v^{\mathrm{op}}(t)^*\big)\big) \\
&= \lim_{t\to\infty} \sum_{i\in\mathcal{S}} \operatorname{tr}\big(\bar{F}_i E\big(e^{\mathrm{op}}(t) e^{\mathrm{op}}(t)^* 1_{\{\theta(t)=i\}}\big) \bar{F}_i^*\big) \\
&= \sum_{i\in\mathcal{S}} \operatorname{tr}\big(\bar{F}_i Y_i \bar{F}_i^*\big).
\end{aligned}$$

Consider any $(\widehat{\mathbf{A}}, \widehat{\mathbf{B}}, \widehat{\mathbf{C}})$ such that the closed-loop system (6.42) is MSS. By Proposition 6.22,

$$E\big(\|v(t)\|^2\big) = \sum_{i\in\mathcal{S}} \operatorname{tr}\big(\Lambda_i P_i(t) \Lambda_i^*\big) \geq \sum_{i\in\mathcal{S}} \operatorname{tr}\big(\bar{F}_i Y_i \bar{F}_i^*\big), \tag{6.66}$$

where $\mathbf{P}(t) = (P_1(t), \dots, P_N(t)) \in \mathbb{H}^{n+}$ is as in (6.50), and $\mathbf{P}(t) \xrightarrow{t\uparrow\infty} \mathbf{P}$ with $\mathbf{P} = (P_1, \dots, P_N) \in \mathbb{H}^{n+}$ the unique solution of the controllability Gramian (6.48), so that

$$\lim_{t\to\infty} E\big(\|v(t)\|^2\big) = \sum_{i\in\mathcal{S}} \operatorname{tr}\big(\Lambda_i P_i \Lambda_i^*\big) \geq \sum_{i\in\mathcal{S}} \operatorname{tr}\big(\bar{F}_i Y_i \bar{F}_i^*\big),$$

proving the desired result. □

6.3.3 The Separation Principle

We next present the main theorem, which establishes the separation principle for H_2 control of MJLS with partial observations. In what follows, we recall that $\|\mathcal{G}\|_2$ represents the H_2-norm of (6.41) under a control law of the form (6.42).

Suppose that there exists $\mathbf{X} = (X_1, \dots, X_N) \in \mathbb{H}^{n+}$, the stabilizing solution (see Definition 4.11) of the optimal control CARE (6.52), and let $\mathbf{F} = (F_1, \dots, F_N)$ be as in (6.53). Set also $\mathbf{R} = (R_1, \dots, R_N)$ with $R_i = D_i^* D_i$ for $i \in \mathcal{S}$. We introduce a change of variable in (6.41) for the control law in the following form:

$$u(t) = v(t) - F_{\theta(t)} x(t), \tag{6.67}$$

where $\nu(t)$ represents the new control variable. By making this change, (6.41) can be rewritten as

$$\begin{cases} \dot{x}(t) = \widetilde{A}_{\theta(t)}x(t) + B_{\theta(t)}\nu(t) + J_{\theta(t)}w(t), \\ z(t) = \widetilde{C}_{\theta(t)}x(t) + D_{\theta(t)}\nu(t), \\ y(t) = H_{\theta(t)}x(t) + G_{\theta(t)}w(t), \end{cases}$$

where $\widetilde{A}_i = A_i - B_i F_i$ and $\widetilde{C}_i = C_i - D_i F_i$. We can decompose the above system so that $x(t) = x_1(t) + x_2(t)$ and $z(t) = z_1(t) + z_2(t)$, where

$$\mathcal{G}_c = \begin{cases} \dot{x}_1(t) = \widetilde{A}_{\theta(t)}x_1(t) + J_{\theta(t)}w(t), \\ z_1(t) = \widetilde{C}_{\theta(t)}x_1(t) \end{cases} \tag{6.68}$$

will be associated to the cost of control, and

$$\mathcal{G}_U = \begin{cases} \dot{x}_2(t) = \widetilde{A}_{\theta(t)}x_2(t) + B_{\theta(t)}R_{\theta(t)}^{-1/2}v(t), \\ z_2(t) = \widetilde{C}_{\theta(t)}x_2(t) + D_{\theta(t)}R_{\theta(t)}^{-1/2}v(t), \end{cases} \tag{6.69}$$

with $v(t) = R_{\theta(t)}^{1/2}\nu(t)$, $v = \{v(t); t \in \mathbb{R}\}$, will be associated to the separation of the cost of estimation with the cost of control. Notice that system $\mathcal{G}_c$ does not depend on the control $u(t)$ and that

$$z(t) = \mathcal{G}_c(w)(t) + \mathcal{G}_U(v)(t).$$

In what follows, $\widetilde{\Phi}(s,t)$ will be as in (3.19) with $\widetilde{A}_{\theta(t)}$ replacing $A_{\theta(t)}$. We can now prove the following result (for the definition of the adjoint operator for the MJLS, see Appendix B).

Proposition 6.25 *Let $\mathcal{G}_c$ and $\mathcal{G}_U$ be as in* (6.68) *and* (6.69), *respectively. Then, for any $w = \{w(t); t \in \mathbb{R}\} \in L_2^r(\Omega, \mathcal{F}, P)$,*

$$\text{(a) } \mathcal{G}_U^*\mathcal{G}_U = I,$$

$$\text{(b) } \mathcal{G}_U^*\mathcal{G}_c(w)(t) = R_{\theta(t)}^{-1/2}B_{\theta(t)}^* \int_t^\infty E(\widetilde{\Phi}(s,t)^* X_{\theta(s)} J_{\theta(s)} w(s) | \mathcal{F}_t)\, ds. \tag{6.70}$$

Proof See Appendix B. □

The main result reads as follows.

Theorem 6.26 *Consider system* (6.41) *and Markov jump stochastically stabilizing controllers as in* (6.42). *Suppose that there exist stabilizing solutions $\mathbf{Y} = (Y_1, \dots, Y_N)$ and $\mathbf{X} = (X_1, \dots, X_N)$ for the filtering and control CARE as in* (6.54) *and* (6.52), *respectively, and let $\mathbf{K}^f = (K_1^f, \dots, K_N^f)$ and $\mathbf{F} = (F_1, \dots, F_N)$ be as in* (6.55) *and* (6.53), *respectively. Then an optimal solution for the H_2 control problem with partial observations is given by $\widehat{\mathbf{A}}^{\mathrm{op}} = (\widehat{A}_1^{\mathrm{op}}, \dots, \widehat{A}_N^{\mathrm{op}})$, $\widehat{\mathbf{B}}^{\mathrm{op}} = (\widehat{B}_1^{\mathrm{op}}, \dots, \widehat{B}_N^{\mathrm{op}})$,* *and* $\widehat{\mathbf{C}}^{\mathrm{op}} = (\widehat{C}_1^{\mathrm{op}}, \ldots, \widehat{C}_N^{\mathrm{op}})$ *as in Eqs.* 6.64), *that is, a Markov optimal controller* $\mathcal{G}_K^{\mathrm{op}}$ *is given by*

$$\mathcal{G}_K^{\mathrm{op}} = \begin{cases} \dot{\widehat{x}}^{\mathrm{op}}(t) = \big(A_{\theta(t)} - K_{\theta(t)}^f H_{\theta(t)} - B_{\theta(t)} F_{\theta(t)}\big)\widehat{x}^{\mathrm{op}}(t) + K_{\theta(t)}^f y(t), \\ u(t) = -F_{\theta(t)}\widehat{x}^{\mathrm{op}}(t). \end{cases}$$

Moreover, the H_2*-norm for this control is*

$$\min_{\mathcal{G}_K} \|\mathcal{G}\|_2^2 = \sum_{i\in\mathcal{S}} \pi_i \operatorname{tr}\big(J_i^* X_i J_i\big) + \sum_{i\in\mathcal{S}} \operatorname{tr}\big(D_i F_i Y_i F_i^* D_i^*\big).$$

Proof From (6.42), (6.56), and (6.67) we have that

$$\mathcal{G}_v = \begin{cases} \dot{\mathbf{v}}(t) = \Gamma_{\theta(t)}\mathbf{v}(t) + \Psi_{\theta(t)} w(t), \\ v(t) = R_{\theta(t)}^{1/2}\begin{bmatrix} F_{\theta(t)} & \widehat{C}_{\theta(t)} \end{bmatrix}\begin{bmatrix} x(t) \\ \widehat{x}(t) \end{bmatrix} = \Lambda_{\theta(t)}\mathbf{v}(t), \end{cases}$$

where $\boldsymbol{\Gamma} = (\Gamma_1, \ldots, \Gamma_N)$ and $\boldsymbol{\Psi} = (\Psi_1, \ldots, \Psi_N)$ are as in (6.44), and $\boldsymbol{\Lambda} = (\Lambda_1, \ldots, \Lambda_N)$, $\Lambda_i = R_i^{1/2}[\, F_i \;\; \widehat{C}_i \,]$, $i \in \mathcal{S}$. We have from (6.68) and (6.69) that

$$z(t) = \mathcal{G}(w)(t) = \mathcal{G}_c(w)(t) + \mathcal{G}_U\big(\mathcal{G}_v(w)\big)(t).$$

The norm of the operator $\mathcal{G}$ applied to w can be written as

$$\begin{aligned} \big\|\mathcal{G}(w)\big\|_2^2 &= \big\langle \mathcal{G}_c(w) + \mathcal{G}_U\big(\mathcal{G}_v(w)\big); \mathcal{G}_c(w) + \mathcal{G}_U\big(\mathcal{G}_v(w)\big)\big\rangle \\ &= \big\|\mathcal{G}_c(w)\big\|_2^2 + \big\langle \mathcal{G}_U^*\mathcal{G}_c(w); \mathcal{G}_v(w)\big\rangle + \big\langle \mathcal{G}_v(w); \mathcal{G}_U^*\mathcal{G}_c(w)\big\rangle \\ &\quad + \big\langle \mathcal{G}_U^*\mathcal{G}_U\mathcal{G}_v(w); \mathcal{G}_v(w)\big\rangle. \end{aligned}$$

We recall from (6.46) that

$$\|\mathcal{G}\|_2^2 = \sum_{k=1}^{r} \big\|\mathcal{G}(w_k)\big\|_2^2,$$

where $w_k(t) = \delta(t)e_k$, and e_k is a vector with 1 at the kth position and zero elsewhere. Notice now that, by (6.70) in Proposition 6.25,

$$\mathcal{G}_U^*\mathcal{G}_c(w_k)(t) = \begin{cases} R_{\theta(t)}^{-1/2} B_{\theta(t)}^* E\big(\widetilde{\Phi}(0,t)^* X_{\theta_0} J_{\theta_0} e_k | \mathcal{F}_t\big), & t \le 0, \\ 0, & t > 0, \end{cases}$$

and since

$$\mathcal{G}_v(w)(t) = \Lambda_{\theta(t)} \int_{-\infty}^{t} \Phi(t,s)\Psi_{\theta(s)} w(s)\, ds,$$

we have that

$$\mathcal{G}_v(w_k)(t) = \begin{cases} \Lambda_{\theta(t)}\Phi(t,0)\Psi_{\theta_0}e_k, & t \geq 0, \\ 0, & t < 0. \end{cases}$$

Thus,

$$\langle \mathcal{G}_U^* \mathcal{G}_c(w_k); \mathcal{G}_v(w_k) \rangle = 0.$$

Furthermore, by Proposition 6.25, $\mathcal{G}_U^* \mathcal{G}_U = I$, and thus,

$$\langle \mathcal{G}_U^* \mathcal{G}_U \mathcal{G}_v(w); \mathcal{G}_v(w) \rangle = \| \mathcal{G}_v(w) \|_2^2.$$

This leads to

$$\|\mathcal{G}\|_2^2 = \sum_{k=1}^{r} \| \mathcal{G}(w_k) \|_2^2 = \|\mathcal{G}_c\|_2^2 + \|\mathcal{G}_v\|_2^2,$$

and since $\mathcal{G}_c$ does not depend on u,

$$\min_{\mathcal{G}_K} \|\mathcal{G}\|_2^2 = \|\mathcal{G}_c\|_2^2 + \min_{\mathcal{G}_K} \|\mathcal{G}_v\|_2^2, \tag{6.71}$$

where $\mathcal{G}_K$, $\mathcal{G}_v$, and $\mathcal{G}_c$ are as in (6.42), (6.56), and (6.68), respectively. But the solution of $\min_{\mathcal{G}_K} \|\mathcal{G}_v\|_2^2$ is as in Theorem 6.24. Therefore, from Theorems 5.4 and 6.24,

$$\|\mathcal{G}_c\|_2^2 = \sum_{i \in \mathcal{S}} \pi_i \operatorname{tr}(J_i^* X_i J_i) \tag{6.72}$$

and

$$\min_{\mathcal{G}_K} \|\mathcal{G}_v\|_2^2 = \sum_{i \in \mathcal{S}} \operatorname{tr}(R_i^{1/2} F_i Y_i F_i^* R_i^{1/2}), \tag{6.73}$$

completing the proof of the theorem. □

Remark 6.27 Notice that as for the deterministic case (see, for instance, [251] or [324]), we have from (6.71) that the H_2-norm can be written as the sum of two H_2-norms; the first one does not depend on the control u and has value given by (6.72), and the second one is equivalent to problem OF and has optimal value given by (6.73).

Remark 6.28 It is worth noticing that from Theorem 6.26 an optimal dynamic MJLS controller for the problem can be obtained from the stabilizing solution for the CARE associated with the filtering problem (6.54), and from the stabilizing solution for the CARE associated with the optimal control (6.52). The controller equations are as in (6.64), and the optimal cost given by the sum of the terms in (6.72) and (6.73).

6.3.4 An LMIs Approach for the H_2 Control Problem

In this subsection we present a formulation for the H_2 control problem with partial observations as posed in Sect. 6.3.1 based on an LMIs optimization problem. As seen in Sect. 6.3.1, the H_2 control problem with partial observations can be written as the optimization Problem 1 on the matrix variables $\widehat{A}_i$, $\widehat{B}_i$, $\widehat{C}_i$, S_i, V_i for $i \in \mathcal{S}$ (recall from (6.44) that Γ_i, Ψ_i, Λ_i depend on $\widehat{A}_i$, $\widehat{B}_i$, $\widehat{C}_i$):

Problem 1

$$\inf \sum_{i\in\mathcal{S}} \pi_i \operatorname{tr}(V_i)$$

subject, for $i \in \mathcal{S}$, to

$$S_i > 0, \tag{6.74}$$

$$V_i > \Psi_i^* S_i \Psi_i, \tag{6.75}$$

$$\Gamma_i^* S_i + S_i \Gamma_i + \sum_{j\in\mathcal{S}} \lambda_{ij} S_j + \Lambda_i^* \Lambda_i < 0. \tag{6.76}$$

Consider now the LMIs optimization Problem 2 on the matrix variables X_i, Y_i, L_i, F_i, V_i for $i \in \mathcal{S}$ (recall in what follows the definition of $\mathcal{R}$ and $\mathcal{D}$ in (3.111) and (3.112), respectively) defined as follows.

Problem 2

$$\inf \sum_{i\in\mathcal{S}} \pi_i \operatorname{tr}(V_i)$$

subject, for $i \in \mathcal{S}$, to

$$\begin{bmatrix} Y_i & I & J_i \\ I & X_i & X_i J_i + L_i G_i \\ J_i^* & J_i^* X_i + G_i^* L_i & V_i \end{bmatrix} > 0, \tag{6.77}$$

$$\begin{bmatrix} A_i Y_i + Y_i A_i^* + B_i F_i + F_i^* B_i^* + \lambda_{ii} Y_i & Y_i C_i^* + F_i^* D_i^* & \mathcal{R}_i(\mathbf{Y}) \\ C_i Y_i + D_i F_i & -I & 0 \\ \mathcal{R}_i^*(\mathbf{Y}) & 0 & -\mathcal{D}_i(\mathbf{Y}) \end{bmatrix} < 0, \tag{6.78}$$

$$A_i^* X_i + X_i A_i + L_i H_i + H_i^* L_i^* + C_i^* C_i + \sum_{j\in\mathcal{S}} \lambda_{ij} X_j < 0. \tag{6.79}$$

We have the following theorem, based on Theorem 4.1 in [101].

Theorem 6.29 *Problem* 1 *and Problem* 2 *are equivalent. Moreover, if* $X_i > 0$, $Y_i > 0$, L_i, F_i, V_i, $i \in \mathcal{S}$, *is an* ϵ-*optimal solution for Problem* 2, *then an* ϵ-*optimal* *solution for Problem* 1 *can be obtained by choosing the same* V_i *and by making, for* $i \in \mathcal{S}$,

$$S_i = \begin{bmatrix} X_i & Y_i^{-1} - X_i \\ Y_i^{-1} - X_i & X_i - Y_i^{-1} \end{bmatrix} > 0 \tag{6.80}$$

and

$$\widehat{A}_i = (X_i - Y_i^{-1})^{-1}\Big(A_i^* + X_i A_i Y_i + X_i B_i F_i + L_i H_i Y_i + C_i^*(C_i Y_i + D_i F_i) + \sum_{j\in\mathcal{S}} \lambda_{ij} Y_j^{-1} Y_i\Big) Y_i^{-1}, \tag{6.81}$$

$$\widehat{B}_i = (Y_i^{-1} - X_i)^{-1} L_i, \tag{6.82}$$

$$\widehat{C}_i = F_i Y_i^{-1}. \tag{6.83}$$

Proof Suppose first that there exist $\widehat{A}_i$, $\widehat{B}_i$, $\widehat{C}_i$, S_i, V_i satisfying (6.74)–(6.76) and consider the following partition for S_i:

$$S_i = \begin{bmatrix} Z_i & U_i \\ U_i^* & \widehat{Z}_i \end{bmatrix}. \tag{6.84}$$

Without loss of generality, suppose further that U_i is nonsingular and define the matrices $\mathcal{Y}_i$, T_i, M_i as in (3.118), (3.119), (3.120), respectively. Multiplying (6.76) to the left by $T_i^* M_i^*$ and to the right by $M_i T_i$, we get, after performing some calculations, that

$$\begin{bmatrix} \mathcal{T}_i & \mathcal{M}_i^* \\ \mathcal{M}_i & \mathcal{B}_i \end{bmatrix} + \sum_{j\in\mathcal{S}} \lambda_{ij} \begin{bmatrix} \mathcal{Y}_i[\mathcal{Y}_j^{-1} + (U_i\widehat{Z}_i^{-1}\widehat{Z}_j - U_j)\widehat{Z}_j^{-1}(U_i\widehat{Z}_i^{-1}\widehat{Z}_j - U_j)^*]\mathcal{Y}_i & 0 \\ 0 & Z_j \end{bmatrix} < 0, \tag{6.85}$$

where

$$\begin{aligned}
\mathcal{T}_i &= A_i\mathcal{Y}_i + \mathcal{Y}_i A_i^* + B_i\mathcal{F}_i + \mathcal{F}_i^* B_i^* + (\mathcal{Y}_i C_i^* + \mathcal{F}_i^* D_i^*)(C_i\mathcal{Y}_i + D_i\mathcal{F}_i),\\
\mathcal{B}_i &= A_i^* Z_i + Z_i A_i + \mathcal{L}_i H_i + H_i^*\mathcal{L}_i^* + C_i^* C_i,\\
\mathcal{F}_i &= -\widehat{C}_i\widehat{Z}_i^{-1} U_i^*\mathcal{Y}_i,\\
\mathcal{L}_i &= U_i\widehat{B}_i,\\
\mathcal{M}_i &= A_i^* + Z_i A_i\mathcal{Y}_i + Z_i B_i\mathcal{F}_i + \mathcal{L}_i H_i\mathcal{Y}_i - U_i\widehat{A}_i\widehat{Z}_i^{-1}U_i^*\mathcal{Y}_i\\
&\quad + C_i^*(C_i\mathcal{Y}_i + D_i\mathcal{F}_i) + \sum_{j\in\mathcal{S}} \lambda_{ij}(Z_j - U_j\widehat{Z}_i^{-1}U_i^*)\mathcal{Y}_i.
\end{aligned} \tag{6.86}$$

From (6.85) by the same reasoning as in (3.124) we get that (6.78) and (6.79) hold with $Y_i = \mathcal{Y}_i$, $X_i = Z_i$, $F_i = \mathcal{F}_i$, and $L_i = \mathcal{L}_i$, $i \in \mathcal{S}$. Moreover, from the same reasoning as in (3.123) we get that $\begin{bmatrix} \mathcal{Y}_i & I \\ I & Z_i \end{bmatrix} > 0$ and that

$$T_i^* M_i^* S_i \Psi_i = \begin{bmatrix} J_i \\ Z_i J_i + \mathcal{L}_i G_i \end{bmatrix},$$

$$T_i^* M_i^* S_i M_i T_i = \begin{bmatrix} \mathcal{Y}_i & I \\ I & Z_i \end{bmatrix}.$$

Therefore, by (6.75),

$$\begin{aligned} V_i > \Psi_i^* S_i \Psi_i &= (\Psi_i^* S_i M_i T_i)(T_i^* M_i^* S_i M_i T_i)^{-1}(T_i^* M_i^* S_i \Psi_i) \\ &= \begin{bmatrix} J_i^* & J_i^* Z_i + G_i^* \mathcal{L}_i^* \end{bmatrix} \begin{bmatrix} \mathcal{Y}_i & I \\ I & Z_i \end{bmatrix}^{-1} \begin{bmatrix} J_i \\ Z_i J_i + \mathcal{L}_i G_i \end{bmatrix}, \end{aligned} \tag{6.87}$$

which shows that (6.77) holds with the above choice for Y_i, X_i, F_i, L_i.

On the other hand, if $X_i > 0$, $Y_i > 0$, L_i, F_i, V_i is an ϵ-optimal solution for Problem 2, we get, by choosing S_i, $\widehat{A}_i$, $\widehat{B}_i$, $\widehat{C}_i$ as in (6.80)–(6.83), respectively, and setting

$$\begin{aligned} T_i &= \begin{bmatrix} Y_i & I \\ Y_i & 0 \end{bmatrix}, \\ \mathcal{J}_i &= A_i Y_i + Y_i A_i^* + B_i F_i + F_i^* B_i^* + (Y_i C_i^* + F_i^* D_i^*)(C_i Y_i + D_i F_i) \\ &\quad + \sum_{j \in \mathcal{S}} \lambda_{ij} Y_i Y_j^{-1} Y_i, \\ \mathcal{H}_i &= A_i^* X_i + X_i A_i + L_i H_i + H_i^* L_i^* + C_i^* C_i + \sum_{j \in \mathcal{S}} \lambda_{ij} X_j, \end{aligned}$$

that

$$T_i^* \Big(\Gamma_i^* P_i + P_i \Gamma_i + \Lambda_i^* \Lambda_i + \sum_{j \in \mathcal{S}} \lambda_{ij} P_j \Big) T_i = \begin{bmatrix} \mathcal{J}_i & 0 \\ 0 & \mathcal{H}_i \end{bmatrix} < 0 \tag{6.88}$$

since, by (6.78), $\mathcal{J}_i < 0$, and, by (6.79), $\mathcal{H}_i < 0$, showing that (6.76) is satisfied. Moreover, from (6.77) and noticing that

$$\begin{aligned} \Psi_i^* S_i \Psi_i &= (\Psi_i^* S_i M_i T_i)(T_i^* M_i^* S_i M_i T_i)^{-1}(T_i^* M_i^* S_i \Psi_i) \\ &= \begin{bmatrix} J_i^* & J_i^* X_i + G_i^* L_i^* \end{bmatrix} \begin{bmatrix} Y_i & I \\ I & X_i \end{bmatrix}^{-1} \begin{bmatrix} J_i \\ X_i J_i + L_i G_i \end{bmatrix} < V_i, \end{aligned}$$

we get that (6.75) holds, completing the proof. □

Remark 6.30 As in Remark 3.44, the choice in (6.80)–(6.83) for $\widehat{A}_i$, $\widehat{B}_i$, $\widehat{C}_i$, S_i corresponds to choosing in (6.84) the trivial solution $U_i = -\widehat{Z}_i = Y_i^{-1} - X_i$ and, in (6.86), $\mathcal{M}_i = 0$.

6.4 Historical Remarks

It is well known that stochastic control problems with partial observation are, in general, very hard problems to deal with. In this scenario, we can certainly assert that separation principles are powerful tools. In this regard, the separation principle, studied by Wonham [302], is a distinctive result, and it has had a crucial bearing on the study of the control problem with partial observations (an early foray in this topic was made in [193]). Its famous version, which is a great achievement for the LQG problem with partial observations, is known in the specialized literature as the certainty equivalence principle (first excursions in this scenario were made in [260] and [277]). An example in which the control has the separation property but not the certainty equivalence is given in [265]. These concepts were used extensively in areas such as adaptive control (the enforced separation approach) and self-tuning controllers, and gave rise to a huge amount of literature on these subjects. Other attempts to advance in the theory of optimal control with partial observations were made, for instance, in [19, 99, 139, 300, 303]. Despite many efforts, control problems with partial observations remain a great challenge, and the LQG problem with partial observations is one of the rare problems in which we can explicitly obtain the optimal control solution and in which the separation principle applies (the certainty equivalence situation).

Regarding MJLS, it is perhaps worth noting here that the partial observation may be associated either with the state variable or with the Markov chain, or yet with both variables, which is of course the hardest problem. For the control problem with partial observations of the Markov chain, the readers are referred, for instance, to [51, 134, 143]. The case with partial information of the state and perfect measurement of the Markov chain (including the H_2 control problem) is treated, for instance, in [80, 89, 101, 125, 153]. The case in which both the state variable and Markov chain are only partially observable was also studied in [125]. Another setting with partial observations, which has a strong relation with robustness, is that in which we have partial information of the transition matrix (or transition probability) of the Markov chain, i.e., the partial observations are related to the uncertainty in the transition matrix. One kind of uncertainty found in the literature is that in which the transition rate matrix belongs to a polytope. For a brief account of the results in this scenario, see, e.g., [305, 306, 316–319].

For those interested in the theory of stochastic control problems with partial observation, we refer, for instance, to the books [25, 97] and [311].

The material of this chapter was drawn, essentially, from [80, 101], and [153].

Chapter 7
Best Linear Filter with Unknown $(x(t), \theta(t))$

7.1 Outline of the Chapter

The aim of this chapter is to derive the best linear mean-square estimator for continuous-time MJLS assuming that only an output $y(t)$ is available. It is important to emphasize that in this chapter we assume that both the state variable $x(t)$ and jump parameter $\theta(t)$ are not known. The idea is to derive a filter that bears those desirable properties of the Kalman filter: a recursive scheme suitable for computer implementation which allows some offline computation that alleviates the computational burden. The filter is derived as a function of the error covariance matrix whose dynamics is governed by two matrix differential equations, one associated with the second moment of the state variable and the other one associated with the second moment of the estimator. The linear filter has dimension Nn (recall that n denotes the dimension of the state vector, and N the number of states of the Markov chain). Both the finite-horizon and infinite horizon cases are considered. A brief outline of the content of this chapter is as follows. In Sect. 7.2 we recall some basic facts on linear filtering, which can be found in [97]. The problem statement for the finite-horizon case is described in Sect. 7.3. The filter equations for this case are derived in Sect. 7.4, and the convergence of the solution of the associated Riccati equation to a stationary value is considered in Sect. 7.5. The stationary filter is studied in Sect. 7.6. An equivalent LMIs formulation to solve the problem, which can be extended to consider uncertainties on the parameters of the possible modes of operation of the system, together with a numerical example, will be analyzed in Chaps. 9 and 10, Sects. 9.5 and 10.5.

7.2 Preliminaries

The purpose of this section is to recall some basic facts on linear filtering for stochastic continuous-time systems, following the approach adopted in [97]. We set $\mathcal{H} := L_2(\Omega, \mathcal{F}, P)$, which represents the Hilbert space of all square-integrable random variables in the probability space $(\Omega, \mathcal{F}, P)$, equipped with inner product $\langle x, y \rangle = E(x^* y)$. Convergence here will be in the *quadratic mean* (*q.m.*) sense,

O.L.V. Costa et al., *Continuous-Time Markov Jump Linear Systems*,
Probability and Its Applications, DOI 10.1007/978-3-642-34100-7_7,

i.e., a sequence $\{x(n)\}$ converges to x if $\|x(n) - x\|_2 \to 0$. We define also $\mathcal{H}_0 = \{x \in \mathcal{H} | Ex = 0\}$, the closed subspace of all centered random variables of $\mathcal{H}$ and therefore a Hilbert space. In addition, $x \in \mathcal{H}$ and $y \in \mathcal{H}$ are said to be *orthogonal* (from now on $x \bot y$) if $\langle x, y\rangle = 0$. Furthermore, for $y \in \mathcal{H}_0$, we consider the subspace $\mathcal{H}_t^y \subset \mathcal{H}_0$ defined by $\mathcal{H}_t^y = \mathfrak{L}\{y(s), 0 \le s \le t\}$ that consists of all *linear combinations* $\sum_i \alpha_i^* y(t_i)$, where $t_i < t$, and q.m. limits of these combinations (a closed subspace), so that $\mathcal{H}_t^y \subset \mathcal{H}_{t'}^y \subset \mathcal{H}^y := \mathcal{H}_\infty^y$ for $t < t'$. We recall that if $\{y(t)\}$ is q.m. continuous, then $\mathcal{H}^y$ is a separable Hilbert space, and, as a fundamental property of a Hilbert space, any $z \in \mathcal{H}_0$ has a unique decomposition (cf. [97], p. 45) $z = \hat{z} + \tilde{z}$ where $\hat{z} = \mathcal{P}_t^y z \in \mathcal{H}_t^y$ and $\tilde{z} \perp \mathcal{H}_t^y$. Here $\mathcal{P}_t^y$ denotes the *projection operator*, which projects each element of $\mathcal{H}_0$ onto $\mathcal{H}_t^y$. Moreover, we have the following properties (cf. [97], p. 45): (i) $\|z - \mathcal{P}_t^y z\| = \min_{v \in \mathcal{H}_t^y} \|z - v\|$, and therefore $\hat{z} = \mathcal{P}_t^y z$ *is the linear least-square estimator* of z given $\mathcal{H}_t^y$, i.e., *the best linear estimator is the projection* of z onto $\mathcal{H}_t^y$; (ii) $\tilde{z} = z - \hat{z} \perp \mathcal{H}_t^y$.

A vector process $\{x(t) = [x_1(t) \ \dots \ x_n(t)]^*; t \in \mathbb{R}^+\} \in \mathbb{R}^n$ has *orthogonal increments* (o.i.) if for all i, j and any nonoverlapping intervals (u, r) and (s, t), $(x_i(t) - x_i(s)) \perp (x_j(r) - x_j(u))$ or, equivalently, $(x_i(t) - x_i(s)) \perp \mathcal{H}_s^x$. For a second-order vector process $\{x(t)\}$, $\text{cov}(x(t))$ will refer to its covariance function.

We recall now some basic facts on filtering theory, the main reference being [97]. Without any loss of generality, we assume in the sequel that all the processes are centered (zero mean).

Proposition 7.1 *A stochastic process $\{x(t)\}$ is q.m. continuous if and only if its covariance function $r(s, t)$ is continuous at the diagonal point (t, t). Furthermore, if $\{x(t)\}$ is q.m. continuous for all t, then $r(\cdot, \cdot)$ is continuous at every point (s, t).*

Proposition 7.2 *There is a measurable version of every q.m. continuous stochastic process.*

Proposition 7.3 *A q.m. continuous stochastic process $\{x(t)\}$ has stationary o.i. if and only if its covariance function is given by $r(t, s) = \Gamma \min\{t, s\}$, where Γ is a positive definite matrix.*

Proposition 7.4 *Let $\{x(t)\}$ be a q.m. continuous stochastic process, and $y \in \mathcal{H}_t^x$ for some t. Then, $\mathcal{P}_s^x y \to y$ as $s \uparrow t$.*

Proposition 7.5 *Let $\{\pi(t)\}$ be a process with stationary o.i., and $\{z(t)\}$ a q.m. continuous second-order process. Then the following assertions are equivalent*:

(a) *For each $t \in \mathbb{R}^+$, $z(t) \in \mathcal{H}_t^\pi$, and for $s \le t$, $(z(t) - z(s)) \perp \mathcal{H}_s^\pi$.*
(b) *There exists a matrix-valued function $G(s)$ such that, for each t,*

$$\sum_{i,j} \int_0^t \left(G_{ij}(s)\right)^2 ds < \infty$$

and $z(t) = \int_0^t G(s)\, d\pi(s)$.

Proposition 7.6 *Let x be a random variable, and $\{\pi(t)\}$ a stationary o.i. process. If $\hat{x}(t) = \mathcal{P}_t^\pi x$, then*

$$\hat{x}(t) = \int_0^t \left(\frac{d}{ds} E\big[x\pi(s)^*\big] \right) d\pi(s). \tag{7.1}$$

7.3 Problem Formulation for the Finite-Horizon Case

On the complete probability space $(\Omega, \mathcal{F}, P)$ carrying its natural filtration $\{\mathcal{F}_t, t \in \mathbb{R}^+\}$, as usual augmented by all null sets in the P-completion of $\mathcal{F}$, consider the dynamical systems modeled by the following MJLS:

$$dx(t) = A_{\theta(t)} x(t)\, dt + J_{\theta(t)}\, dw_0(t), \quad t \in \mathbb{R}^+, \qquad x(0) = x_0, \tag{7.2}$$

$$dy(t) = H_{\theta(t)} x(t)\, dt + G_{\theta(t)}\, dw(t), \qquad y(0) = 0, \tag{7.3}$$

where, as before, $\{x(t)\}$ denotes the state vector in $\mathbb{R}^n$, and $\{y(t)\}$ the output process in $\mathbb{R}^q$, which generates the observational information that is available at time t. As before, we assume that $\mathbf{A} = (A_1, \ldots, A_N) \in \mathbb{H}^n$, $\mathbf{J} = (J_1, \ldots, J_N) \in \mathbb{H}^{r,n}$, $\mathbf{H} = (H_1, \ldots, H_N) \in \mathbb{H}^{n,q}$, and $\mathbf{G} = (G_1, \ldots, G_N) \in \mathbb{H}^{r,q}$. The results here also apply for the case in which the above matrices are of class **PC** (see Definition 2.3), but for easiness of notation, we dropped the dependence on the time variable t. We also assume that:

(A.1) $W = \{(w(t), \mathcal{F}_t), t \in \mathbb{R}^+\}$ and $W_0 = \{(w_0(t), \mathcal{F}_t), t \in \mathbb{R}^+\}$ are independent standard Wiener process in $\mathbb{R}^p$ and $\mathbb{R}^r$, respectively.
(A.2) For each $i \in \mathcal{S}$, $x_0 1_{\{\theta_0 = i\}}$ is a second-order random variable such that $E(x_0 1_{\{\theta_0 = i\}}) = \mu_i$ and $E(x_0 x_0^* 1_{\{\theta_0 = i\}}) = V_i$.
(A.3) x_0 and $\{\theta(t)\}$ are independent of $\{w_0(t)\}$ and $\{w(t)\}$.
(A.4) $G_i G_i^* > 0$ for each $i \in \mathcal{S}$.

We set $\widehat{r}(t) := \mathcal{P}_t^y(r)$, $z_i(t) = x(t) 1_{\{\theta(t) = i\}}$, and $Q_i(t) = E(z_i(t) z_i(t)^*)$, $i \in \mathcal{S}$, in $\mathbb{R}^n$ and $\mathbb{B}(\mathbb{R}^n)^+$, respectively. We also set

$$J_t^\theta := \operatorname{diag}(J_i 1_{\{\theta(t) = i\}}),$$

$$G_t^\theta := \big[1_{\{\theta(t)=1\}} G_1 \quad \ldots \quad 1_{\{\theta(t)=N\}} G_N \big],$$

$$d\overline{w}_0(t) := \big[dw_0(t)^* \quad \ldots \quad dw_0(t)^* \big]^*,$$

$$d\overline{w}(t) := \big[dw(t)^* \quad \ldots \quad dw(t)^* \big]^*.$$

Furthermore, by making $z(t) = [z_1(t)^* \ \ldots \ z_N(t)^*]^* \in \mathbb{R}^{Nn}$, we consider the following additional notation (some of which already presented in Chap. 3):

$$Z(t) := E\big(z(t) z(t)^*\big), \qquad \hat{Z}(t) := E\big(\hat{z}(t) \hat{z}(t)^*\big), \qquad \tilde{Z}(t) := E\big(\tilde{z}(t) \tilde{z}(t)^*\big),$$

with $Z(t)$, $\hat{Z}(t)$, and $\tilde{Z}(t) \in \mathbb{B}(\mathbb{R}^{Nn})^+$, and

$$\mathcal{J} := \Pi' \otimes I_n \in \mathbb{B}(\mathbb{R}^{Nn}), \tag{7.4}$$

$$F = \Pi' \otimes I_n + \operatorname{diag}(A_i) \in \mathbb{B}(\mathbb{R}^{Nn}), \tag{7.5}$$

$$H := \begin{bmatrix} H_1 & \dots & H_N \end{bmatrix} \in \mathbb{B}(\mathbb{R}^{Nn}, \mathbb{R}^m), \tag{7.6}$$

$$G_t^p := \begin{bmatrix} \sqrt{p_1(t)}G_1 & \dots & \sqrt{p_N(t)}G_N \end{bmatrix} \in \mathbb{B}(\mathbb{R}^{Np}, \mathbb{R}^m), \tag{7.7}$$

$$J_t^p := \operatorname{diag}\big(\sqrt{p_i(t)}J_i\big) \in \mathbb{B}(\mathbb{R}^{Np}, \mathbb{R}^{Nn}). \tag{7.8}$$

The *linear filtering problem* consists in finding the best linear mean-square estimator of $\{x(t)\}$ given $\mathcal{H}_t^y$, i.e., finding $\hat{x}(t)$ such that it minimizes the mean-square error $\|\tilde{x}(t)\|_2^2$. With the techniques of Hilbert spaces underpinning our geometric approach, in conjunction with stationary o.i. processes, a filter is defined via

$$\mathfrak{L}_K(t) = \int_0^t K(t,s)\, d\pi(s),$$

where $\{\pi(t)\}$ is an adequate stationary o.i. process, and to find the best linear mean-square filter is tantamount to identifying the kernel $K(t,s)$ that minimizes

$$J\big(K(t,s)\big) = E\big(\big[x(t) - \mathfrak{L}_K(t)\big]^*\big[x(t) - \mathfrak{L}_K(t)\big]\big).$$

As in [65], the idea here is, instead of estimating directly $x(t)$, to estimate $z_i(t) = x(t)1_{\{\theta(t)=i\}}$, taking advantage of its Markovian property. Via $\hat{z}_i(t)$ we get easily $\hat{x}(t)$, since $x(t) = \sum_{i\in\mathcal{S}} x(t)1_{\{\theta(t)=i\}}$ and therefore $\hat{x}(t) = \sum_{i\in\mathcal{S}} \hat{z}_i(t)$.

We conclude this section with some auxiliary results. For system (7.2), it has been shown in Proposition 3.28 (see also [152]) that for $t \in \mathbb{R}^+$ and $i \in \mathcal{S}$,

$$\dot{Q}_i(t) = A_i Q_i(t) + Q_i(t)A_i^* + \sum_{j\in\mathcal{S}} \lambda_{ji} Z_j(t) + J_i J_i^* p_i(t). \tag{7.9}$$

In what follows, we set $d\mathcal{D}_t = [x(t)^* d(1_{\{\theta(t)=1\}}) \ \dots \ x(t)^* d(1_{\{\theta(t)=N\}})]^* \in \mathbb{R}^{Nn}$. We have the following result.

Lemma 7.7 *We have that*

$$dz(t) = Fz(t)\,dt - \mathcal{J}z(t)\,dt + J_t^\theta\, d\overline{w}_0(t) + d\mathcal{D}_t, \tag{7.10}$$

$$\mathcal{P}_t^y(d\mathcal{D}_t) = \mathcal{J}\hat{z}(t)\,dt. \tag{7.11}$$

Proof First notice that

$$\begin{aligned} dz_i(t) &= 1_{\{\theta(t)=i\}}\, dx(t) + x(t)\, d(1_{\{\theta(t)=i\}}) \\ &= A_i z_i(t)\, dt + J_i 1_{\{\theta(t)=i\}}\, dw_0(t) + x(t)\, d(1_{\{\theta(t)=i\}}) \end{aligned} \tag{7.12}$$

and thus

$$\begin{aligned}dz(t) &= \mathrm{diag}(A_i)z(t)\,dt + J_t^\theta\, d\overline{w}_0(t) + d\mathcal{D}_t\\ &= \bigl[\mathrm{diag}(A_i) + \Pi' \otimes I_n\bigr]z(t)\,dt - \bigl[\Pi' \otimes I_n\bigr]z(t)\,dt + J_t^\theta\, d\overline{w}_0(t) + d\mathcal{D}_t\\ &= Fz(t)\,dt - \bigl[\Pi' \otimes I_n\bigr]z(t)\,dt + J_t^\theta\, d\overline{w}_0(t) + d\mathcal{D}_t\\ &= Fz(t)\,dt - \mathcal{J}z(t)\,dt + J_t^\theta\, d\overline{w}_0(t) + d\mathcal{D}_t,\end{aligned}$$

showing (7.10). Now,

$$\begin{aligned}&\mathcal{P}_t^y\bigl(x(t)\bigl(1_{\{\theta(t+h)=i\}} - 1_{\{\theta(t)=i\}}\bigr)\bigr)\\ &\quad= \mathcal{P}_t^y\bigl[E\bigl(x(t)(1_{\{\theta(t+h)=i\}} - 1_{\{\theta(t)=i\}})|\mathcal{F}_t\bigr)\bigr]\\ &\quad= \mathcal{P}_t^y\bigl[E\bigl(x(t)1_{\{\theta(t+h)=i\}}|\mathcal{F}_t\bigr) - x(t)1_{\{\theta(t)=i\}}\bigr]\\ &\quad= \mathcal{P}_t^y\Bigl[\sum_{j\in\mathcal{S}} E\bigl(x(t)1_{\{\theta(t+h)=i\}}1_{\{\theta(t)=j\}}|\mathcal{F}_t\bigr)\Bigr] - \hat{z}_i(t)\\ &\quad= \sum_{j\in\mathcal{S},j\neq i} \mathcal{P}_t^y\bigl[x(t)1_{\{\theta(t)=j\}}\lambda_{ji}\bigr]h\\ &\qquad+ \mathcal{P}_t^y\bigl[x(t)1_{\{\theta(t)=i\}}(1+\lambda_{ii}h)\bigr] - \hat{z}_i(t) + o(h)\\ &\quad= \sum_{j\in\mathcal{S}} \lambda_{ji}\hat{z}_j(t)h + o(h),\end{aligned} \tag{7.13}$$

showing (7.11). □

Remark 7.8 Notice that

$$dy(t) = Hz(t)\,dt + G_t^\theta\, d\overline{w}(t), \tag{7.14}$$

and from (7.14), the assumptions above, and (7.9) it easily follows that $\{z(t)\}$ and $\{y(t)\}$ are second-order q.m. continuous processes.

7.4 Main Result for the Finite-Horizon Case

The main result for the finite-horizon case reads as follows:

Theorem 7.9 *For system* (7.2)–(7.3) *with assumptions* (A.1)–(A.5), *the best linear mean-square estimator* $\hat{x}(t)$ *is given by the following filter*:

$$\hat{x}(t) = \sum_{i\in\mathcal{S}} \hat{z}_i(t),$$

where

$$d\hat{z}(t) = F\hat{z}(t)\,dt + \tilde{Z}(t)H^*\big(G_t^p G_t^{p*}\big)^{-1}\big(dy(t) - H\hat{z}(t)\,dt\big), \tag{7.15}$$

$$\hat{z}(0) = \mu := \begin{bmatrix}\mu_1^* & \dots & \mu_N^*\end{bmatrix}^*, \tag{7.16}$$

with

$$\tilde{Z}(t) = Z(t) - \hat{Z}(t), \tag{7.17}$$

and $Z(t) = \operatorname{diag}(Q_i(t))$ *with*

$$\dot{Q}_i(t) = A_i Z_i + Z_i A_i^* + \sum_{j\in\mathcal{S}} \lambda_{ji} Z_j + J_i J_i^* p_i(t), \tag{7.18}$$

$$Z_i(0) = V_i, \quad i \in \mathcal{S}, \tag{7.19}$$

and

$$\dot{\hat{Z}}(t) = \bar{F}(t)\hat{Z}(t) + \hat{Z}(t)\bar{F}^*(t) + \hat{Z}(t)H^*\big(G_t^p G_t^{p*}\big)^{-1} H\hat{Z}(t) + Z(t)H^*\big(G_t^p G_t^{p*}\big)^{-1} HZ(t), \tag{7.20}$$

$$\bar{F}(t) = F - Z(t)H^*\big(G_t^p G_t^{p*}\big)^{-1} H, \tag{7.21}$$

$$\hat{Z}(0) = \mu\mu^*. \tag{7.22}$$

Remark 7.10 Notice that for the case with no jumps ($N = 1$), suppressing the superscript and subscript, it follows from (7.17)–(7.22) that $\dot{\tilde{Z}}(t) = A\tilde{Z}(t) + \tilde{Z}(t)A^* + JJ^* - \tilde{Z}(t)H^*(GG^*)^{-1}H\tilde{Z}(t)$, which coincides with the standard filtering Riccati differential equation, and thus (7.15), (7.16) reduce to the standard Kalman filter.

Before going into the proof of this result, let us consider some remarks and results which play a central role in the arguments of the proof. We begin by assuming that $\mu_i = 0$, $i \in \mathcal{S}$, in (A.3), and show later on that the results remain true for the nonzero case. In this case, under assumptions (A.1) and (A.4), all the random variables have zero mean. Defining the innovation process $\{\nu(t)\}$ as

$$d\nu(t) := dy(t) - H\hat{z}(t)\,dt, \tag{7.23}$$

we have from (7.14) that

$$d\nu(t) = H\tilde{z}(t)\,dt + G_{\theta(t)}\,dw(t) = H\tilde{z}(t)\,dt + G_t^\theta\,d\overline{w}(t). \tag{7.24}$$

Lemma 7.11 *For* $t > s$, *we have that*:

(a) $\nu(t) - \nu(s) \perp \mathcal{H}_s^y$,
(b) $\operatorname{cov}(\nu(t)) = \int_0^t G_u^p G_u^{p*}\,du$.

Proof (a) Let $x \in \mathcal{H}_s^y$ and denote by $\nu^i(t)$ the ith component of $\nu(t)$, by H^i the ith row of H and define $G_{\theta(t)} = [g_\theta^{ij}(t)]$. Then from (7.24) we have

$$\begin{aligned}\langle x, \nu^i(t) - \nu^i(s)\rangle &= E\left(x\left(\int_s^t H^i \tilde{z}(u)\,du + \sum_{j=1}^p \int_s^t g_\theta^{ij}(t)\,dw^j(t)\right)\right)\\ &= E\left(x\int_s^t H^i \tilde{z}(u)\,du\right) + \sum_{j=1}^p E\left(\int_s^t x g_\theta^{ij}(t)\,dw^j(t)\right).\end{aligned}$$

The last term is obviously zero. Furthermore, as $H^i \tilde{z}(u) \perp \mathcal{H}_u^y \supset \mathcal{H}_s^y$, we have that

$$E\left(x\int_s^t H^i \tilde{z}(u)\,du\right) = \int_s^t E\big(xH^i \tilde{z}(u)\,du\big) = 0.$$

(b) From (7.24) and the Itô's rule, $d\nu(t)\,d\nu(t)^* = G_t^\theta\,d\overline{w}(t)\,d\overline{w}(t)^* G_t^{\theta*}$. Thus, from the Itô's differential rule,

$$d\big(\nu(t)\nu(t)^*\big) = \big(d\nu(t)\big)\nu(t)^* + \nu(t)\big(d\nu(t)^*\big) + G_t^\theta\,d\overline{w}(t)\,d\overline{w}(t)^* G_t^{\theta*}.$$

From (7.23) and $\nu(t) \in \mathcal{H}_t^y$, bearing in mind part (a), we have that $E(d\nu(t)\nu(t)^*) = 0$. Taking the expectation operator on both sides, we have

$$\begin{aligned}dE\big(\nu(t)\nu(t)^*\big) &= E\big(d\nu(t)\,\nu(t)^*\big) + E\big(\nu(t)\,d\nu(t)^*\big) + E\big(G_t^\theta\,d\overline{w}(t)\,d\overline{w}(t)^* G_t^{\theta*}\big)\\ &= E\big(G_t^\theta G_t^{\theta*}\big)\,dt = G_t^p G_t^{p*}\,dt,\end{aligned}$$

showing the desired result. □

Since by hypothesis $G_i G_i{}^* > 0$, we have $\Delta(t) := (G_t^p G_t^{p*})^{-1/2} > 0$. Set

$$\pi(t) := \int_0^t \Delta(s)\,d\nu(s), \tag{7.25}$$

so that, by Proposition 7.3, $\{\pi(t)\}$ *has stationary orthogonal increments.*

Lemma 7.12 *For $\nu(t)$ and $\pi(t)$ defined as in* (7.23) *and* (7.25), *we have that $\mathcal{H}_t^\pi = \mathcal{H}_t^\nu = \mathcal{H}_t^y$.*

Proof It follows from the same arguments as in Theorem 4.3.4, p. 127, in [97]. □

Now since $\mathcal{H}_t^\pi = \mathcal{H}_t^\nu = \mathcal{H}_t^y$, we have from Proposition 7.6 that

$$\hat{z}(t) = \int_0^t k(t,s)\,d\pi(s), \tag{7.26}$$

where the kernel $k(t,s)$ is given by

$$k(t,s) = \frac{d}{ds} E\big(z(t)\pi(s)^*\big). \tag{7.27}$$

Set

$$q(t) := \hat{z}(t) - \hat{z}(0) - \int_0^t F\hat{z}(u)\,du. \tag{7.28}$$

Recalling that $\{y(t)\}$ is a q.m. continuous process, from Remark 7.8 and from Proposition 7.4 it is not difficult to prove that $\{\hat{z}(t)\}$ is also a q.m. continuous process, and by Proposition 7.2, it has a measurable version, which implies that the integral above is well defined. The next result shows that $\{q(t)\}$ is an o.i. process.

Lemma 7.13 *For $s < t$, we have that $(q(t) - q(s)) \perp \mathcal{H}_s^y$.*

Proof First notice that $\mathcal{P}_s^y \hat{z}(t) = \mathcal{P}_s^y \mathcal{P}_t^y z(t) = \mathcal{P}_s^y z(t)$, by the property of the projection operator, and

$$\mathcal{P}_s^y \int_s^t F\hat{z}(u)\,du = \int_s^t F\mathcal{P}_s^y \mathcal{P}_u^y z(u)\,du = \int_s^t F\mathcal{P}_s^y z(u)\,du = \mathcal{P}_s^y \int_s^t Fz(u)\,du. \tag{7.29}$$

Bearing in mind equations (7.10), (7.28), and (7.29), we have

$$\begin{aligned}
\mathcal{P}_s^y\big(q(t) - q(s)\big) &= \mathcal{P}_s^y\left(z(t) - z(s) - \int_s^t Fz(u)\,du\right) \\
&= \mathcal{P}_s^y \int_s^t \big(-\mathcal{J}z(u)\,du + J_u^\theta\,d\overline{w}_0(u) + d\mathcal{D}_u\big) \\
&= \mathcal{P}_s^y \int_s^t \big(-\mathcal{J}\hat{z}(u)\,du + J_u^\theta\,d\overline{w}_0(u) + d\mathcal{D}_u\big).
\end{aligned} \tag{7.30}$$

We have from assumptions (A.1) and (A.4) that $\mathcal{P}_s^y(\int_s^t J_u^\theta\,d\overline{w}_0(u)) = 0$. As shown in (7.11), $\mathcal{P}_u^y(d\mathcal{D}_u) = \mathcal{J}\hat{z}(u)\,du$, proving that (7.30) equals zero. □

Since for each $t \in \mathbb{R}^+$, $q(t) \in \mathcal{H}_t^y$ and $(q(t) - q(s)) \perp \mathcal{H}_s^y$ for $s < t$, it follows that $\{q(t)\}$ is an o.i. process, and by Proposition 7.5 there exists a matrix-valued function $G(s)$ such that, for each t, $\sum_{i,j}\int_0^t (G_{ij}(s))^2\,ds < \infty$ and $q(t) = \int_0^t G(s)\,d\pi(s)$. In view of (7.28) with $\hat{z}(0) = 0$, we can write

$$d\hat{z}(t) = F\hat{z}(t)\,dt + G(t)\,d\pi(t). \tag{7.31}$$

We are now ready to prove the main result of this section.

Proof of Theorem 7.9 Equation (7.18) follows from (7.9). In view of (7.31), the problem of getting (7.15) boils down now to identify $G(t)$. For this, notice that, since $\{\pi(t)\}$ is a stationary o.i. process, (7.31) has the solution $\hat{z}(t) =$

$\int_0^t \Phi(t,s)G(s)\,d\pi(s)$, where $\Phi(t,s)$ is the transition matrix of F (i.e., $\Phi(t,s) = e^{F(t-s)}$). Comparing this with (7.26), we have that $\Phi(t,s)G(s) = k(t,s)$, and evaluating this at $t = s$ gives $G(t) = k(t,t)$. In order to calculate $k(t,t)$, we first derive an adequate expression for (7.27). This is achieved by first noticing that

$$\pi(s) = \int_0^s \Delta(u)\,d\nu(u) = \int_0^s \Delta(u)H\tilde{z}(u)\,du + \int_0^s \Delta(u)G_u^\theta\,d\overline{w}(u). \tag{7.32}$$

From assumptions (A.1) and (A.4) we have that

$$E\big(z(t)\pi(s)^*\big) = \int_0^s E\big(z(t)\tilde{z}^*(u)\big)H^*\Delta(u)\,du. \tag{7.33}$$

We now prove that $E(z(t)\tilde{z}(u)^*) = \Phi(t,u)\tilde{Z}(u)$. Defining

$$\Psi(t,u) := E\big(z(t)\tilde{z}(u)^*\big),$$

we have from (7.10) that

$$\begin{aligned} d\Psi(t,u) = F\Psi(t,u)\,dt - \mathcal{J}\Psi(t,u)\,dt + E\big(J_t^\theta\,d\overline{w}_0(t)\tilde{z}(u)^*\big) \\ + E\big(d\mathcal{D}_t\tilde{z}(u)^*\big). \end{aligned} \tag{7.34}$$

Let us now analyze the terms in (7.34). Recalling that J_t^θ and $\tilde{z}(u)^*$, with $u \le t$, are $\mathcal{F}_t$-measurable and $d\overline{w}_0(t)$ is independent of $\mathcal{F}_t$, we have that

$$\begin{aligned} E\big(J_t^\theta\,d\overline{w}_0(t)\tilde{z}(u)^*\big) &= E\big(E\big(J_t^\theta\,d\overline{w}_0(t)\tilde{z}(u)^*\big|\mathcal{F}_t\big)\big) = E\big(\mathcal{J}_t^\theta E\big(d\overline{w}_0(t)|\mathcal{F}_t\big)\tilde{z}(u)^*\big) \\ &= E\big(J_t^\theta E\big(d\overline{w}_0(t)\big)\tilde{z}(u)^*\big) = 0 \end{aligned}$$

since $E(d\overline{w}_0(t)) = 0$. By using the same arguments as in (7.13) we get that $E(d\mathcal{D}_t\tilde{z}(u)^*) = \mathcal{J}\Psi(t,u)\,dt$, and, therefore, the matrix differential equation (7.34) becomes

$$\frac{d}{dt}\Psi(t,u) = F\Psi(t,u). \tag{7.35}$$

Noticing that

$$\Psi(u,u) = E\big(\tilde{z}(u)\tilde{z}(u)^*\big) + E\big(\hat{z}(u)\tilde{z}(u)^*\big) = \tilde{Z}(u) + E\big(\hat{z}(u)\tilde{z}(u)^*\big) = \tilde{Z}(u),$$

since $\hat{z}(u) \in \mathcal{H}_u^y$ and $\tilde{z}(u)^* \perp \mathcal{H}_u^y$, it follows from (7.35) that $\Psi(t,u) = e^{F(t-u)}\tilde{Z}(u)$. Consequently,

$$E\big(z(t)\pi(s)^*\big) = \int_0^s \Phi(t,u)\tilde{Z}(u)H^*\Delta(u)\,du,$$

and, bearing in mind (7.27), we get

$$k(t,s) = \frac{d}{ds}E\big(z(t)\pi(s)^*\big) = \Phi(t,s)\tilde{Z}(s)H^*\Delta(s),$$

and therefore $G(t) = k(t,t) = \tilde{Z}(t)H^*\Delta(t)$, showing (7.15). Defining $M_t := H^*(G_t^p G_t^{p*})^{-1}$, we have from (7.31) and Itô's rule that

$$d\big(\hat{z}(t)\hat{z}(t)^*\big) = F\hat{z}(t)\hat{z}(t)^*\,dt + \tilde{Z}(t)M_t\,d\nu(t)\hat{z}(t)^* + \hat{z}(t)\hat{z}(t)^*F^*\,dt \\ + \hat{z}(t)\,d\nu(t)^*M_t^*\tilde{Z}(t) + \tilde{Z}(t)M_t\,d\nu(t)\,d\nu(t)^*M_t^*\tilde{Z}(t).$$

Taking now the expectation operator on both sides, and bearing in mind Lemma 7.11(a), we have

$$\begin{aligned} d\hat{Z}(t) &= F\hat{Z}(t)\,dt + \tilde{Z}(t)M_tE\big(d\nu(t)\hat{z}(t)^*\big) + \hat{Z}(t)F^*\,dt \\ &\quad + E\big(\hat{z}(t)\,d\nu(t)^*\big)M_t^*\tilde{Z}(t) + \tilde{Z}(t)M_tE\big(d\nu(t)\,d\nu(t)^*\big)M_t^*\tilde{Z}(t) \\ &= F\hat{Z}(t)\,dt + \hat{Z}(t)F^*\,dt + \tilde{Z}(t)M_t\big(G_t^pG_t^{p*}\big)M_t^*\tilde{Z}(t)\,dt \\ &= \big[F - Z(t)H^*\big(G_t^pG_t^{p*}\big)^{-1}H\big]\hat{Z}(t)\,dt \\ &\quad + \hat{Z}(t)\big[F^* - H^*\big(G_t^pG_t^{p*}\big)^{-1}HZ(t)\big]\,dt \\ &\quad + Z(t)H^*\big(G_t^pG_t^{p*}\big)^{-1}HZ(t)\,dt + \hat{Z}(t)H^*\big(G_t^pG_t^{p*}\big)^{-1}H\hat{Z}(t)\,dt \\ &= \bar{F}(t)\hat{Z}(t)\,dt + \hat{Z}(t)\bar{F}^*(t)\,dt + Z(t)H^*\big(G_t^pG_t^{p*}\big)^{-1}HZ(t)\,dt \\ &\quad + \hat{Z}(t)H^*\big(G_t^pG_t^{p*}\big)^{-1}H\hat{Z}(t)\,dt, \end{aligned}$$

and (7.20) follows. Finally, consider now that $\mu_i \neq 0,\ i \in \mathcal{S}$, and define $\mu(t) = [\mu_1(t)^* \ \dots \ \mu_N(t)^*]^*$ with $\mu_i(t) := E(z_i(t))$. Then $\dot{\mu}(t) = F\mu(t)$, $\mu(0) = \mu = [\mu_1^* \ \cdots \ \mu_N^*]^*$. We have shown that for the centered process $z^c(t) = z(t) - \mu(t)$, $\hat{z}^c(t)$ satisfies (7.15) with $\hat{z}^c(0) = 0$. Now, from (7.24) and the equation for $\mu(t)$ above we can easily show that $\hat{z}(t) = \hat{z}^c(t) + \mu(t)$ satisfies (7.15) with $\mu(0) = \mu$, and the proof is concluded. □

7.5 Stationary Solution for the Algebraic Riccati Equation

The goal of this section is to obtain the convergence of the error covariance matrix $\tilde{Z}(t)$ defined in (7.17) to a stationary value under the assumption of mean-square stability of the MJLS (7.2) and ergodicity of the associated Markov chain $\theta(t)$. It is shown that there exists a unique solution for the associated stationary Riccati filter equation, and this solution is the limit of the error covariance matrix of the filter. This result will be used in Sect. 7.6 in order to derive the stationary filter for (7.15)–(7.19).

We start by deriving a matrix Riccati differential equation for $\tilde{Z}(t)$ in a more direct way than in (7.17). We recall that $\tilde{z}_i(t) = z_i(t) - \hat{z}_i(t)$, $\tilde{z}(t) = [\tilde{z}_1(t)^* \ \dots \ \tilde{z}_N(t)^*]^* = z(t) - \hat{z}(t)$, $\tilde{Z}(t) = E[\tilde{z}(t)\tilde{z}(t)^*]$, and $\tilde{x}(t) = x(t) - \hat{x}(t) =$

$\sum_{i\in\mathcal{S}}[z_i(t)-\hat{z}_i(t)]=\sum_{i\in\mathcal{S}}\tilde{z}_i(t)$. In addition, we recall that

$$0 < G_t^p G_t^{p*} = \sum_{j\in\mathcal{S}} G_j G_j^* p_j(t) \in \mathbb{B}(\mathbb{R}^p)$$

and that the innovation process $\{\nu(t)\}$, defined in (7.23), is written as

$$\nu(t) = y(t) - \int_0^t \hat{m}(s)\,ds,$$

where $\hat{m}(t)=\mathcal{P}_t^y[H_{\theta(t)}x(t)]=\sum_{i\in\mathcal{S}} H_i\hat{z}_i(t)$, or

$$d\nu(t) = dy(t) - H\hat{z}(t)\,dt.$$

Notice that in (7.17) the term $\tilde{Z}(t)$ is obtained as the difference of two terms $(Z(t)-\hat{Z}(t))$ derived via (7.18) and (7.20). Our first step then is to obtain a Riccati differential equation for $\tilde{Z}(t)$ as follows.

Lemma 7.14 *$\tilde{Z}(t)$ satisfies the following matrix Riccati differential equation*:

$$\begin{aligned}\dot{\tilde{Z}}(t) &= F\tilde{Z}(t)+\tilde{Z}(t)F^* \\ &\quad - \tilde{Z}(t)H^*\big(G_t^pG_t^{p*}\big)^{-1}H\tilde{Z}(t) + J_t^pJ_t^{p*} + \mathcal{V}\big(\mathbf{Q}(t)\big), \qquad (7.36)\\ \tilde{Z}(0) &= \operatorname{diag}(V_i) - \mu\mu^* \ge 0,\end{aligned}$$

with $\mathbf{Q}(t)=(Q_1(t),\dots,Q_N(t))$, where $Q_i(t)$ is the solution of (7.18), *and $\mathcal{V}(\mathbf{Q}(t))$ is defined by*

$$\begin{aligned}\mathcal{V}\big(\mathbf{Q}(t)\big) &:= \begin{bmatrix}\sum_{j\in\mathcal{S}} Q_j(t)\lambda_{j1} & \cdots & 0\\ \vdots & \ddots & \vdots \\ 0 & \cdots & \sum_{j\in\mathcal{S}} Q_j(t)\lambda_{jN}\end{bmatrix}\\ &\quad - \big(\Pi'\otimes I_n\big)Z(t) - Z(t)\big(\Pi'\otimes I_n\big)^*. \qquad (7.37)\end{aligned}$$

Moreover, $\mathcal{V}$ is a linear operator, and $\mathcal{V}(\mathbf{Q}(t))\ge 0$.

Proof Let us first prove that $\mathcal{V}$ is a linear operator and $\mathcal{V}(\mathbf{Q}(t))\ge 0$. For any $\mathbf{Q}=(Q_1,\dots,Q_N)\in\mathbb{H}^{n+}$, $\mathbf{R}=(R_1,\dots,R_N)\in\mathbb{H}^{n+}$, α and $\beta\in\mathbb{R}$, it is straightforward to show that $\mathcal{V}(\alpha\mathbf{Q}+\beta\mathbf{R})=\alpha\mathcal{V}(\mathbf{Q})+\beta\mathcal{V}(\mathbf{R})$. Recalling that $d\mathcal{D}_t=[x(t)^*d(1_{\{\theta(t)=1\}})\ \dots\ x(t)^*d(1_{\{\theta(t)=N\}})]^*\in\mathbb{R}^{Nn}$, by using Lemma 3.6 one can show that $E[d\mathcal{D}_t d\mathcal{D}_t^*]=\mathcal{V}(\mathbf{Q}(t))\,dt$. Then, for any constant vector $v\in\mathbb{R}^{Nn}$,

$$\begin{aligned}v^*\mathcal{V}\big(\mathbf{Q}(t)\big)v\,dt &= v^*E\big(d\mathcal{D}_t d\mathcal{D}_t^*\big)v\\ &= E\big(\big\|d\mathcal{D}_t^*v\big\|^2\big)\ge 0,\end{aligned}$$

and therefore, $\mathcal{V}(\mathbf{Q}(t)) \geq 0$. By the definition of $Z(t)$, we have that

$$\dot{Z}(t) = \operatorname{diag}(A_i)Z(t) + Z(t)\operatorname{diag}(A_i)^* + J_t^p J_t^{p*} + \begin{bmatrix} \sum_{j\in\mathcal{S}} Q_j(t)\lambda_{j1} & \dots & 0 \\ \vdots & \ddots & \vdots \\ 0 & \dots & \sum_{j\in\mathcal{S}} Q_j(t)\lambda_{jN} \end{bmatrix},$$

and bearing in mind the definitions of $\mathcal{V}(\mathbf{Q}(t))$ and F, it follows that

$$\dot{Z}(t) = FZ(t) + Z(t)F^* + J_t^p J_t^{p*} + \mathcal{V}\big(\mathbf{Q}(t)\big).$$

From (7.20) and (7.21) we have that

$$\dot{\hat{Z}}(t) = F\hat{Z}(t) + \hat{Z}(t)F^* + \big[Z(t) - \hat{Z}(t)\big]H^*\big(G_t^p G_t^{p*}\big)^{-1}H\big[Z(t) - \hat{Z}(t)\big].$$

Finally, bearing in mind that $\tilde{Z}(t) = Z(t) - \hat{Z}(t)$, we get that

$$\dot{\tilde{Z}}(t) = F\tilde{Z}(t) + \tilde{Z}(t)F^* - \tilde{Z}(t)H^*\big(G_t^p G_t^{p*}\big)^{-1}H\tilde{Z}(t) + J_t^p J_t^{p*} + \mathcal{V}\big(\mathbf{Q}(t)\big),$$

which proves the lemma. □

We obtain next the main result concerning the asymptotic behavior of the matrix Riccati differential equation (7.36). We show that, under the assumption of irreducibility of the process $\{\theta(t)\}$ and of mean-square stability of (7.2), the unique solution of the matrix Riccati differential equation (7.36) converges to the unique solution of an algebraic Riccati equation.

The main result reads as follows.

Theorem 7.15 *Assume that system* (7.2) *is MSS and* $\{\theta(t)\}$ *is ergodic. Then for any* $\hat{\mathbf{Q}}(0) = (Q_1(0), \dots, Q_N(0)) \in \mathbb{H}^{n+}$ *and* $\tilde{Z}_0 \geq 0$, *we have that* $\tilde{Z}(t) \to \tilde{Z}$ *exponentially with* $\tilde{Z}$ *the unique positive semi-definite solution of the algebraic Riccati equation* (*ARE*):

$$F\tilde{Z} + \tilde{Z}F^* - \tilde{Z}H^*\big(G^p G^{p*}\big)^{-1}H\tilde{Z} + J^p J^{p*} + \mathcal{V}(\mathbf{Q}) = 0, \tag{7.38}$$

where $F - \tilde{Z}H^*(G^p G^{p*})^{-1}H$ *is a stable matrix,*

$$J^p := \operatorname{diag}(\sqrt{\pi_i}J_i), \tag{7.39}$$

$$G^p := \big[\sqrt{\pi_1}G_1 \quad \dots \quad \sqrt{\pi_N}G_N\big], \tag{7.40}$$

with $\{\pi_i; i \in \mathcal{S}\}$ *the limit distribution of* $\theta(t)$, $\mathbf{Q} := (Q_1, \dots, Q_N)$ *with* Q_i *the limit of* $Q_i(t)$, *in the sense that* $\|Q_i(t) - Q_i\| \to 0$ *as* $t \to \infty$, *where* $Q_i = -\hat{\varphi}_i^{-1}(\mathcal{A}^{-1}\hat{\varphi}(\mathbf{R}))$ (*recalling that* $\mathcal{A} = \Pi' \otimes I_{n^2} + \operatorname{diag}(A_i \oplus A_i)$) *and* $R_i = J_i J_i^* \pi_i$, $\mathbf{R} = (R_1, \dots, R_N)$.

Proof See Sect. A.7 in the Appendix A. □

Remark 7.16 For the case in which there is no jump ($N = 1$), we have that $\mathcal{V}(\mathbf{Q}) = 0$, so that in this case the above filter reduces to the Kalman filter.

7.6 Stationary Filter

The main goal of this section is to derive the stationary filter for (7.15)–(7.19). The idea is to design a time-invariant version of the dynamic linear filter (7.15) such that the joint system obtained from (7.2)–(7.3) and the filter is mean-square stable and optimal in the sense that it minimizes the stationary expected value of the square estimation error. It is shown that a solution to this optimal filter problem is obtained from the stationary solution associated to the ARE analyzed in Sect. 7.5. This section is divided into two subsections. In Sect. 7.6.1 several auxiliary results are presented in order to formalize the infinite-horizon filtering problem. The solution of the stationary filtering problem based on the associated filtering ARE obtained in Sect. 7.5 is derived in Sect. 7.6.2. As mentioned in the outline of the chapter, an equivalent LMIs formulation to solve the problem, which can be extended to consider uncertainties on the parameters of the possible modes of operation of the system, together with a numerical example, will be analyzed in Chaps. 9 and 10, Sects. 9.5 and 10.5.

7.6.1 Auxiliary Results and Problem Formulation

As before, we consider a hybrid dynamical system modeled by the following MJLS:

$$dx(t) = A_{\theta(t)}x(t)\,dt + J_{\theta(t)}\,dw(t), \tag{7.41}$$

$$dy(t) = H_{\theta(t)}x(t)\,dt + G_{\theta(t)}\,dw(t), \tag{7.42}$$

$$v(t) = L_{\theta(t)}x(t), \tag{7.43}$$

under the same hypotheses as in Sect. 7.3. Again $\{x(t)\}$ denotes the state vector in $\mathbb{R}^n$, $\{y(t)\}$ the output process in $\mathbb{R}^q$, which generates the observational information that is available at time t, and $\{v(t)\}$ denotes the vector in $\mathbb{R}^r$ that it is desired to estimate.

We recall from Theorem 3.25 that if $\mathrm{Re}\{\lambda(\mathcal{L})\} < 0$, then for any $\mathbf{V} = (V_1, \ldots, V_N) \in \mathbb{H}^n$, there exists a unique $\mathbf{U} = (U_1, \ldots, U_N) \in \mathbb{H}^n$ such that

$$\mathcal{L}(\mathbf{U}) + \mathbf{V} = 0, \tag{7.44}$$

and furthermore, if $\mathbf{V} \geq 0$ (respectively $\mathbf{V} > 0$), then $\mathbf{U} \geq 0$ (respectively $\mathbf{U} > 0$). We will need a matrix representation of (7.44) that will yield a convenient LMIs formulation for the filter problem. To this end, we consider the following notation:

$$L := \begin{bmatrix} L_1 & \dots & L_N \end{bmatrix} \in \mathbb{B}(\mathbb{R}^{Nn}, \mathbb{R}^r). \tag{7.45}$$

The following identity can be easily established for $\mathbf{Z} = (Z_1, \dots, Z_N) \in \mathbb{H}^n$ and $Z = \operatorname{diag}(Z_i)$ (recall the definition of $\mathcal{V}$ in (7.37)):

$$\operatorname{diag}(\mathcal{L}_i(\mathbf{Z})) = FZ + ZF^* + \mathcal{V}(\mathbf{Z}). \tag{7.46}$$

Recall that $Z(t) = E(z(t)z(t)^*) = \operatorname{diag}(Q_i(t))$, $Q_i(t) = E(z_i(t)z_i(t)^*)$, and $\mathbf{Q}(t) = (Q_1(t), \dots, Q_N(t))$. From (3.71) and (7.46) it can be easily shown the following matrix equation for $Z(t)$:

$$\dot{Z}(t) = FZ(t) + Z(t)F^* + \mathcal{V}(\mathbf{Q}(t)) + J_t^p J_t^{p*}. \tag{7.47}$$

It is desired to design a dynamic estimator $\widehat{v}(t)$ for $v(t)$ given in (7.43) of the following form:

$$d\widehat{z}(t) = A_f \widehat{z}(t)\,dt + B_f\,dy(t), \tag{7.48}$$

$$\widehat{v}(t) = L_f \widehat{z}(t), \tag{7.49}$$

$$e(t) = v(t) - \widehat{v}(t), \tag{7.50}$$

where $A_f \in \mathbb{B}(\mathbb{R}^{n_f}, \mathbb{R}^{n_f})$, $B_f \in \mathbb{B}(\mathbb{R}^p, \mathbb{R}^{n_f})$, $L_f \in \mathbb{B}(\mathbb{R}^{n_f}, \mathbb{R}^r)$, and $e(t)$ denotes the estimation error. Defining $x_e(t)^* = [x(t)^* \ \widehat{z}(t)^*]$, we have from (7.41), (7.42), and (7.48)–(7.50) that

$$\begin{aligned} dx_e(t) &= \begin{bmatrix} A_{\theta(t)} & 0 \\ B_f H_{\theta(t)} & A_f \end{bmatrix} x_e(t)\,dt + \begin{bmatrix} J_{\theta(t)} \\ B_f G_{\theta(t)} \end{bmatrix} dw(t), \\ e(t) &= [L_{\theta(t)} - L_f] x_e(t), \end{aligned} \tag{7.51}$$

which is a continuous-time Markov jump linear system. We will be interested in filters such that (7.51) is mean-square stable. The next results aim at providing necessary and sufficient conditions for the mean-square stability of (7.51). Define

$$\widehat{Z}(t) := E(\widehat{z}(t)\widehat{z}(t)^*),$$

$$U(t) := \begin{bmatrix} U_1(t) \\ \vdots \\ U_N(t) \end{bmatrix}, \qquad U_i(t) := E(z_i(t)\widehat{z}(t)^*),$$

$$P(t) := E\left(\begin{bmatrix} z(t) \\ \widehat{z}(t) \end{bmatrix} \begin{bmatrix} z(t)^* & \widehat{z}(t)^* \end{bmatrix} \right) = \begin{bmatrix} Z(t) & U(t) \\ U(t)^* & \widehat{Z}(t) \end{bmatrix}, \tag{7.52}$$

$$\widetilde{L} := \begin{bmatrix} L & -L_f \end{bmatrix}, \qquad \widetilde{F} := \begin{bmatrix} F & 0 \\ B_f H & A_f \end{bmatrix}, \tag{7.53}$$

$$\widetilde{J} := \begin{bmatrix} J^p \\ B_f G^p \end{bmatrix}, \qquad \widetilde{J}_t := \begin{bmatrix} J_t^p \\ B_f G_t^p \end{bmatrix}. \tag{7.54}$$

Proposition 7.17 *For* $t \in \mathbb{R}^+$,

$$\dot{P}(t) = \widetilde{F} P(t) + P(t)\widetilde{F}^* + \begin{bmatrix} \mathcal{V}(\mathbf{Q}(t)) & 0 \\ 0 & 0 \end{bmatrix} + \widetilde{J}_t \widetilde{J}_t^*. \tag{7.55}$$

Proof By Itô's calculus, noting that $B_f H_{\theta(t)} x(t) = B_f H z(t)$, from (7.51) we have

$$\dot{\widehat{Z}}(t) = B_f H U(t) + A_f \widehat{Z}(t) + U(t)^* H^* B_f^* + \widehat{Z}(t) A_f^* + B_f G_t^p G_t^{p*} B_f^*. \tag{7.56}$$

Similarly, by Itô's calculus,

$$\dot{U}(t) = F U(t) + Z(t) H^* B_f^* + U(t) A_f^* + J_t^p G_t^{p*} B_f^*. \tag{7.57}$$

By combining (7.47), (7.56), and (7.57), we get (7.55). □

We now want to rewrite (7.47) so that the term $\mathcal{V}(\mathbf{Q}(t))$ can be decomposed as a sum of matrices. To this end, we first define $\Gamma_\ell := [\varphi_{\ell,i,j}] \in \mathbb{B}(\mathbb{R}^N)$, $\ell \in \mathcal{S}$, where

$$\varphi_{\ell,i,j} = \begin{cases} |\lambda_{\ell i}|, & i = j, \\ -\lambda_{\ell i}, & i \neq j,\ j = \ell, \\ -\lambda_{\ell j}, & i \neq j,\ i = \ell, \\ 0 & \text{otherwise}. \end{cases}$$

After some straightforward calculations we have that (7.37) can be rewritten, for $\mathbf{Q} = (Q_1, \dots, Q_N)$, as

$$\mathcal{V}(\mathbf{Q}) = \sum_{\ell \in \mathcal{S}} (\Gamma_\ell \otimes I_n)\, \mathrm{dg}(Q_\ell)$$

where, for $S_i \in \mathbb{B}(\mathbb{R}^n)$, $i \in \mathcal{S}$, $\mathrm{dg}(S_j)$ stands for an $Nn \times Nn$ matrix where only S_j is put together corner-to-corner diagonally with all other entries being zero. We have the following result.

Proposition 7.18 $\Gamma_\ell \geq 0$.

Proof For any vector $v^* = [v_1 \ \dots \ v_N]$, we have that

$$\begin{aligned} v^* \Gamma_\ell v &= \sum_{i \in \mathcal{S}} \sum_{j \in \mathcal{S}} \varphi_{\ell,i,j} v_i v_j \\ &= |\lambda_{\ell\ell}| v_\ell^2 + \sum_{i \neq \ell} \lambda_{\ell i} v_i^2 - v_\ell \Big(\sum_{j \neq \ell} \lambda_{\ell j} v_j \Big) - \Big(\sum_{i \neq \ell} v_i \lambda_{\ell i} \Big) v_\ell. \end{aligned}$$

But recalling that $-\lambda_{\ell\ell} = \sum_{j\neq\ell} \lambda_{\ell j}$, we get that

$$\begin{aligned} v^* \Gamma_\ell v &= \sum_{i\neq\ell} \lambda_{\ell i} v_\ell^2 + \sum_{i\neq\ell} \lambda_{\ell i} v_i^2 - 2\Big(\sum_{i\neq\ell} \lambda_{\ell i} v_i v_\ell\Big) \\ &= \sum_{i\neq\ell} \lambda_{\ell i}\big(v_\ell^2 + v_i^2 - 2 v_i v_\ell\big) = \sum_{i\neq\ell} \lambda_{\ell i}(v_\ell - v_i)^2 \geq 0, \end{aligned}$$

showing the desired result. □

It follows from Proposition 7.18 that for $\mathbf{Q} = (Q_1, \ldots, Q_N) \geq 0$, we have that

$$\begin{aligned} \mathcal{V}(\mathbf{Q}) &= \sum_{\ell\in\mathcal{S}} \big(\Gamma_\ell^{1/2} \otimes I_n\big)\big(\Gamma_\ell^{1/2} \otimes I_n\big) \operatorname{dg}(Q_\ell) \\ &= \sum_{\ell\in\mathcal{S}} \big(\Gamma_\ell^{1/2} \otimes I_n\big) \operatorname{dg}(Q_\ell)\big(\Gamma_\ell^{1/2} \otimes I_n\big) \geq 0. \end{aligned} \tag{7.58}$$

Therefore, writing for $\ell \in \mathcal{S}$,

$$\Psi_\ell := \Gamma_\ell^{1/2} \otimes I_n, \tag{7.59}$$

$$\Upsilon_\ell := \begin{bmatrix} \Psi_\ell \\ 0 \end{bmatrix}, \tag{7.60}$$

we have from (7.55) that

$$\dot{P}(t) = \widetilde{F} P(t) + P(t) \widetilde{F}^* + \sum_{\ell\in\mathcal{S}} \Upsilon_\ell \operatorname{dg}\big(Q_\ell(t)\big) \Upsilon_\ell^* + \widetilde{J}_t \widetilde{J}_t^*. \tag{7.61}$$

We next present some conditions to get the mean-square stability of system (7.51). First, we present a necessary and sufficient condition based on the spectra of the operator $\mathcal{L}$ and matrix A_f.

Proposition 7.19 *System* (7.51) *is MSS if and only if* $\operatorname{Re}\{\lambda(\mathcal{L})\} < 0$ *and* A_f *is a stable matrix.*

Proof (⇒) If (7.51) is MSS, then, considering $w(t) = 0$ in (7.51), we have for any initial condition $x_e(0)$, θ_0, that

$$E\big(\|x_e(t)\|^2\big) = E\big(\|x(t)\|^2\big) + E\big(\|\widehat{z}(t)\|^2\big) \overset{t\to\infty}{\longrightarrow} 0,$$

that is, $E(\|x(t)\|^2) \overset{t\to\infty}{\longrightarrow} 0$ and $E(\|\widehat{z}(t)\|^2) \overset{t\to\infty}{\longrightarrow} 0$. By Theorem 3.25, $\operatorname{Re}\{\lambda(\mathcal{L})\} < 0$. Consider now an initial condition $x_e(0) = [0^* \ \widehat{z}(0)^*]^*$. Then, clearly $\widehat{z}(t) = e^{A_f t}\widehat{z}(0)$, and since $\widehat{z}(t) \overset{t\to\infty}{\longrightarrow} 0$ for any initial condition $\widehat{z}(0)$, it follows that A_f is a stable matrix.

($\Leftarrow$) Suppose now that $\mathrm{Re}\{\lambda(\mathcal{L})\} < 0$ and A_f is a stable matrix. By Proposition 2.23, $\mathrm{Re}\{\lambda(\mathcal{L})\} < 0$ implies that $\|e^{\mathcal{L}t}\| \le ae^{-bt}$ for some $a > 0$, $b > 0$, and all $t \in \mathbb{R}^+$. From (7.51) with $w(t) = 0$ and according to Theorem 3.25, $E(\|x(t)\|^2) \le ce^{-bt}\|x(0)\|^2$ for some $c > 0$. Moreover,

$$d\widehat{z}(t) = A_f\widehat{z}(t)\,dt + B_f H_{\theta(t)}x(t)\,dt,$$

and thus,

$$\widehat{z}(t) = e^{A_f t}\widehat{z}(0) + \int_0^t e^{A_f(t-s)}B_f H_{\theta(s)}x(s)\,ds.$$

From the triangular inequality we have

$$\begin{aligned}
E\bigl(\|\widehat{z}(t)\|^2\bigr)^{1/2} &\le E\bigl(\|e^{A_f t}\widehat{z}(0)\|^2\bigr)^{1/2} + \int_0^t E\bigl(\|e^{A_f(t-s)}B_f H_{\theta(s)}x(s)\|^2\bigr)^{1/2}\,ds \\
&\le \|e^{A_f t}\|E\bigl(\|\widehat{z}(0)\|^2\bigr)^{1/2} \\
&\quad + \|B_f\|\|\mathbf{H}\|_{\max}\int_0^t \|e^{A_f(t-s)}\|E\bigl(\|x(s)\|^2\bigr)^{1/2}\,ds,
\end{aligned}$$

where we recall that $\|\mathbf{H}\|_{\max} = \max\{\|H_i\|, i \in \mathcal{S}\}$. Since A_f is stable, we can find $a' > 0$ and $b' > 0$ such that $\|e^{A_f t}\| \le a'e^{-b't}$. Then, for some $\bar{a} > 0$ and $\bar{b} > 0$,

$$E\bigl(\|\widehat{z}(t)\|^2\bigr)^{1/2} \le \bar{a}\bigl(e^{-b't}\|\widehat{z}(0)\| + e^{-\bar{b}t}t\|x(0)\|\bigr),$$

showing that $E(\|x_e(t)\|^2) = E(\|x(t)\|^2) + E(\|\widehat{z}(t)\|^2) \stackrel{t\to\infty}{\longrightarrow} 0$ for any initial condition $x_e(0)$, θ_0, so that by Theorem 3.25 system (7.51) is MSS. □

The next proposition presents a necessary and sufficient condition for mean-square stability of system (7.51) based on an LMIs representation.

Proposition 7.20 *System* (7.51) *is MSS if and only if there exists $P > 0$ with*

$$P = \begin{bmatrix} \mathrm{diag}(Q_i) & U \\ U^* & \widehat{Z} \end{bmatrix}$$

such that

$$\widetilde{F}P + P\widetilde{F}^* + \sum_{\ell\in\mathcal{S}} \Upsilon_\ell\,\mathrm{dg}(Q_\ell)\Upsilon_\ell^* < 0. \tag{7.62}$$

Proof ($\Rightarrow$) Consider the operator $\widetilde{\mathcal{L}} \in \mathbb{B}(\mathbb{H}^{n+n_f})$ as follows: for $\mathbf{V} = (V_1, \dots, V_N) \in \mathbb{H}^{n+n_f}$, $\widetilde{\mathcal{L}}(\mathbf{V}) = (\widetilde{\mathcal{L}}_1(\mathbf{V}), \dots, \widetilde{\mathcal{L}}_N(\mathbf{V}))$ is given by

$$\widetilde{\mathcal{L}}_j(\mathbf{V}) := \widetilde{A}_j V_j + V_j\widetilde{A}_j^* + \sum_{i\in\mathcal{S}} \lambda_{ij}V_i, \tag{7.63}$$

where

$$\widetilde{A}_i = \begin{bmatrix} A_i & 0 \\ B_f H_i & A_f \end{bmatrix}.$$

Consider model (7.51) with $w(t) = 0$, and for $t \in \mathbb{R}^+$,

$$\widetilde{P}_i(t) = \begin{bmatrix} E(z_i(t)z_i(t)^*) & E(z_i(t)\widehat{z}(t)^*) \\ E(\widehat{z}(t)z_i(t)^*) & E(\widehat{z}(t)\widehat{z}(t)^* 1_{\{\theta(t)=i\}}) \end{bmatrix} \tag{7.64}$$

and $\widetilde{\mathbf{P}}(t) = (\widetilde{P}_1(t), \dots, \widetilde{P}_N(t)) \in \mathbb{H}^{n+n_f}$. By Proposition 7.9, $\dot{\widetilde{\mathbf{P}}}(t) = \widetilde{\mathcal{L}}(\widetilde{\mathbf{P}}(t))$, and by Theorem 3.25 system (7.51) is MSS if and only if there exists $\widetilde{\mathbf{P}} = (\widetilde{P}_1, \dots, \widetilde{P}_N) \in \mathbb{H}^{n+n_f}$, $\widetilde{P}_i > 0$, $i \in \mathcal{S}$, such that

$$\widetilde{\mathcal{L}}_j(\widetilde{\mathbf{P}}) < 0, \quad j \in \mathcal{S}. \tag{7.65}$$

Partitionate $\widetilde{P}_j$ as follows:

$$\widetilde{P}_j = \begin{bmatrix} Q_j & U_j \\ U_j^* & \widehat{Z}_j \end{bmatrix}, \quad j \in \mathcal{S},$$

where $Q_j \in \mathbb{B}(\mathbb{R}^n)$ and $\widehat{Z}_j \in \mathbb{B}(\mathbb{R}^{n_f})$, and define $\widehat{Z} = \sum_{j\in\mathcal{S}} \widehat{Z}_j$, $Z = \operatorname{diag}(Q_i)$, and

$$U = \begin{bmatrix} U_1 \\ \vdots \\ U_N \end{bmatrix}, \qquad P = \begin{bmatrix} Z & U \\ U^* & \widehat{Z} \end{bmatrix}.$$

Notice that $P > 0$. Indeed since for each $j \in \mathcal{S}$, $\widetilde{P}_j > 0$, it follows from Schur's complement (Lemma 2.26) that $\widehat{Z}_j > U_j^* Z_j^{-1} U_j$ for $j \in \mathcal{S}$, so that

$$\widehat{Z} = \sum_{j\in\mathcal{S}} \widehat{Z}_j > \sum_{j\in\mathcal{S}} U_j^* Q_j^{-1} U_j = U^* Z^{-1} U.$$

From Schur's complement again it follows that

$$\begin{bmatrix} Z & U \\ U^* & \widehat{Z} \end{bmatrix} = P > 0.$$

Reorganizing (7.65) and using (7.58), we obtain that (7.62) holds.

($\Leftarrow$) From (7.62) and (7.58) it follows that $\mathcal{L}(\mathbf{Q}) < 0$, where $\mathbf{Q} = (Q_1, \dots, Q_N) \in \mathbb{H}^n$, which means, by Theorem 3.25, that $\operatorname{Re}\{\lambda(\mathcal{L})\} < 0$. From the Lyapunov equation (7.62) it is easy to see that $\widetilde{F}$ is stable, and thus in particular A_f is stable. Since $\operatorname{Re}\{\lambda(\mathcal{L})\} < 0$ and A_f is stable, it follows from Proposition 7.19 that system (7.51) is MSS. □

The next result guarantees the convergence of $P(t)$ defined in (7.52), (7.55) to a $P \geq 0$ as $t \to \infty$.

Proposition 7.21 *Consider* $P(t)$ *given by* (7.52), (7.55) *and assume that* $\operatorname{Re}\{\lambda(\mathcal{L})\} < 0$ *and that* A_f *is stable. Then* $P(t) \overset{t\to\infty}{\longrightarrow} P \geq 0$, *where* P *is of the following form*:

$$P = \begin{bmatrix} Z & U \\ U^* & \widehat{Z} \end{bmatrix}, \quad Z = \operatorname{diag}(Q_i) \geq 0.$$

Moreover, P *is the only solution of the following equation in* V:

$$0 = \widetilde{F}V + V\widetilde{F}^* + \sum_{\ell \in \mathcal{S}} \Upsilon_\ell \operatorname{dg}(X_\ell)\Upsilon_\ell^* + J^p J^{p*}, \tag{7.66}$$

$$V = \begin{bmatrix} X & R \\ R^* & \widehat{X} \end{bmatrix}, \quad X = \operatorname{diag}(X_i). \tag{7.67}$$

Furthermore, if V *satisfies*

$$\widetilde{F}V + V\widetilde{F}^* + \sum_{\ell \in \mathcal{S}} \Upsilon_\ell \operatorname{dg}(X_\ell)\Upsilon_\ell^* + \widetilde{J}\widetilde{J}^* \leq 0 \quad (\textit{respectively} < 0), \tag{7.68}$$

then $V \geq P$ (*respectively* $V > P$).

Proof Consider the operator $\widetilde{\mathcal{L}}$ as in (7.63) and $\widetilde{P}_i(t)$ as in (7.64). As shown in Corollary 3.31,

$$\dot{\widetilde{P}}_j(t) = \widetilde{\mathcal{L}}_j\big(\widetilde{\mathbf{P}}(t)\big) + p_j(t)\begin{bmatrix} J_j J_j^* & J_j G_j^* B_f^* \\ B_f G_j J_j^* & B_f G_j G_j^* B_f^* \end{bmatrix},$$

and $\widetilde{P}_j(t) \overset{t\to\infty}{\longrightarrow} \widetilde{P}_j \geq 0$, where $\widetilde{\mathbf{P}} = (\widetilde{P}_1, \dots, \widetilde{P}_N)$ satisfies

$$0 = \widetilde{\mathcal{L}}_j(\widetilde{\mathbf{P}}) + \pi_i \begin{bmatrix} J_j J_j^* & J_j G_j^* B_f^* \\ B_f G_j J_j^* & B_f G_j G_j^* B_f^* \end{bmatrix}. \tag{7.69}$$

Note that

$$\widetilde{P}_i(t) = \begin{bmatrix} Q_i(t) & U_i(t) \\ U_i^*(t) & \widehat{Q}_i(t) \end{bmatrix} \overset{t\to\infty}{\longrightarrow} \begin{bmatrix} Q_i & U_i \\ U_i^* & \widehat{Q}_i \end{bmatrix},$$

where

$$\widehat{Q}_i(t) = E\big(\widehat{z}(t)\widehat{z}(t)^* 1_{\{\theta(t)=i\}}\big).$$

Moreover,

$$\widehat{Z}(t) = \sum_{i \in \mathcal{S}} \widehat{Q}_i(t) \overset{t\to\infty}{\longrightarrow} \sum_{i \in \mathcal{S}} \widehat{Q}_i = \widehat{Z}.$$

Defining

$$P := \begin{bmatrix} Z & U \\ U^* & \widehat{Z} \end{bmatrix}, \quad U := \begin{bmatrix} U_1 \\ \vdots \\ U_N \end{bmatrix}$$

and recalling that $Z = \text{diag}(Q_i)$, it follows that $P(t) \overset{t\to\infty}{\longrightarrow} P$. Furthermore, from (7.69) we have that P satisfies (7.66), (7.67). Suppose that V also satisfies (7.66), (7.67) and set $\mathbf{Q} = (Q_1, \ldots, Q_N)$, $\mathbf{X} = (X_1, \ldots, X_N)$. Then $\mathcal{L}(\mathbf{X}) = 0$, $\mathcal{L}(\mathbf{Q}) = 0$, and since $\text{Re}\{\lambda(\mathcal{L})\} < 0$, we have from Corollary 3.31 that $Q_j = X_j$, $j \in \mathcal{S}$. This yields that $0 = \widetilde{F}(V - P) + (V - P)\widetilde{F}^*$. By Proposition 3.13, $\text{Re}\{\lambda(\mathcal{L})\} < 0$ implies that F is stable and thus that the block diagonal matrix $\widetilde{F}$ is also stable, yielding that $V = P$. Finally, suppose that V is such that (7.67), (7.68) are satisfied. Then $\mathcal{L}(\mathbf{X}) \leq 0$ (respectively $\mathcal{L}(\mathbf{X}) < 0$), and it follows that $\mathcal{L}(\mathbf{X} - \mathbf{Q}) \leq 0$ (respectively $\mathcal{L}(\mathbf{X} - \mathbf{Q}) < 0$). This implies by Theorem 3.21 that $\mathbf{X} \geq \mathbf{Q}$ (respectively $\mathbf{X} > \mathbf{Q}$). Using this fact, we conclude that $\widetilde{F}(V - P) + (V - P)\widetilde{F}^* \leq 0$ (respectively $\widetilde{F}(V - P) + (V - P)\widetilde{F}^* < 0$), and again from stability of $\widetilde{F}$ it follows that $V - P \geq 0$ (respectively $V - P > 0$). □

We conclude this subsection presenting the formulation of the stationary filtering problem. We assume from now on the following hypothesis:

Assumption 7.22 $\text{Re}\{\lambda(\mathcal{L})\} < 0$ (or equivalently $\text{Re}\{\lambda(\mathcal{A})\} < 0$, see Theorem 3.21).

In what follows, recall that $v(t) = Lz(t)$ and that $e(t) = v(t) - \widehat{v}(t) = Lz(t) - L_f\widehat{z}(t)$.

Definition 7.23 (The stationary filtering problem) Find (A_f, B_f, L_f) such that A_f is stable and minimizes $\lim_{t\to\infty} E(\|e(t)\|^2)$, that is, minimizes

$$\begin{aligned} \lim_{t\to\infty} E\big(\|e(t)\|^2\big) &= \text{tr}\Big(\lim_{t\to\infty} E\big(e(t)e(t)^*\big)\Big) \\ &= \text{tr}\Big(\widetilde{L} \lim_{t\to\infty} P(t)\widetilde{L}^*\Big) = \text{tr}\big(\widetilde{L}P\widetilde{L}^*\big), \end{aligned} \tag{7.70}$$

where the last equality follows from Proposition 7.21 and

$$P = \begin{bmatrix} Z & U \\ U^* & \widehat{Z} \end{bmatrix}, \quad Z = \text{diag}(Q_i), \tag{7.71}$$

in which, according to Proposition 7.21, P is the unique solution of the equation

$$\widetilde{F}P + P\widetilde{F}^* + \sum_{\ell\in\mathcal{S}} \Upsilon_\ell \, \text{dg}(Q_\ell)\Upsilon_\ell^* + \widetilde{J}\widetilde{J}^* = 0. \tag{7.72}$$

7.6.2 Solution for the Stationary Filtering Problem via the ARE

In this subsection we present a solution for the stationary filtering problem posed in Definition 7.23 based on the associated filtering ARE derived in Sect. 7.5. To keep the same notation as in Sects. 7.4 and 7.5, we assume here that $J_i G_i^* = 0$ for each $i \in \mathcal{S}$ (although this assumption could be easily removed). We recall the assumption that for each $i \in \mathcal{S}$, $G_i G_i^* > 0$, so that from this hypothesis we have that $G_t^p G_t^{p*} = \sum_{i\in\mathcal{S}} p_i(t) G_i G_i^* > 0$ and similarly $G^p G^{p*} = \sum_{i\in\mathcal{S}} \pi_i G_i G_i^* > 0$. By Corollary 3.31, $\mathrm{Re}\{\lambda(\mathcal{L})\} < 0$ implies that there exists a unique solution $\mathbf{Q} = (Q_1, \dots, Q_N)$ for (3.72). We recall that $Z = \mathrm{diag}(Q_i)$. Define

$$T(U) = U H^* \bigl(G^p G^{p*}\bigr)^{-1}. \tag{7.73}$$

We have the following theorem.

Theorem 7.24 *Consider* $\mathbf{Q} = (Q_1, \dots, Q_N) = -\mathcal{L}^{-1}(\mathbf{R})$ *as in* (3.72) *and let* $\widetilde{Z}$ *be the unique positive semi-definite solution of the algebraic Riccati equation* (7.38). *An optimal solution for the stationary filtering problem posed in Definition* 7.23 *is*

$$A_{f,\mathrm{op}} = F - T(\widetilde{Z})H, \qquad B_{f,\mathrm{op}} = T(\widetilde{Z}), \qquad L_{f,\mathrm{op}} = L \tag{7.74}$$

with the optimal cost given by $\mathrm{tr}(L\widetilde{Z}L^*)$.

Proof Consider any A_f, B_f, L_f such that A_f is stable and write

$$\begin{aligned} d\widehat{z}(t) &= A_f \widehat{z}(t)\,dt + B_f\,dy(t), \\ \widehat{v}(t) &= L_f \widehat{z}(t). \end{aligned}$$

Consider $\widehat{v}_{\mathrm{op}}(t) = L\widehat{z}_{\mathrm{op}}(t)$, where $\widehat{z}_{\mathrm{op}}(t)$ is given by (7.15), (7.16), and $e(t) = v(t) - \widehat{v}(t)$. Noticing that

$$\begin{aligned} e(t) &= Lz(t) - L_f\widehat{z}(t) = Lz(t) - L\widehat{z}_{\mathrm{op}}(t) + L\widehat{z}_{\mathrm{op}}(t) - L_f\widehat{z}(t) \\ &= L\widetilde{z}_{\mathrm{op}}(t) + L\widehat{z}_{\mathrm{op}}(t) - L_f\widehat{z}(t) = L\widetilde{z}_{\mathrm{op}}(t) + \widehat{v}_{\mathrm{op}}(t) - \widehat{v}(t), \end{aligned}$$

from the orthogonality (see Sect. 7.4) between $\widetilde{z}_{\mathrm{op}}(t)$ and $\widehat{v}_{\mathrm{op}}(t) - \widehat{v}(t)$ we have that

$$\begin{aligned} E\bigl(\|e(t)\|^2\bigr) &= \mathrm{tr}\Bigl(E\bigl(\bigl(L\widetilde{z}_{\mathrm{op}}(t) + \widehat{v}_{\mathrm{op}}(t) - \widehat{v}(t)\bigr)\bigl(L\widetilde{z}_{\mathrm{op}}(t) + \widehat{v}_{\mathrm{op}}(t) - \widehat{v}(t)\bigr)^*\bigr)\Bigr) \\ &= \mathrm{tr}\bigl(L\widetilde{Z}(t)L^*\bigr) + E\bigl(\|\widehat{v}_{\mathrm{op}}(t) - \widehat{v}(t)\|^2\bigr) \\ &\ge \mathrm{tr}\bigl(L\widetilde{Z}(t)L^*\bigr) \stackrel{t\to\infty}{\longrightarrow} \mathrm{tr}\bigl(L\widetilde{Z}L^*\bigr) \end{aligned}$$

since, as stated in Theorem 7.9, $\widetilde{Z}(t) \stackrel{t\to\infty}{\longrightarrow} \widetilde{Z}$. Combining this with (7.70) yields that

$$\lim_{t\to\infty} E\bigl(\|e(t)\|^2\bigr) = \mathrm{tr}\bigl(\widetilde{L}P\widetilde{L}^*\bigr) \ge \mathrm{tr}\bigl(L\widetilde{Z}L^*\bigr). \tag{7.75}$$

On the other hand, consider $A_f = A_{f,\mathrm{op}}$, $B_f = B_{f,\mathrm{op}}$, $L_f = L_{f,\mathrm{op}}$ as in (7.74). We want to show that in this case the unique solution of (7.72), denoted by

$$P_{\mathrm{op}} = \begin{bmatrix} Z_{\mathrm{op}} & U_{\mathrm{op}} \\ U_{\mathrm{op}}^* & \widehat{Z}_{\mathrm{op}} \end{bmatrix}, \qquad Z_{\mathrm{op}} = \mathrm{diag}(Z_{\mathrm{op},i}), \tag{7.76}$$

is given in (7.76) by $Z_{\mathrm{op}} = Z$, $U_{\mathrm{op}} = \widehat{Z}_{\mathrm{op}} = Z - \widetilde{Z}$, that is, we have the following equations satisfied:

$$0 = B_{f,\mathrm{op}} H U_{\mathrm{op}} + A_{f,\mathrm{op}} \widehat{Z}_{\mathrm{op}} + U_{\mathrm{op}}^* H^* B_{f,\mathrm{op}}^* + \widehat{Z}_{\mathrm{op}} A_{f,\mathrm{op}}^* + B_{f,\mathrm{op}} G^p G^{p*} B_{f,\mathrm{op}}^*, \tag{7.77}$$

$$0 = F U_{\mathrm{op}} + Z_{\mathrm{op}} H_{\mathrm{op}}^* B_{f,\mathrm{op}}^* + U_{\mathrm{op}} A_{f,\mathrm{op}}^*. \tag{7.78}$$

Notice that

$$0 = FZ + ZF^* + \mathcal{V}(\mathbf{Q}) + J^p J^{p*}, \tag{7.79}$$

and from (7.79) and (7.38) we have that

$$F(Z - \widetilde{Z}) + (Z - \widetilde{Z})F^* + \widetilde{Z} H^* \big(G^p G^{p*}\big)^{-1} H \widetilde{Z} = 0. \tag{7.80}$$

For (7.77) with $U_{\mathrm{op}} = \widehat{Z} = Z - \widetilde{Z}$, we have from (7.80) that

$$\begin{aligned}
&B_{f,\mathrm{op}} H U_{\mathrm{op}} + A_{f,\mathrm{op}} \widehat{Z}_{\mathrm{op}} + U_{\mathrm{op}}^* H^* B_{f,\mathrm{op}}^* + \widehat{Z}_{\mathrm{op}} A_{f,\mathrm{op}}^* + B_{f,\mathrm{op}} G^p G^{p*} B_{f,\mathrm{op}}^* \\
&\quad = T(\widetilde{Z}) H \widehat{Z} + \big(F - T(\widetilde{Z}) H\big) \widehat{Z} + \widehat{Z} H^* T(\widetilde{Z})^* + \widehat{Z}\big(F - T(\widetilde{Z}) H\big)^* \\
&\qquad + T(\widetilde{Z}) G^p G^{p*} T(\widetilde{Z})^* \\
&\quad = F\widehat{Z} + \widehat{Z} F^* + \widetilde{Z} H^* \big(G^p G^{p*}\big)^{-1} H \widetilde{Z} \\
&\quad = 0
\end{aligned}$$

and similarly from (7.78) that

$$\begin{aligned}
&F U_{\mathrm{op}} + Z_{\mathrm{op}} H_{\mathrm{op}}^* B_{f,\mathrm{op}}^* + U_{\mathrm{op}} A_{f,\mathrm{op}}^* \\
&\quad = F\widehat{Z} + Z H^* T(\widetilde{Z})^* + \widehat{Z}\big(F - T(\widetilde{Z}) H\big)^* \\
&\quad = F\widehat{Z} + \widehat{Z} F^* + (Z - \widehat{Z}) H^* \big(G^p G^{p*}\big)^{-1} H \widetilde{Z} \\
&\quad = F\widehat{Z} + \widehat{Z} F^* + \widetilde{Z} \big(G^p G^{p*}\big)^{-1} H \widetilde{Z} = 0.
\end{aligned}$$

This shows that for the choice of $A_{f,\mathrm{op}}$, $B_{f,\mathrm{op}}$, $L_{f,\mathrm{op}}$ as in (7.74), the unique solution P_{op} of (7.72) is given in (7.76) by $Z_{\mathrm{op}} = Z$, $U_{\mathrm{op}} = \widehat{Z} = Z - \widetilde{Z}$. It is easy to see that in this case $\widetilde{L} P_{\mathrm{op}} \widetilde{L}^* = L \widetilde{Z} L^*$, which shows by (7.75) that (7.74) yields the minimal cost $\mathrm{tr}(L\widetilde{Z}L^*)$. □

The next corollary is an immediate consequence of the above theorem.

Corollary 7.25 *An optimal stationary filter for the stationary filtering problem is*

$$d\widehat{z}_{\mathrm{op}}(t) = F\widehat{z}_{\mathrm{op}}(t)\,dt + \widetilde{Z}H^*\big(G^pG^{p*}\big)^{-1}\big(dy(t) - H\widehat{z}_{\mathrm{op}}(t)\,dt\big),$$
$$\widehat{z}_{\mathrm{op}}(0) = E\big(z(0)\big),$$

where $\widetilde{Z}$ *is a unique positive semi-definite solution of the algebraic Riccati equation* (7.38) *with* Z *satisfying* (7.79).

7.7 Historical Remarks

Filtering theory has been widely celebrated as a great achievement in stochastic system theory and is of fundamental importance in applications. The Wiener–Kolmogorov filter was the first systematic attempt to have a filtering theory for the linear case (the time-invariant case). The crowning achievement in this setting is the Kalman filter, which has spawned a widespread range of applications. The climax of the theory is the Fujisaki–Kallianpur–Kunita equation for the nonlinear setting. Although the theoretical machinery available to deal with nonlinear estimation problems is by now rather considerable, there are yet many challenging questions in this area. One of these is the fact that the description of the optimal nonlinear filter can rarely be given in terms of a closed finite system of stochastic differential equations (the so-called finite filters), i.e., the optimal nonlinear filter is not computable via a finite computation (the filter is infinite-dimensional). This is what happens with the optimal nonlinear filter for the MJLS model in the case in which both x and θ are not accessible (even in the case in which the state space of the Markov chain is finite countable). This, in turn, has engendered a great amount of literature dealing with this scenario, which is the setting treated in this book.

In the MJLS framework, the filtering problem has three possible settings: (i) the Markov chain is observed, but not the state x; (ii) the state is observed, but not the chain; and finally (iii) none of them is observed. In the first case, the optimal filter is a Kalman-like filter, and the result can be found, for instance, in [223]. In this scenario, H_∞ filtering and *robust* Kalman filtering for uncertain MJLS (including the time-lag case) have been studied, for instance, in [105–107, 205, 217, 257]. For iterative and sampling algorithms, see, e.g., [115] and [116]. An optimal finite nonlinear filter for the second case was derived in [301]. In a general setting, as the one considered in this book, the optimal nonlinear filter for (iii) is infinite-dimensional, which makes the optimal approach impractical from an application point of view (see, e.g., [30, 31, 123, 132, 214, 308]). Due to this, some particular situations and suboptimal approaches have been studied. For instance, with the simplification that the observation process is not fed by the state, in [30] a finite filter is derived. This problem has been also studied in [320] for a *small noise* observation scenario (asymptotic optimality is proved). In [155], an optimal linear mean-square filter was obtained, and the associated stationary filter was derived in [160] (see also [65] and [84] for the discrete-time case). A robust version of this filter was analyzed in [82] (see also [83]). For a sample of suboptimal filters, we mention, for instance, [7, 22, 37, 58, 123, 125, 131, 214, 287, 288, 308, 321, 322].

For an introduction to the linear filtering theory, the readers are referred, for instance, to [97] and [100] (see also [194] for an authoritative account of the nonlinear filtering theory and [98] for a nice introduction). Regarding MJLS, we recommend, for instance, [22] and [271].

The main sources for this chapter were, essentially, [82, 154, 155, 160].

Chapter 8
H_∞ Control

8.1 Outline of the Chapter

This chapter is devoted to the H_∞ control of continuous-time MJLS in the infinite-horizon setting. The statement of a bounded real lemma is the starting point toward a complete LMIs characterization of static state feedback, as well as of full-order dynamic output feedback stabilizing controllers that guarantee that a prescribed closed-loop H_∞ performance is attained. This chapter is organized as follows. Section 8.2 presents a description of the problems. In order to study the H_∞ control problem, we start by deriving a bounded real lemma for MJLS in Sect. 8.3. This allows us to characterize, in terms of the feasibility of LMIs, whether a system is mean-square stable with prescribed degree of performance. In the sequence we proceed to study in Sects. 8.5 and 8.6 the disturbance attenuation problem. This, roughly speaking, consists of determining whether a given system can be stabilized and driven to a desired H_∞ performance level, by the action of control, and how such a controller can be designed. Two cases are considered, the static state feedback (Sect. 8.5) and dynamic output feedback (Sect. 8.6) control problems. The main results are presented through explicit formulas with the corresponding design algorithms for obtaining the controllers of interest.

8.2 Description of the Problem

Our concern in this chapter is with MJLS described as

$$\dot{x}(t) = A_{\theta(t)}x(t) + B_{\theta(t)}u(t) + J_{\theta(t)}w(t), \tag{8.1}$$

which maps input disturbances w into a to-be-controlled output z,

$$z(t) = C_{\theta(t)}x(t) + D_{\theta(t)}u(t) + L_{\theta(t)}w(t), \tag{8.2}$$

O.L.V. Costa et al., *Continuous-Time Markov Jump Linear Systems*,
Probability and Its Applications, DOI 10.1007/978-3-642-34100-7_8,

that should be made as close to zero as possible, in an appropriate sense, under the action of the control signal u. We wish to characterize and to design mode-dependent controllers that, on the basis of partial measurements of the state, of the form

$$y(t) = H_{\theta(t)}x(t) + G_{\theta(t)}w(t), \tag{8.3}$$

are capable of stabilizing the system with prescribed degree of H_∞ *performance.* Roughly speaking, such performance requirement will be intimately related to a quantity similar to (the precise definition is given in (8.15))

$$\sup_{w\in L_2^r(\Omega,\mathcal{F},P),\|w\|_2\neq 0} \left(\frac{\int_0^\infty E[\|z(t)\|^2]\,dt}{\int_0^\infty E[\|w(t)\|^2]\,dt} \right)^{1/2}, \tag{8.4}$$

which measures to which extent *finite-energy* input disturbances are amplified (or attenuated) by the system, in a worst-case scenario. In other words, the larger this norm, the greater the effect of finite-energy input disturbances on the system's controlled output. The two basic problems considered here are the one of *static state feedback*, in which the controller is of the form

$$u(t) = F_{\theta(t)}x(t), \tag{8.5}$$

and that of *dynamic output feedback*, in which the control signal is extracted from the output of a dynamic controller driven by the measurement, such as

$$\begin{cases} \dot{\widehat{x}}(t) = \widehat{A}_{\theta(t)}\widehat{x}(t) + \widehat{B}_{\theta(t)}y(t), \\ u(t) = \widehat{C}_{\theta(t)}\widehat{x}(t) + \widehat{D}_{\theta(t)}y(t). \end{cases} \tag{8.6}$$

The state feedback approach is interesting whenever $H_i \equiv I$ and $G_i \equiv 0$, so that the complete state (x,θ) is available to the controller. In the general case of incomplete information, on the other hand, the output feedback approach is the choice adopted here. The scenario in which θ is not available will not be addressed.

As it will be shown more precisely later, whenever either (8.5) or (8.6) is plugged in the above system, we will end up with a closed-loop system of the form

$$\mathcal{G}_w^{\mathrm{cl}} = \begin{cases} \dot{x}_{\mathrm{cl}}(t) = A^{\mathrm{cl}}_{\theta(t)}x_{\mathrm{cl}}(t) + J^{\mathrm{cl}}_{\theta(t)}w(t), \\ z(t) = C^{\mathrm{cl}}_{\theta(t)}x_{\mathrm{cl}}(t) + L^{\mathrm{cl}}_{\theta(t)}w(t), \end{cases} \tag{8.7}$$

whose main H_∞ analysis tool comes in the form of a *bounded real lemma*. This result, presented in Lemma 8.2, states that system (8.7) is mean-square stable, with performance $\|\mathcal{G}_w^{\mathrm{cl}}\|_\infty < \gamma$, whenever a specific set of linear matrix inequalities is satisfied. Besides the interest in its own right, such LMIs characterization of performance makes it possible for us to: (i) tell whether there exists a stabilizing controller such as (8.5) or (8.6) which guarantees a desired level of disturbance attenuation, and (ii) design such a controller. These two problems are respectively solved in Theorems 8.13 and 8.15, as well as in Algorithms 8.20 and 8.21–8.22.

8.3 The Bounded Real Lemma

8.3.1 Problem Formulation and Main Result

Throughout this section, we denote by $L_2^r(\Omega, \mathcal{F}, P, [0, T])$ the space of all random processes $w = \{w(t); t \in \mathbb{R}^+\}$ in $\mathbb{R}^r$ such that the norm

$$\|w\|_{2,T} := \left(\int_0^T E\left(\|w(t)\|^2\right) dt \right)^{1/2} \tag{8.8}$$

is finite. Letting $T \to \infty$, this obviously yields that $\|\cdot\|_{2,\infty} \equiv \|\cdot\|_2$, as defined in (2.3) in Chap. 2. Moreover, notice that $L_2^n(\Omega, \mathcal{F}, P, [0, T]) \supset L_2^n(\Omega, \mathcal{F}, P)$ for every $T > 0$.

The basic H_∞-analysis problem we address here is related to a system of the form

$$\mathcal{G}_w = \begin{cases} \dot{x}(t) = A_{\theta(t)} x(t) + J_{\theta(t)} w(t), \\ z(t) = C_{\theta(t)} x(t) + L_{\theta(t)} w(t), \\ \vartheta_0 = (x_0, \theta_0), \qquad P(\theta_0 = i) = \nu_i, \quad i \in \mathcal{S}, \end{cases} \tag{8.9}$$

with $\mathbf{A} = (A_1, \ldots, A_N) \in \mathbb{H}^n$, $\mathbf{J} = (J_1, \ldots, J_N) \in \mathbb{H}^{r,n}$, $\mathbf{C} = (C_1, \ldots, C_N) \in \mathbb{H}^{n,p}$, and $\mathbf{L} = (L_1, \ldots, L_N) \in \mathbb{H}^{r,p}$. Here, $x = \{x(t); t \in \mathbb{R}^+\}$ in $\mathbb{R}^n$ denotes the state, $w = \{w(t); t \in \mathbb{R}^+\} \in L_2^r(\Omega, \mathcal{F}, P)$ is a finite-energy stochastic disturbance, and $z = \{z(t); t \in \mathbb{R}^+\}$ in $\mathbb{R}^p$ is the system output.

The notion of stability considered in this chapter is defined as follows.

Definition 8.1 System (8.9) is said to be internally mean-square stable (internally MSS) if, for an arbitrary initial condition ϑ_0, we have

$$\lim_{t \to \infty} E\left(\|x_{\text{zi}}(t)\|^2\right) = 0, \tag{8.10}$$

in which $x_{\text{zi}} = \{x_{\text{zi}}(t); t \in \mathbb{R}^+\}$, termed the zero-input response of system (8.9), stands for the unique solution to the equation

$$\dot{x}_{\text{zi}}(t) = A_{\theta(t)} x_{\text{zi}}(t), \quad x_{\text{zi}}(0) = x_0. \tag{8.11}$$

Throughout this section, a very convenient decomposition for the state of system (8.9) will be such as

$$x(t) = x_{\text{zi}}(t) + x_{\text{zs}}(t), \tag{8.12}$$

in which $x_{\text{zs}} = \{x_{\text{zs}}(t); t \in \mathbb{R}^+\}$, termed the *zero-state response* of system (8.9), stands for the unique solution to the equation

$$\dot{x}_{\text{zs}}(t) = A_{\theta(t)} x_{\text{zs}}(t) + J_{\theta(t)} w(t), \quad x_{\text{zs}}(0) = 0. \tag{8.13}$$

The importance of (8.12) is due to the fact that it provides us with a clear distinction between the response to initial conditions and the response to the external input signal. In our approach, we consider that only the latter is relevant to H_∞ analysis. Bearing this in mind, we define the *perturbation operator* as

$$\mathbb{L}w(t) = C_{\theta(t)}x_{zs}(t) + L_{\theta(t)}w(t), \quad t \in \mathbb{R}^+, \tag{8.14}$$

which, whenever internal MSS holds, is well-defined (bearing in mind Theorem 3.33) as $\mathbb{L}: L_2^r(\Omega, \mathcal{F}, P) \to L_2^p(\Omega, \mathcal{F}, P)$, a bounded linear operator. Since, in our framework of H_∞ analysis, $\mathbb{L}$ entirely describes how input disturbances affect the output of system (8.9), the induced norm

$$\|\mathbb{L}\| = \sup\left\{\frac{\|\mathbb{L}w\|_2}{\|w\|_2}; w \in L_2^r(\Omega, \mathcal{F}, P), \|w\|_2 \neq 0\right\} \tag{8.15}$$

measures precisely the worst-case effect that finite-energy disturbances may cause at the output. Due to this, throughout this section we interchangeably write that

$$\|\mathcal{G}_w\|_\infty = \|\mathbb{L}\| \tag{8.16}$$

and refer to it as the H_∞-*norm* of system (8.9). As mentioned earlier, the larger this norm, the greater the worst-case effect of the unknown disturbance w on the system output, z. Thus, (8.15) may be seen as a measure of *performance* for system (8.9). An LMIs criterion for the assessment of the H_∞-norm is provided by the main result of this section, which goes as follows.

Lemma 8.2 (Bounded real lemma) *Given $\gamma > 0$, the following statements are equivalent*:

(i) *System* (8.9) *is internally MSS with $\|\mathcal{G}_w\|_\infty = \|\mathbb{L}\| < \gamma$.*
(ii) *There is $\mathbf{P} = (P_1, \ldots, P_N) > 0$ in $\mathbb{H}^{n+}$ such that $M^\gamma(\mathbf{P}) = (M_1^\gamma(\mathbf{P}), \ldots, M_N^\gamma(\mathbf{P})) < 0$ in $\mathbb{H}^m$, where $m := n + r$, and, for each $i \in \mathcal{S}$,*

$$M_i^\gamma(\mathbf{P}) = \begin{bmatrix} \mathcal{T}_i(\mathbf{P}) + C_i^* C_i & \mathcal{U}_i(\mathbf{P}) \\ \mathcal{U}_i(\mathbf{P})^* & \mathcal{V}_i^\gamma \end{bmatrix}, \tag{8.17}$$

with $\mathcal{T}_i(\mathbf{P}) = A_i^* P_i + P_i A_i + \sum_{j\in\mathcal{S}} \lambda_{ij} P_j$ as in (3.21) *and*

$$\mathcal{U}_i(\mathbf{P}) = P_i J_i + C_i^* L_i, \qquad \mathcal{V}_i^\gamma = L_i^* L_i - \gamma^2 I_r. \tag{8.18}$$

(iii) *There is $\mathbf{R} = (R_1, \ldots, R_N) > 0$ in $\mathbb{H}^{n+}$ such that, for each $i \in \mathcal{S}$,*

$$\begin{bmatrix} A_i^* R_i + R_i A_i + \sum_{j\in\mathcal{S}} \lambda_{ij} R_j & R_i J_i & C_i^* \\ J_i^* R_i & -\gamma I_r & L_i^* \\ C_i & L_i & -\gamma I_p \end{bmatrix} < 0. \tag{8.19}$$

The sufficiency part of the proof of Lemma 8.2 will rely on the following result, to be proved in a more generalized form in Proposition 8.6 of Sect. 8.3.2.

Proposition 8.3 *For arbitrary θ_0, all $w \in L_2^r(\Omega, \mathcal{F}, P)$, and every $\mathbf{P} = (P_1, \ldots, P_N) \in \mathbb{H}^{n*}$, we have*

$$\begin{aligned}\int_0^T & E\big(\|\mathbb{L}w(t)\|^2 - \gamma^2\|w(t)\|^2\big)\,dt \\ &= -E\big(\langle x_{zs}(T); P_{\theta(T)}x_{zs}(T)\rangle\big) \\ &\quad + \int_0^T E\left(\left\langle \begin{bmatrix} x_{zs}(t) \\ w(t) \end{bmatrix}; M^\gamma_{\theta(t)}(\mathbf{P}) \begin{bmatrix} x_{zs}(t) \\ w(t) \end{bmatrix}\right\rangle\right) dt \end{aligned} \tag{8.20}$$

for all $T > 0$.

Proof See Proposition 8.6 in Sect. 8.3.2 below. □

Conversely, the necessity in Lemma 8.2 is a consequence of the following lemma, which will be proved in Sect. 8.3.2.

Lemma 8.4 *Suppose that system* (8.9) *is internally MSS with $\|\mathbb{L}\| < \gamma$. Then there is $\mathbf{P} = (P_1, \ldots, P_N)$ in $\mathbb{H}^{n+}$ such that the coupled algebraic Riccati equations*

$$\mathcal{T}_i(\mathbf{P}) + C_i^* C_i - \mathcal{U}_i(\mathbf{P})\big(\mathcal{V}_i^\gamma\big)^{-1}\mathcal{U}_i(\mathbf{P})^* = 0, \quad i \in \mathcal{S}, \tag{8.21}$$

are satisfied.

Bearing in mind the two preceding results, the proof of the bounded real lemma goes as follows.

Proof of Lemma 8.2 The equivalence between (ii) and (iii) is easily obtained with the aid of Schur complements, Lemma 2.26, by letting $\mathbf{R} = \gamma^{-1}\mathbf{P}$. Thus, it suffices to prove that (i) and (ii) are equivalent.

Given $\gamma > 0$, suppose that there is $\mathbf{P} = (P_1, \ldots, P_N) > 0$ such that $M_i^\gamma(\mathbf{P}) < 0$ for all $i \in \mathcal{S}$, so that (ii) is true. Then, it is immediate that

$$\mathcal{T}_i(\mathbf{P}) \le \mathcal{T}_i(\mathbf{P}) + C_i^* C_i < 0, \quad i \in \mathcal{S},$$

which, by Theorem 3.21, guarantees that system (8.9) is internally MSS as in Definition 8.1. Next, we have that for such $\mathbf{P}$ and γ, there is $\delta > 0$ such that, according to Proposition 8.3,

$$\begin{aligned}\|\mathbb{L}w\|_{2,T}^2 - \gamma^2\|w\|_{2,T}^2 &= -E\big(\langle x_{zs}(T); P_{\theta(T)}x_{zs}(T)\rangle\big) \\ &\quad + \int_0^T E\left(\left\langle \begin{bmatrix} x_{zs}(t) \\ w(t) \end{bmatrix}; M^\gamma_{\theta(t)}(\mathbf{P}) \begin{bmatrix} x_{zs}(t) \\ w(t) \end{bmatrix}\right\rangle\right) dt \\ &\le 0 + \int_0^T E\left(\left\langle \begin{bmatrix} x_{zs}(t) \\ w(t) \end{bmatrix}; (-\delta I) \begin{bmatrix} x_{zs}(t) \\ w(t) \end{bmatrix}\right\rangle\right) dt \\ &\le \delta\|w\|_{2,T}^2. \end{aligned}$$

Thus, letting $T \to \infty$, we get that, for every $w \in L_2^r(\Omega, \mathcal{F}, P)$ such that $\|w\|_2 \neq 0$,

$$\|\mathbb{L}w\|_2^2 - \gamma^2\|w\|_2^2 \leq \delta^2\|w\|_2^2 < 0, \tag{8.22}$$

which, due to (8.15), yields $\|\mathbb{L}\| < \gamma$.

Conversely, let $\widetilde{C}_i^\varepsilon := \left[\begin{smallmatrix} C_i \\ \varepsilon I \end{smallmatrix}\right]$, $\widetilde{L}_i := \left[\begin{smallmatrix} L_i \\ 0 \end{smallmatrix}\right]$, and notice that $\|\mathbb{L}\| < \gamma$ implies $\|\mathbb{L}^\varepsilon\| < \gamma$ if $\varepsilon > 0$ is small enough, where

$$\mathbb{L}^\varepsilon w(t) = \widetilde{C}_{\theta(t)}^\varepsilon x_{\mathrm{zs}}(t) + \widetilde{L}_{\theta(t)} w(t) \tag{8.23}$$

by definition (whose norm is defined analogously to (8.15)). In fact, since $\|\mathbb{L}^\varepsilon w(t)\|^2 = \|\mathbb{L}w(t)\|^2 + \varepsilon^2\|w(t)\|^2$, it follows from (3.62) that there is $c > 0$ such that, for all $w \in L_2^r(\Omega, \mathcal{F}, P)$,

$$\left\|\mathbb{L}^\varepsilon w\right\|_2^2 \leq \|\mathbb{L}w\|_2^2 + c\varepsilon^2\|w\|_2^2,$$

so that $\|\mathbb{L}^\varepsilon\|^2 \leq \|\mathbb{L}\|^2 + c\varepsilon^2$. Hence, whenever (ii) is true, we have, due to Lemma 8.4, that

$$\mathcal{T}_i(\mathbf{P}) + \left(\widetilde{C}_i^\varepsilon\right)^* \widetilde{C}_i^\varepsilon - \widetilde{\mathcal{U}}_i^\varepsilon(\mathbf{P})\left(\widetilde{\mathcal{V}}_i^\gamma\right)^{-1} \widetilde{\mathcal{U}}_i^\varepsilon(\mathbf{P})^* = 0 \tag{8.24}$$

is satisfied for some $\mathbf{P} = (P_1, \ldots, P_N) > 0$ and all $i \in \mathcal{S}$, where

$$\begin{aligned} \left(\widetilde{C}_i^\varepsilon\right)^* \widetilde{C}_i^\varepsilon &= C_i^* C_i + \varepsilon^2 I, \\ \widetilde{\mathcal{U}}_i^\varepsilon(\mathbf{P}) &:= P_i J_i + \left(\widetilde{C}_i^\varepsilon\right)^* \widetilde{L}_i = P_i J_i + C_i^* L_i = \mathcal{U}_i(\mathbf{P}), \\ \widetilde{\mathcal{V}}_i^\gamma &:= \widetilde{L}_i^* \widetilde{L}_i - \gamma^2 I = L_i^* L_i - \gamma^2 I = \mathcal{V}_i^\gamma, \end{aligned}$$

so that, for all $i \in \mathcal{S}$,

$$\mathcal{T}_i(\mathbf{P}) + \left\{\varepsilon^2 I + C_i^* C_i - \mathcal{U}_i(\mathbf{P})\left(\mathcal{V}_i^\gamma\right)^{-1} \mathcal{U}_i(\mathbf{P})^*\right\} = 0, \tag{8.25}$$

which, from Theorem 3.21, yields $\mathbf{P} > 0$. Furthermore, this may be rewritten in the form

$$\mathcal{T}_i(\mathbf{P}) + C_i^* C_i - \mathcal{U}_i(\mathbf{P})\left(\mathcal{V}_i^\gamma\right)^{-1} \mathcal{U}_i(\mathbf{P})^* = -\varepsilon^2 I < 0, \tag{8.26}$$

and the result follows from Schur complements (Lemma 2.26). □

8.3.2 Proof of Proposition 8.3 and Lemma 8.4

The purpose of this subsection is to prove Proposition 8.3 and Lemma 8.4. Our approach relies on maximizing the cost functional

$$w \mapsto \mathcal{J}_T^\gamma(\chi, i, w) = \int_0^T E\left(\left\|z(t)\right\|^2 - \gamma^2\left\|w(t)\right\|^2 \middle| \theta_0 = i\right) dt, \tag{8.27}$$

in which $T > 0$ is an arbitrary time horizon, $z = \{z(t), 0 \leq t < T\}$ is the output of system (8.9) with respect to $x(0) = \chi \in \mathbb{R}^n$ and given $\theta_0 = i \in \mathcal{S}$, and the input $w \in L_2^r(\Omega, \mathcal{F}, P) \subset L_2^r(\Omega, \mathcal{F}, P, [0, T])$. For later use, let us also define the cost associated with an arbitrary distribution $\nu = \{\nu_i, i \in \mathcal{S}\}$ to θ_0,

$$w \mapsto \mathcal{J}_T^\gamma(\chi, w) = \int_0^T E\left(\|z(t)\|^2 - \gamma^2\|w(t)\|^2\right) dt, \tag{8.28}$$

which is linked with (8.27) by means of

$$\mathcal{J}_T^\gamma(\chi, w) = \sum_{i\in\mathcal{S}} \mathcal{J}_T^\gamma(\chi, i, w) P(\theta_0 = i) = \sum_{i\in\mathcal{S}} \mathcal{J}_T^\gamma(\chi, i, w)\nu_i. \tag{8.29}$$

The relevance of (8.28) is largely due to the fact that it generalizes the cost at the left-hand side of (8.20), which was already shown to play a central role in the proof of Lemma 8.2. Roughly speaking, there we used the fact that, whenever system (8.9) is internally MSS as in Definition 8.1, then

$$\lim_{T\to\infty} \mathcal{J}_T^\gamma(0, w) = \|\mathbb{L}w\|_2^2 - \gamma^2\|w\|_2^2 < 0 \quad \Longleftrightarrow \quad \frac{\|\mathbb{L}w\|_2}{\|w\|_2} < \gamma \tag{8.30}$$

for any given $\gamma > 0$ and $w \in L_2^r(\Omega, \mathcal{F}, P)$. Besides, noticing that

$$\frac{\|\mathbb{L}v\|_{2,T}}{\|v\|_{2,T}} \leq \gamma \leq \frac{\|\mathbb{L}w\|_{2,T}}{\|w\|_{2,T}} \quad \Longrightarrow \quad \mathcal{J}_T^\gamma(0, v) \leq \mathcal{J}_T^\gamma(0, w), \tag{8.31}$$

we get that a candidate for worst-case disturbance in the right-hand side of (8.15) should maximize the cost $w \mapsto \|\mathbb{L}w\|_2^2 - \gamma^2\|w\|_2^2$. In order to study the more general problem of maximizing the cost in (8.27), we will need the following auxiliary result regarding a certain coupled differential Riccati equation which will subsequently come up.

Proposition 8.5 *There is* $\mathbf{Y} = (Y_1, \ldots, Y_N)$ *mapping* $(0, T)$ *into* $\mathbb{H}^{n*}$ *such that, for every* $i \in \mathcal{S}$ *and all* $T \in \mathbb{R}^+$,

$$\dot{Y}_i + \mathcal{T}_i(\mathbf{Y}) + C_i^* C_i - \mathcal{U}_i(\mathbf{Y}) V_i^{-1} \mathcal{U}_i(\mathbf{Y})^* = 0, \quad Y_i(T) = 0, \tag{8.32}$$

for any given $\mathbf{V} = (V_1, \ldots, V_N) < 0$ *in* $\mathbb{H}^r$. *Moreover,* $\mathbf{Y}$ *is unique, continuous, and continuously differentiable.*

Proof This is a consequence of Theorem A.1 in Appendix A. □

Proceeding further, we have the following generalization of Proposition 8.3 for the case of time-varying $\mathbf{P} : [0, T] \to \mathbb{H}^{n*}$ with given $x(0) = \chi \in \mathbb{R}^n$.

Proposition 8.6 *For any initial condition* $x(0) = \chi \in \mathbb{R}^n$, $\theta(0) = i \in \mathcal{S}$, *all* $w \in L_2^r(\Omega, \mathcal{F}, P, [0,T])$, $T > 0$, *and every* $\mathbf{P} = (P_1, \dots, P_N) : [0,T] \to \mathbb{H}^{n*}$ *continuously differentiable, the cost functional defined in* (8.27) *can be written as*

$$\begin{aligned}
&\mathcal{J}_T^\gamma(\chi, i, w)\\
&\quad = \langle \chi; P_i(0)\chi\rangle - E\big(\langle x(T); P_{\theta(T)}(T)x(T)\rangle \big| \theta_0 = i\big)\\
&\qquad + \int_0^T E\left(\langle x(t); \dot{P}_{\theta(t)}(t)x(t)\rangle + \left\langle \begin{bmatrix} x(t)\\ w(t)\end{bmatrix}; M_{\theta(t)}^\gamma\big(\mathbf{P}(t)\big)\begin{bmatrix} x(t)\\ w(t)\end{bmatrix}\right\rangle \middle| \theta_0 = i\right) dt,
\end{aligned} \tag{8.33}$$

with $M_i^\gamma(\mathbf{P})$ *given by* (8.17) *for each* $i \in \mathcal{S}$. *In particular, this guarantees the validity of* (8.20).

Proof In order to prove this result, let us show that

$$\begin{aligned}
&F(T) - F(0)\\
&\quad = \int_0^T E\left(\left\langle \begin{bmatrix} x(t)\\ w(t)\end{bmatrix}; \begin{bmatrix} \dot{P}_{\theta(t)} + \mathcal{T}_{\theta(t)}(\mathbf{P}) & P_{\theta(t)}J_{\theta(t)}\\ J_{\theta(t)}^* P_{\theta(t)} & 0\end{bmatrix}\begin{bmatrix} x(t)\\ w(t)\end{bmatrix}\right\rangle \middle| \theta_0 = i\right) dt,
\end{aligned} \tag{8.34}$$

where

$$F(t) := E\big(\langle x(t); P_{\theta(t)}(t)x(t)\rangle \big| \theta_0 = i\big). \tag{8.35}$$

To see why (8.34) is true, notice that $F(t) = \sum_{j=1}^N E(\phi_j(t)1_{\{\theta(t)=j\}}|\theta_0 = i)$ with $\phi_j(t) := \langle x(t), P_j(t)x(t)\rangle$. Hence, by Itô's rule, we have

$$\begin{aligned}
dF(t) &= \sum_{j\in\mathcal{S}} E\big(d\big[\phi_j(t)1_{\{\theta(t)=j\}}\big]|\theta_0 = i\big)\\
&= \sum_{j\in\mathcal{S}} E\big(d\phi_j(t)1_{\{\theta(t)=j\}} + \phi_j(t)\,d1_{\{\theta(t)=j\}}|\theta_0 = i\big),
\end{aligned} \tag{8.36}$$

where $d\phi_j(t) = \langle x(t); \dot{P}_j(t)x(t)\rangle\,dt + \langle dx(t); P_j(t)x(t)\rangle + \langle x(t); P_j(t)\,dx(t)\rangle$, so that, letting $\mathbf{x}(t) := \big[\begin{smallmatrix} x(t)\\ w(t)\end{smallmatrix}\big]$, we have

$$\begin{aligned}
&\sum_{j\in\mathcal{S}} E\big(d\phi_j(t)1_{\{\theta(t)=j\}}|\theta_0 = i\big)\\
&\quad = E\left(\left\langle \mathbf{x}(t); \begin{bmatrix} \dot{P}_{\theta(t)} + A_{\theta(t)}^* P_{\theta(t)} + P_{\theta(t)}A_{\theta(t)} & P_{\theta(t)}J_{\theta(t)}\\ J_{\theta(t)}^* P_{\theta(t)} & 0\end{bmatrix}\mathbf{x}(t)\right\rangle \middle| \theta_0 = i\right) dt.
\end{aligned}$$

On the other hand, bearing in mind that $E(\cdot|\theta_0 = i)$ is by itself a mathematical expectation, we have from Lemma 3.6 that

$$\begin{aligned}
&\sum_{j\in\mathcal{S}} E(\phi_j(t)d1_{\{\theta(t)=j\}}|\theta_0 = i) \\
&\quad = \sum_{j\in\mathcal{S}} E\Big(\phi_j(t)\Big(\sum_{k\in\mathcal{S}}\lambda_{kj}1_{\{\theta(t)=k\}}\Big)\Big|\theta_0 = i\Big)\,dt \\
&\quad = \sum_{j\in\mathcal{S}} E(\phi_j(t)\lambda_{\theta(t)j}|\theta_0 = i)\,dt \\
&\quad = E\Big(\Big\langle x(t); \Big(\sum_{j\in\mathcal{S}}\lambda_{\theta(t)j}P_j\Big)x(t)\Big\rangle\Big|\theta_0 = i\Big)\,dt.
\end{aligned}$$

Proceeding further, notice that (8.34) is obtained after integrating both sides of (8.36). This, when combined with the easily verifiable fact that

$$\begin{aligned}
\mathcal{J}_T^\gamma(\chi, i, w) &= \int_0^T E(\langle z(t); z(t)\rangle - \gamma^2\langle x(t); x(t)\rangle|\theta_0 = i)\,dt \\
&= \int_0^T E\left(\left\langle \mathbf{x}(t); \begin{bmatrix} C_{\theta(t)}^* C_{\theta(t)} & C_{\theta(t)}^* L_{\theta(t)} \\ L_{\theta(t)}^* C_{\theta(t)} & L_{\theta(t)}^* L_{\theta(t)} - \gamma^2 I \end{bmatrix} \mathbf{x}(t)\right\rangle\middle|\theta_0 = i\right) dt
\end{aligned}$$

because of $z(t) = C_{\theta(t)}x(t) + L_{\theta(t)}w(t)$ and $\mathbf{x}(t) = [x(t)^* \; w(t)^*]^*$, immediately yields (8.33). Finally, bearing in mind (8.29), we obtain

$$\begin{aligned}
\mathcal{J}_T^\gamma(0, w) &= \sum_{i\in\mathcal{S}} \mathcal{J}_T^\gamma(0, i, w)P(\theta(0) = i) \\
&= -\sum_{i\in\mathcal{S}} E(\langle x_{\mathrm{zs}}(T); P_{\theta(T)}x_{\mathrm{zs}}(T)\rangle 1_{\{\theta(0)=i\}}) \\
&\quad + \int_0^T \sum_{i\in\mathcal{S}} E\left(\left\langle \begin{bmatrix} x_{\mathrm{zs}}(t) \\ w(t) \end{bmatrix}; M_{\theta(t)}^\gamma(\mathbf{P}) \begin{bmatrix} x_{\mathrm{zs}}(t) \\ w(t) \end{bmatrix}\right\rangle 1_{\{\theta(0)=i\}}\right) dt \qquad (8.37)
\end{aligned}$$

for any $\mathbf{P} \in \mathbb{H}^{n*}$ independent of t, which is just (8.20). □

Notice that, in the same vein as (8.37), the preceding proposition yields

$$\begin{aligned}
\mathcal{J}_T^\gamma(\chi, w) &= E(\langle \chi; P_{\theta_0}(0)\chi\rangle) - E(\langle x(T); P_{\theta(T)}(T)x(T)\rangle) \\
&\quad + \int_0^T E\left(\langle x(t); \dot{P}_{\theta(t)}(t)x(t)\rangle + \left\langle \begin{bmatrix} x(t) \\ w(t) \end{bmatrix}; M_{\theta(t)}^\gamma(\mathbf{P}(t)) \begin{bmatrix} x(t) \\ w(t) \end{bmatrix}\right\rangle\right) dt,
\end{aligned} \qquad (8.38)$$

a fact that will be employed in the proof of the next proposition, which states that the cost associated to any disturbance in $L_2^r(\Omega, \mathcal{F}, P)$ and fixed initial conditions can be uniformly bounded from above.

Proposition 8.7 *Suppose that system* (8.9) *is internally MSS with* $\|\mathbb{L}\| < \gamma$. *Then there is* $h \in (0, \infty)$ *such that, for all* $T > 0$, $\chi \in \mathbb{R}^n$, $i \in \mathcal{S}$, *and* $w \in L_2^r(\Omega, \mathcal{F}, P, [0, T])$,

$$\mathcal{J}_T^\gamma(\chi, i, w) \le h\|\chi\|^2. \tag{8.39}$$

Proof Bearing in mind Lemma A.2, notice that there is $\mathbf{X}^T(t) = (X_1^T(t), \dots, X_N^T(t)) \in \mathbb{H}^{n*}$ that satisfies, for all $t \in [0, T]$,

$$\dot{X}_i^T(t) + \mathcal{T}_i(\mathbf{X}^T(t)) + C_i^* C_i = 0, \quad X_i^T(T) = 0, \tag{8.40}$$

along with $a \ge 0$ such that

$$\|X_i^T(t)\| \le a \quad \text{for all } i \in \mathcal{S},\ t \in [0, T]. \tag{8.41}$$

Thus, letting $\mathbf{P}(t) \equiv \mathbf{X}^T(t)$ in (8.38), we obtain that

$$\begin{aligned}\mathcal{J}_T^\gamma(\chi, w) = E\big(\langle\chi; X_{\theta_0}^T(0)\chi\rangle\big) + \int_0^T E\big(\langle\mathcal{U}_{\theta(t)}(\mathbf{X}^T(t))^* x(t); w(t)\rangle \\ + \langle w(t); \mathcal{U}_{\theta(t)}(\mathbf{X}^T(t))^* x(t)\rangle + \langle w(t); \mathcal{V}_{\theta(t)}^\gamma w(t)\rangle\big)\, dt,\end{aligned} \tag{8.42}$$

or, bearing in mind (8.12),

$$\begin{aligned}\mathcal{J}_T^\gamma(\chi, w) = \mathcal{J}_T^\gamma(0, w) + E\big(\langle\chi; X_{\theta_0}^T(0)\chi\rangle\big) + \int_0^T E\big(\langle\mathcal{U}_{\theta(t)}(\mathbf{X}^T(t))^* x_{\mathrm{zi}}(t); w(t)\rangle \\ + \langle w(t); \mathcal{U}_{\theta(t)}(\mathbf{X}^T(t))^* x_{\mathrm{zi}}(t)\rangle\big)\, dt.\end{aligned} \tag{8.43}$$

Next, let us denote the zero extension of w to $t > T$ as

$$w_0(t) = \begin{cases} w(t), & t \le T, \\ 0, & t > T, \end{cases}$$

so that, for any $0 < \varepsilon < \sqrt{\gamma^2 - \|\mathbb{L}\|^2}$,

$$\begin{aligned}\mathcal{J}_T^\gamma(0, w) &= \int_0^T E\big(\|\mathbb{L}w_0(t)\|^2 - \gamma^2\|w_0(t)\|^2\big)\, dt \\ &\le \int_0^\infty E\big(\|\mathbb{L}w_0(t)\|^2 - \gamma^2\|w_0(t)\|^2\big)\, dt \\ &\le \big(\|\mathbb{L}\|^2 - \gamma^2\big)\int_0^\infty E\big(\|w_0(t)\|^2\big)\, dt \\ &\le -\varepsilon^2\int_0^\infty E\big(\|w_0(t)\|^2\big)\, dt = -\int_0^T E\big(\langle\varepsilon w(t); \varepsilon w(t)\rangle\big)\, dt,\end{aligned}$$

and thus, back to (8.43), we obtain:

$$\begin{aligned}
\mathcal{J}_T^{\gamma}(\chi, w) &\le E\big(\langle \chi; X_{\theta_0}^T(0)\chi\rangle\big) - \int_0^T E\big(\langle \varepsilon w(t); \varepsilon w(t)\rangle \\
&\quad - \big\langle \varepsilon^{-1}\mathcal{U}_{\theta(t)}\big(\mathbf{X}^T(t)\big)^* x_{\mathrm{zi}}(t); \varepsilon w(t)\big\rangle \\
&\quad - \big\langle \varepsilon w(t); \varepsilon^{-1}\mathcal{U}_{\theta(t)}\big(\mathbf{X}^T(t)\big)^* x_{\mathrm{zi}}(t)\big\rangle \\
&\quad + \big\|\varepsilon^{-1}\mathcal{U}_{\theta(t)}\big(\mathbf{X}^T(t)\big)^* x_{\mathrm{zi}}(t)\big\|^2 - \big\|\varepsilon^{-1}\mathcal{U}_{\theta(t)}\big(\mathbf{X}^T(t)\big)^* x_{\mathrm{zi}}(t)\big\|^2\big)\, dt \\
&= E\big(\langle \chi; X_{\theta_0}^T(0)\chi\rangle\big) + \int_0^T E\big(\varepsilon^{-2}\big\|\mathcal{U}_{\theta(t)}\big(\mathbf{X}^T(t)\big)^* x_{\mathrm{zi}}(t)\big\|^2 \\
&\quad - \big\|\varepsilon w(t) - \varepsilon^{-1}\mathcal{U}_{\theta(t)}\big(\mathbf{X}^T(t)\big)^* x_{\mathrm{zi}}(t)\big\|^2\big)\, dt \\
&\le E\big(\langle \chi; X_{\theta_0}^T(0)\chi\rangle\big) + \varepsilon^{-2}\int_0^T E\big(\big\|\mathcal{U}_{\theta(t)}\big(\mathbf{X}^T(t)\big)^* x_{\mathrm{zi}}(t)\big\|^2\big)\, dt.
\end{aligned}$$

Bearing in mind (8.41) and (3.38), we thus obtain for a sufficiently large $\bar{a}$ that

$$\begin{aligned}
\mathcal{J}_T^{\gamma}(\chi, w) &\le \Big(\sup_{0\le t\le T}\big\|\mathbf{X}^T(t)\big\|_{\max}\Big)\|\chi\|^2 \\
&\quad + \varepsilon^{-2}\Big(\sup_{0\le t\le T}\max_{i\in\mathcal{S}}\big\|X_i^T(t)J_i + C_i^* L_i\big\|\Big)^2 \int_0^T E\big(\big\|x_{\mathrm{zi}}(t)\big\|^2\big)\, dt \\
&\le \bar{a}\|\chi\|^2 + \varepsilon^{-2}\big(\bar{a}\|\mathbf{J}\|_{\max} + \|\mathbf{C}\|_{\max}\|\mathbf{L}\|_{\max}\big)^2 \int_0^{\infty} \bar{a}e^{-bt}\|\chi\|^2\, dt \\
&= h\|\chi\|^2 \qquad (8.44)
\end{aligned}$$

for $h := \bar{a} + \varepsilon^{-2}(\bar{a}\|\mathbf{J}\|_{\max} + \|\mathbf{C}\|_{\max}\|\mathbf{L}\|_{\max})^2\bar{a}/b$. Finally, bearing in mind (8.29) and the fact that the right-hand side of (8.44) is independent of the distribution of θ_0, we get (8.39). □

In the following lemma we prove that, whenever system (8.9) is internally MSS with $\|\mathbb{L}\| < \gamma$, then the second diagonal block of $M^{\gamma}(\mathbf{P})$ in (8.17) is negative definite.

Proposition 8.8 *Suppose that system* (8.9) *is internally MSS with* $\|\mathbb{L}\| < \gamma$. *Then, bearing in mind* (8.18), *we have that* $\mathcal{\mathbf{V}}^{\gamma} = (\mathcal{V}_1^{\gamma}, \dots, \mathcal{V}_N^{\gamma}) < 0$ *in* $\mathbb{H}^r$.

Proof Assume, by contradiction, that there is $\varphi \in \mathbb{R}^r$ such that $\langle \varphi; \mathcal{V}_i^{\gamma}\varphi\rangle \ge \alpha$ for some $i \in \mathcal{S}$ and $\alpha > 0$. Also, for fixed $T > \delta > 0$, consider $w^{\delta} = \{w^{\delta}(t); t \in \mathbb{R}^+\} \in L_2^r(\Omega, \mathcal{F}, P)$ given by

$$w^{\delta}(t) := \begin{cases} -\varphi 1_{\{\theta(t)=i\}}, & 0 \le t < \delta, \\ 0 & \text{otherwise}. \end{cases} \qquad (8.45)$$

Then, in the spirit of (8.42), if we introduce $\mathcal{U} := \mathcal{U}(\mathbf{X}^T(t))$ with $\mathbf{X}^T(t) = (X_1^T(t), \ldots, X_N^T(t))$ satisfying (8.40) for all $t \in [0, T]$, it follows that

$$\begin{aligned}
\mathcal{J}_T^\gamma(0, i, w^\delta) &= \int_0^T E\big(2\,\mathrm{Re}\langle x_{zs}(t); \mathcal{U}_{\theta(t)} w^\delta(t)\rangle + \langle w^\delta(t); \mathcal{V}_{\theta(t)}^\gamma w^\delta(t)\rangle \big| \theta_0 = i\big)\, dt \\
&= \int_0^\delta E\big(\big[-2\,\mathrm{Re}\langle x_{zs}(t); \mathcal{U}_{\theta(t)} \varphi\rangle + \langle \varphi; \mathcal{V}_{\theta(t)}^\gamma \varphi\rangle\big] 1_{\{\theta(t)=i\}} \big| \theta_0 = i\big)\, dt \\
&\ge \int_0^\delta E\big(\big[-2\,\mathrm{Re}\langle x_{zs}(t); \mathcal{U}_i \varphi\rangle + \alpha\big] 1_{\{\theta(t)=i\}} \big| \theta_0 = i\big)\, dt \\
&\ge -2 \int_0^\delta E\big(\big|\langle x_{zs}(t); \mathcal{U}_i \varphi\rangle\big| 1_{\{\theta(t)=i\}} \big| \theta_0 = i\big)\, dt \\
&\quad + \alpha \int_0^\delta P\big(\theta(t) = i | \theta_0 = i\big)\, dt \\
&\ge -2\|\varphi\| \Big\{ \sup_{0 \le t < \delta} \big\|\mathcal{U}\big(\mathbf{X}^T(t)\big)\big\|_{\max} \Big\} \int_0^\delta E\big(\|x_{zs}(t)\| 1_{\{\theta(t)=i\}} \big| \theta_0 = i\big)\, dt \\
&\quad + \alpha \int_0^\delta P\big(\theta(t) = i | \theta_0 = i\big)\, dt \\
&\ge \int_0^\delta \big(-\beta E\big(\|x_{zs}(t)\| \big| \theta_0 = i\big) + \alpha P\big(\theta(t) = i | \theta_0 = i\big)\big)\, dt
\end{aligned} \tag{8.46}$$

for some constant $\beta > 0$, similarly to (8.44) and due to the fact that $1_{\{\cdot\}} \in \{0, 1\}$. However, since $x_{zs}(t)$ is continuous to the right of $t = 0$ with probability one, so that $\lim_{t \downarrow 0} P(x_{zs}(t) = 0) = 1$, and because of $\lim_{t \downarrow 0} P(\theta(t) = i | \theta_0 = i) = 1$, it follows that the right-hand side of (8.46) is *positive* if we take $\delta > 0$ sufficiently small. But this contradicts (8.39) in Proposition 8.7, from which we conclude that $\mathcal{V}_i^\gamma \le 0$ for any $i \in \mathcal{S}$.

Now consider $0 < \varepsilon < \gamma$ and $\gamma_\varepsilon := (\gamma^2 - \varepsilon^2)^{1/2}$, so that $\|\mathbb{L}\| < \gamma_\varepsilon < \gamma$. Repeating the previous steps right from the start with γ_ε in place of γ, we conclude that $\mathcal{V}_i^{\gamma_\varepsilon} := L_i^* L_i - (\gamma^2 - \varepsilon^2) I_r \le 0$ for all $i \in \mathcal{S}$, so that

$$\mathcal{V}_i^\gamma = L_i^* L_i - \gamma^2 I_r \le -\varepsilon^2 I_r < 0 \tag{8.47}$$

follows immediately. □

With the auxiliary results stated so far, we are now able to present the maximum property of $w \mapsto \mathcal{J}_T^\gamma(\chi, i, w)$ in (8.28).

Lemma 8.9 *Suppose that system* (8.9) *is internally MSS with* $\|\mathbb{L}\| < \gamma$, *and denote by* $\mathbf{P}^T$ *the solution of* (8.32) *with* $\mathcal{V}^\gamma$ *in place of* $\mathbf{V}$. *Then, for* $K^\gamma(\mathbf{P}^T) =$

$(K_1^{\gamma}(\mathbf{P}^T), \ldots, K_N^{\gamma}(\mathbf{P}^T)) \in \mathbb{H}^{n,r}$ *with*

$$K_i^{\gamma}(\mathbf{P}^T) := -(\mathcal{V}_i^{\gamma})^{-1}\mathcal{U}_i(\mathbf{P}^T(t))^*, \quad i \in \mathcal{S}, \tag{8.48}$$

the following statements are true.

(i) *The feedback disturbance* $w_T^{\gamma}(t) := K_{\theta(t)}^{\gamma}(\mathbf{P}^T)x(t)$, $t \in [0, T]$, *is such that*

$$\mathcal{J}_T^{\gamma}(\chi, i, w_T^{\gamma}) \geq \mathcal{J}_T^{\gamma}(\chi, i, w) \quad \textit{for all } w \in L_2^r(\Omega, \mathcal{F}, P, [0, T]). \tag{8.49}$$

(ii) *The associated maximal cost is given by*

$$\mathcal{J}_T^{\gamma}(\chi, i, w_T^{\gamma}) = \langle \chi; P_i^T(0)\chi \rangle. \tag{8.50}$$

Proof By Proposition 8.6 it is not difficult to verify that, for any $\mathbf{P} : [0, T] \to \mathbb{H}^{n*}$,

$$\begin{aligned}\mathcal{J}_T^{\gamma}(\chi, i, w) &= \langle \chi; P_i(0)\chi \rangle - E\big(\langle x(T); P_{\theta(T)}(T)x(T)\rangle \big| \theta_0 = i\big) \\ &\quad + \int_0^T \big(\langle x(t); Z_{\theta(t)}^{\gamma}(\mathbf{P}(t))x(t)\rangle - \Psi_{\theta(t)}^{\gamma}(\mathbf{P}(t)) \big| \theta_0 = i\big)\, dt, \end{aligned} \tag{8.51}$$

where, for all $i \in \mathcal{S}$,

$$Z_i^{\gamma}(\mathbf{P}(t)) := \dot{P}_i(t) + \mathcal{T}_i(\mathbf{P}(t)) + C_i^* C_i - \mathcal{U}_i(\mathbf{P}(t))(\mathcal{V}_i^{\gamma})^{-1}\mathcal{U}_i(\mathbf{P}(t))^*$$

and

$$\Psi_i^{\gamma}(\mathbf{P}(t)) := \langle w(t) - K_i^{\gamma}(\mathbf{P}(t))x(t); (-\mathcal{V}_i^{\gamma})(w(t) - K_i^{\gamma}(\mathbf{P}(t))x(t))\rangle.$$

But since the cost associated to some set of arguments is, by its very definition, independent of any particular choice on $\mathbf{P}$, we may consider (8.51) with $\mathbf{P}^T$ (the solution of (8.32) with $\mathbf{R} = \mathcal{V}^{\gamma}$) in place of $\mathbf{P}$. Thus, since $Z_i^{\gamma}(\mathbf{P}^T) \equiv 0$ and $\mathbf{P}^T(T) = 0$, it follows that (bearing in mind Proposition 8.8)

$$\mathcal{J}_T^{\gamma}(\chi, i, w) = \langle \chi; P_i^T(0)\chi \rangle - \int_0^T E\big(\Psi_{\theta(t)}^{\gamma}(\mathbf{P}^T) \big| \theta_0 = i\big)\, dt \leq \langle \chi; P_i^T(0)\chi \rangle,$$

and equality holds whenever $w(t) = w_T^{\gamma}(t) = K_{\theta(t)}^{\gamma}(\mathbf{P}^T(t))x(t)$ for all $t \in (0, T)$, from which (i) and (ii) follow immediately. □

Finally, the following proposition guarantees the validity of Lemma 8.4.

Proposition 8.10 *Suppose that system* (8.9) *is internally MSS with* $\|\mathbb{L}\| < \gamma$, *and let* $\mathbf{P}^T : [0, T] \to \mathbb{H}^{n*}$ *denote the solution of* (8.32) *with* $\mathbf{V} = \mathcal{V}^{\gamma}$. *Then the following statements are true*:

(i) *There exists* $h \in (0, \infty)$, *independent of* T, *such that*

$$0 \leq P_i^T(t) \leq h I_n \quad \textit{for all } 0 \leq t < T,\ i \in \mathcal{S}. \tag{8.52}$$

(ii) *For every* $0 < \tilde{T} < T$ *and* $i \in \mathcal{S}$, *we have*

$$P_i^{\tilde{T}}(t) \le P_i^T(t) \quad \textit{for any } 0 \le t < \tilde{T}. \tag{8.53}$$

(iii) *There is* $\mathbf{P} \in \mathbb{H}^{n+}$ *such that, for every* $t, \tilde{t} \in \mathbb{R}^+$,

$$\mathbf{P} = \lim_{T\to\infty} \mathbf{P}^T(t) = \lim_{T\to\infty} \mathbf{P}^T(\tilde{t}). \tag{8.54}$$

Moreover, $\mathbf{P}$ *satisfies* (8.21).

Proof In this proof we benefit from the optimality result obtained in Lemma 8.9. First notice that, for any $\chi \in \mathbb{R}^n$, $i \in \mathcal{S}$, and $0 \le t \le T$, we have

$$\begin{aligned}\langle \chi; P_i^T(t)\chi\rangle &= \langle \chi; P_i^{T-t}(0)\chi\rangle = \mathcal{J}_{T-t}^{\gamma}(\chi, i, w_{T-t}^{\gamma}) \\ &\ge \mathcal{J}_{T-t}^{\gamma}(\chi, i, 0) = \int_0^{T-t} E(\|z(s)\|^2 | \theta_0 = i)\, ds \\ &\ge 0,\end{aligned}$$

in addition to

$$\langle \chi; P_i^{T-t}(0)\chi\rangle = \mathcal{J}_{T-t}^{\gamma}(\chi, i, w_{T-t}^{\gamma}) \le h\|\chi\|^2 = \langle \chi; (hI_n)\chi\rangle$$

due to (8.39) in Proposition 8.7, which immediately yields the validity of (i).

To prove (ii), consider $\tilde{w} = \{\tilde{w}(s); s \in \mathbb{R}^+\}$ in $L_2^r(\Omega, \mathcal{F}, P)$ given by

$$\tilde{w}(s) = \begin{cases} w_{\tilde{T}-t}^{\gamma}(s), & 0 \le s < \tilde{T} - t, \\ 0 & \text{otherwise.}\end{cases}$$

Then, for any $\chi \in \mathbb{R}^n$, $i \in \mathcal{S}$, and $0 \le t \le \tilde{T} < T$, we have that

$$\begin{aligned}\langle \chi; P_i^T(t)\chi\rangle &= \langle \chi; P_i^{T-t}(0)\chi\rangle = \mathcal{J}_{T-t}^{\gamma}(\chi, i, w_{T-t}^{\gamma}) \\ &\ge \mathcal{J}_{T-t}^{\gamma}(\chi, i, \tilde{w}) = \mathcal{J}_{\tilde{T}-t}^{\gamma}(\chi, i, w_{\tilde{T}-t}^{\gamma}) + \int_{\tilde{T}-t}^{T-t} E(\|z(s)\|^2 | \theta_0 = i)\, ds \\ &\ge \mathcal{J}_{\tilde{T}-t}^{\gamma}(\chi, i, w_{\tilde{T}-t}^{\gamma}) = \langle \chi; P_i^{\tilde{T}}(t)\chi\rangle,\end{aligned}$$

from which (ii) follows.

Finally, for any given $t, \tilde{t} \in \mathbb{R}^+$, we have

$$\lim_{T\to\infty} P_i^T(t) = \lim_{T\to\infty} P_i^{T-t}(0) = \lim_{T\to\infty} P_i^{T-\tilde{t}}(0) = \lim_{T\to\infty} P_i^T(0), \tag{8.55}$$

for each $i \in \mathcal{S}$. The existence of the limit $P_i := \lim_{T\to\infty} P_i^T(0)$ is then a consequence of (i), (ii), and Lemma 2.17 (p. 24). □

8.4 The H_∞ Control Problem

In this section, for given $\mathbf{A} = (A_1, \dots, A_N) \in \mathbb{H}^n$, $\mathbf{B} = (B_1, \dots, B_N) \in \mathbb{H}^{m,n}$, $\mathbf{J} = (J_1, \dots, J_N) \in \mathbb{H}^{r,n}$, $\mathbf{C} = (C_1, \dots, C_N) \in \mathbb{H}^{n,p}$, $\mathbf{D} = (D_1, \dots, D_N) \in \mathbb{H}^{m,p}$, $\mathbf{L} = (L_1, \dots, L_N) \in \mathbb{H}^{r,p}$, $\mathbf{H} = (H_1, \dots, H_N) \in \mathbb{H}^{n,q}$, and $\mathbf{G} = (G_1, \dots, G_N) \in \mathbb{H}^{r,q}$, we study the control system

$$\mathcal{G}_u = \begin{cases} \dot{x}(t) = A_{\theta(t)}x(t) + B_{\theta(t)}u(t) + J_{\theta(t)}w(t), \\ y(t) = H_{\theta(t)}x(t) + G_{\theta(t)}w(t), \\ z(t) = C_{\theta(t)}x(t) + D_{\theta(t)}u(t) + L_{\theta(t)}w(t), \end{cases} \tag{8.56}$$

where $x = \{x(t); t \in \mathbb{R}^+\}$ in $\mathbb{R}^n$ denotes the state, $w = \{w(t); t \in \mathbb{R}^+\} \in L_2^r(\Omega, \mathcal{F}, P)$ is a finite-energy stochastic disturbance affecting the system, $y = \{y(t); t \in \mathbb{R}^+\}$ in $\mathbb{R}^q$ is the measured output, and $z = \{z(t); t \in \mathbb{R}^+\}$ in $\mathbb{R}^p$ is the system's controlled output. The initial condition $\vartheta_0 = (x_0, \theta_0)$ satisfies $E[\|x_0\|^2] < \infty$ with arbitrary distribution $\nu = \{\nu_i, i \in \mathcal{S}\}$ for θ_0.

Following the same lines of the preliminary problem addressed in the preceding section, the objective of the control signal will be to stabilize the system, while at the same time guaranteeing that the worst-case energy gain of the channel $w \mapsto z$ is smaller than a prespecified level. Analogously to (8.15), such a performance measure will correspond to the quantity

$$\|\mathbb{L}^{\mathrm{cl}}\| = \sup\left\{\frac{\|\mathbb{L}^{\mathrm{cl}}w\|_2}{\|w\|_2}; w \in L_2^r(\Omega, \mathcal{F}, P), \|w\|_2 \neq 0\right\}, \tag{8.57}$$

in which $\mathbb{L}^{\mathrm{cl}} : w(t) \mapsto C^{\mathrm{cl}}_{\theta(t)}x_{\mathrm{cl}}(t) + L^{\mathrm{cl}}_{\theta(t)}w(t)$ stands for the input–output map of the closed-loop system (8.7), when $x_{\mathrm{cl}}(0)$ is set to zero. Our starting point here is the following consequence of Lemma 8.2.

Corollary 8.11 *Given $\gamma > 0$, the closed-loop system (with x_{cl} in $\mathbb{R}^{n_{\mathrm{cl}}}$)*

$$\mathcal{G}^{\mathrm{cl}}_w = \begin{cases} \dot{x}_{\mathrm{cl}}(t) = A^{\mathrm{cl}}_{\theta(t)}x_{\mathrm{cl}}(t) + J^{\mathrm{cl}}_{\theta(t)}w(t), \\ z(t) = C^{\mathrm{cl}}_{\theta(t)}x_{\mathrm{cl}}(t) + L^{\mathrm{cl}}_{\theta(t)}w(t) \end{cases}$$

is internally MSS as in Definition 8.1, *with $\|\mathcal{G}^{\mathrm{cl}}_w\|_\infty = \|\mathbb{L}^{\mathrm{cl}}\| < \gamma$, if and only if there is $\mathbf{P} = (P_1, \dots, P_N) > 0$ in $\mathbb{H}^{n_{\mathrm{cl}}}$ such that the LMIs*

$$\begin{bmatrix} (A^{\mathrm{cl}}_i)^* P_i + P_i A^{\mathrm{cl}}_i + \sum_{j=1}^N \lambda_{ij} P_j & P_i J^{\mathrm{cl}}_i & (C^{\mathrm{cl}}_i)^* \\ (J^{\mathrm{cl}}_i)^* P_i & -\gamma I & (L^{\mathrm{cl}}_i)^* \\ C^{\mathrm{cl}}_i & L^{\mathrm{cl}}_i & -\gamma I \end{bmatrix} < 0 \tag{8.58}$$

are satisfied for all $i \in \mathcal{S}$.

For later use, let us introduce the notation

$$\mathcal{R}_i(\mathbf{R}) := \left[\lambda_{i1}^{1/2} R_i \quad \dots \quad \lambda_{i(i-1)}^{1/2} R_i \quad \lambda_{i(i+1)}^{1/2} R_i \quad \dots \quad \lambda_{iN}^{1/2} R_i\right], \tag{8.59}$$

together with $\mathcal{D} : \mathbb{H}^n \to \mathbb{H}^{(N-1)n}$, which maps $\mathbf{R} = (R_1, \dots, R_N) \in \mathbb{H}^n$ into $\mathcal{D}(\mathbf{R}) = (\mathcal{D}_1(\mathbf{R}), \dots, \mathcal{D}_N(\mathbf{R}))$, with

$$\mathcal{D}_i(\mathbf{R}) := \operatorname{diag}(R_1, \dots, R_{i-1}, R_{i+1}, \dots R_N), \quad i \in \mathcal{S}. \tag{8.60}$$

8.5 Static State Feedback

In the *static state feedback* scenario, we assume that $H_i = I_n$ and $L_i = 0$ for every $i \in \mathcal{S}$, so that system (8.56) becomes

$$\mathcal{G}_u = \begin{cases} \dot{x}(t) = A_{\theta(t)} x(t) + B_{\theta(t)} u(t) + J_{\theta(t)} w(t), \\ z(t) = C_{\theta(t)} x(t) + D_{\theta(t)} u(t) + L_{\theta(t)} w(t), \end{cases} \tag{8.61}$$

with $y(t) \equiv x(t)$. The control law of interest in this case is such as

$$u(t) = F_{\theta(t)} x(t), \tag{8.62}$$

which, when plugged in (8.61), provides us with the closed-loop system of the form

$$\mathcal{G}_w^F = \begin{cases} \dot{x}(t) = (A_{\theta(t)} + B_{\theta(t)} F_{\theta(t)}) x(t) + J_{\theta(t)} w(t), \\ z(t) = (C_{\theta(t)} + D_{\theta(t)} F_{\theta(t)}) x(t) + L_{\theta(t)} w(t). \end{cases} \tag{8.63}$$

In this case, we have the following result (in what follows, we recall from Chap. 2 that for a square matrix U, $\operatorname{Her}(U) = U + U^*$).

Theorem 8.12 *Given $\gamma > 0$, there is a controller such as* (8.62) *which guarantees that system* (8.63) *is internally MSS with $\|\mathcal{G}_w^F\|_\infty < \gamma$, if and only if there are $\mathbf{R} = (R_1, \dots, R_N) > 0$ in $\mathbb{H}^n$ and $\mathbf{W} = (W_1, \dots, W_N) \in \mathbb{H}^{n,m}$, such that*

$$\begin{bmatrix} \operatorname{Her}(A_i R_i + G_i W_i) + \lambda_{ii} R_i & J_i & (C_i R_i + D_i W_i)^* & \mathcal{R}_i(\mathbf{R}) \\ J_i^* & -\gamma I_r & L_i^* & 0 \\ C_i R_i + D_i W_i & L_i & -\gamma I_p & 0 \\ \mathcal{R}_i(\mathbf{R})^* & 0 & 0 & -\mathcal{D}_i(\mathbf{R}) \end{bmatrix} < 0 \tag{8.64}$$

for all $i \in \mathcal{S}$. Moreover, one such controller is given by $u(t) \equiv W_{\theta(t)} R_{\theta(t)}^{-1} x(t)$.

Proof By Corollary 8.11, there is a controller of the form (8.62) if and only if there are $\mathbf{P} = (P_1, \dots, P_N) > 0$ in $\mathbb{H}^n$ and $\mathbf{F} = (F_1, \dots, F_N) \in \mathbb{H}^{n,m}$ such that, for all

$i \in \mathcal{S}$,

$$\begin{bmatrix} (A_i + B_i F_i)^* P_i + P_i(A_i + B_i F_i) + \sum_{j=1}^N \lambda_{ij} P_j & P_i J_i & (C_i + D_i F_i)^* \\ J_i^* P_i & -\gamma I_r & L_i^* \\ C_i + D_i F_i & L_i & -\gamma I_p \end{bmatrix} < 0,$$

which, whenever pre- and post-multiplied by $\begin{bmatrix} P_i^{-1} & 0 & 0 \\ 0 & I & 0 \\ 0 & 0 & I \end{bmatrix}$, yields

$$\begin{aligned} 0 &> \begin{bmatrix} \operatorname{Her}(A_i R_i + B_i W_i) + \sum_{j=1}^N \lambda_{ij} R_i R_j^{-1} R_i & J_i & (C_i R_i + D_i W_i)^* \\ J_i^* & -\gamma I & L_i^* \\ C_i R_i + D_i W_i & L_i & -\gamma I \end{bmatrix} \\ &= \begin{bmatrix} \operatorname{Her}(A_i R_i + B_i W_i) + \lambda_{ii} R_i & * & * \\ J_i^* & -\gamma I & * \\ C_i R_i + D_i W_i & L_i & -\gamma I \end{bmatrix} \\ &\quad + \begin{bmatrix} \mathcal{R}_i(\mathbf{R}) \\ 0 \\ 0 \end{bmatrix} \mathcal{D}_i(\mathbf{R})^{-1} \begin{bmatrix} \mathcal{R}_i(\mathbf{R})^* \\ 0 \\ 0 \end{bmatrix}^*, \end{aligned}$$

in which $R_i := P_i^{-1}$, $W_i := F_i P_i^{-1}$ for all such i. Thus, the result follows easily from Schur complements (Lemma 2.26). □

8.6 Dynamic Output Feedback

In this section we consider again the controlled MJLS (8.56) but, in this case, with dynamic compensators given as

$$\begin{cases} \dot{\widehat{x}}(t) = \widehat{A}_{\theta(t)} \widehat{x}(t) + \widehat{B}_{\theta(t)} y(t), \\ u(t) = \widehat{C}_{\theta(t)} \widehat{x}(t) + \widehat{D}_{\theta(t)} y(t), \end{cases} \tag{8.65}$$

with $\widehat{x}(0) = \widehat{x}_0$. We restrict our attention to *full-order* controllers, i.e., ones such that $\widehat{x} = \{\widehat{x}(t); t \in \mathbb{R}^+\}$ takes values in $\mathbb{R}^n$ and thus has the same number of entries as the state of the to-be-controlled system.

Letting $\mathbf{v}(t) := [x(t)^* \ \widehat{x}(t)^*]^*$, it follows that the closed-loop of (8.56) and (8.65) is of the form

$$\mathcal{G}_w^{\mathcal{K}} = \begin{cases} \dot{\mathbf{v}}(t) = \Gamma_{\theta(t)} \mathbf{v}(t) + \Psi_{\theta(t)} w(t), \\ z(t) = \Lambda_{\theta(t)} \mathbf{v}(t) + \Phi_{\theta(t)} w(t), \end{cases} \tag{8.66}$$

with $\mathbf{v}(0) = (x_0, \widehat{x}_0) \in \mathbb{R}^{2n}$, with the closed-loop data given by

$$\Gamma_i = \begin{bmatrix} A_i + B_i \widehat{D}_i H_i & B_i \widehat{C}_i \\ \widehat{B}_i H_i & \widehat{A}_i \end{bmatrix}, \qquad \Psi_i = \begin{bmatrix} J_i + B_i \widehat{D}_i G_i \\ \widehat{B}_i G_i \end{bmatrix},$$
$$\Lambda_i = \begin{bmatrix} C_i + D_i \widehat{D}_i H_i & D_i \widehat{C}_i \end{bmatrix}, \qquad \Phi_i = L_i + D_i \widehat{D}_i G_i. \tag{8.67}$$

Equivalently, if we define

$$\mathcal{K} = (\mathcal{K}_1, \dots, \mathcal{K}_N) \in \mathbb{H}^{(n+r),(n+m)}, \quad \mathcal{K}_i = \begin{bmatrix} \widehat{A}_i & \widehat{B}_i \\ \widehat{C}_i & \widehat{D}_i \end{bmatrix}, \ i \in \mathcal{S}, \tag{8.68}$$

then it also follows that

$$\Gamma_i = \mathsf{A}_i + \mathsf{B}_i \mathcal{K}_i \mathsf{H}_i, \qquad \Psi_i = \mathsf{J}_i + \mathsf{B}_i \mathcal{K}_i \mathsf{G}_i,$$
$$\Lambda_i = \mathsf{C}_i + \mathsf{D}_i \mathcal{K}_i \mathsf{H}_i, \qquad \Phi_i = \mathsf{L}_i + \mathsf{D}_i \mathcal{K}_i \mathsf{G}_i, \tag{8.69}$$

with

$$\begin{bmatrix} \mathsf{A}_i & \mathsf{J}_i & \mathsf{B}_i \\ \mathsf{C}_i & \mathsf{L}_i & \mathsf{D}_i \\ \mathsf{H}_i & \mathsf{G}_i & \star \end{bmatrix} = \left[\begin{array}{cc|c|cc} A_i & 0 & J_i & 0 & B_i \\ 0 & 0_n & 0 & I_n & 0 \\ \hline C_i & 0 & L_i & 0 & D_i \\ \hline 0 & I_n & 0 & & \\ H_i & 0 & G_i & & \star \end{array}\right]. \tag{8.70}$$

8.6.1 Main Results

In this subsection we present the main results for the dynamic output feedback H_∞ control problem. The first main result will show that the existence of internally stabilizing controllers such as (8.65), which guarantee a desired performance $\|\mathcal{G}_w^{\mathcal{K}}\|_\infty < \gamma$ for the closed-loop system (8.66), is *equivalent* to the feasibility of the coupled LMIs problem (in what follows, we recall from Chap. 2 that $\mathcal{N}$ and $\mathcal{R}$ represent respectively the kernel and the range of a matrix):

$$\begin{bmatrix} A_i^* X_i + X_i A_i + \sum_{j=1}^N \lambda_{ij} X_j & X_i J_i & C_i^* \\ J_i^* X_i & -\gamma I_r & L_i^* \\ C_i & L_i & -\gamma I_p \end{bmatrix} < 0$$
$$\text{on } \mathcal{N}\begin{bmatrix} H_i & G_i & 0_{q\times p} \end{bmatrix}, \tag{8.71a}$$

$$\begin{bmatrix} Y_i A_i^* + A_i Y_i + \lambda_{ii} Y_i & J_i & Y_i C_i^* & \mathcal{R}_i(\mathbf{Y}) \\ J_i^* & -\gamma I_r & L_i^* & 0 \\ C_i Y_i & L_i & -\gamma I_p & 0 \\ \mathcal{R}_i(\mathbf{Y})^* & 0 & 0 & -\mathcal{D}_i(\mathbf{Y}) \end{bmatrix} < 0$$
$$\text{on } \mathcal{N}\begin{bmatrix} B_i^* & 0_{m\times r} & D_i^* & 0_{m\times n(N-1)} \end{bmatrix}, \tag{8.71b}$$

$$\begin{bmatrix} X_i & I \\ I & Y_i \end{bmatrix} > 0 \tag{8.71c}$$

for $i \in \mathcal{S}$, or, alternatively, to the feasibility of the LMIs

$$\begin{bmatrix} \mathrm{Her}(X_i A_i + U_i H_i) + \sum_{j=1}^{N} \lambda_{ij} X_j & * & * \\ (X_i J_i + U_i G_i)^* & -\gamma I_r & * \\ C_i + D_i W_i H_i & L_i + D_i W_i G_i & -\gamma I_p \end{bmatrix} < 0, \tag{8.72a}$$

$$\begin{bmatrix} \mathrm{Her}(A_i Y_i + B_i V_i) + \lambda_{ii} Y_i & * & * & * \\ (J_i + B_i W_i G_i)^* & -\gamma I_r & * & * \\ C_i Y_i + H_i V_i & L_i + D_i W_i G_i & -\gamma I_p & * \\ \mathcal{R}_i(\mathbf{Y})^* & 0 & 0 & -\mathcal{D}_i(\mathbf{Y}) \end{bmatrix} < 0, \tag{8.72b}$$

$$\begin{bmatrix} X_i & I \\ I & Y_i \end{bmatrix} > 0. \tag{8.72c}$$

The precise result, to be proved in Sect. 8.6.2, is as follows.

Theorem 8.13 *Given* $\gamma > 0$, *the following statements are equivalent*:

(i) *There exists a dynamic compensator* $\mathcal{K}$ *such as* (8.65)–(8.68) *for which system* (8.66) *is internally MSS with* $\|\mathcal{G}_w^{\mathcal{K}}\|_\infty < \gamma$.
(ii) *There exist* $\mathbf{X} = (X_1, \ldots, X_N) \in \mathbb{H}^n$ *and* $\mathbf{Y} = (Y_1, \ldots, Y_N) \in \mathbb{H}^n$ *such that* (8.71a)–(8.71c) *is satisfied for all* $i \in \mathcal{S}$.
(iii) *There exist suitable* $\mathbf{X} = (X_1, \ldots, X_N) \in \mathbb{H}^n$, $\mathbf{Y} = (Y_1, \ldots, Y_N) \in \mathbb{H}^n$, $\mathbf{U} = (U_1, \ldots, U_N) \in \mathbb{H}^{q,n}$, $\mathbf{V} = (V_1, \ldots, V_N) \in \mathbb{H}^{n,m}$, *and* $\mathbf{W} = (W_1, \ldots, W_N) \in \mathbb{H}^{q,m}$ *such that* (8.72a)–(8.72c) *is satisfied for all* $i \in \mathcal{S}$.

Moreover, given any $\mathbf{X}$, $\mathbf{Y}$ *satisfying* (ii), *there always exist suitable* $\mathbf{U}$, $\mathbf{V}$, *and* $\mathbf{W}$ *such that* (iii) *is satisfied.*

In addition, notice that, due to Finsler's lemma (see [49, Chap. 2]), we may rewrite the projection constraints in (8.71a)–(8.71c) as follows.

Corollary 8.14 *Given* $\gamma > 0$, *there exists a dynamic compensator* $\mathcal{K}$ *such as* (8.65)–(8.68) *for which system* (8.66) *is internally MSS with* $\|\mathcal{G}_w^{\mathcal{K}}\|_\infty < \gamma$ *if and only if there are* $\mathbf{X} = (X_1, \ldots, X_N)$ *and* $\mathbf{Y} = (Y_1, \ldots, Y_N)$ *in* $\mathbb{H}^n$, *together with* $\boldsymbol{\alpha} = (\alpha_1, \ldots, \alpha_N)$ *and* $\boldsymbol{\beta} = (\beta_1, \ldots, \beta_N)$ *in* $\mathbb{H}^1$, *such that, for all* $i \in \mathcal{S}$, *the LMIs*

$$\begin{bmatrix} A_i^* X_i + X_i A_i + \sum_{j=1}^{N} \lambda_{ij} X_j + \alpha_i H_i^* H_i & * & * \\ J_i^* X_i + \alpha_i G_i^* G_i & \alpha_i G_i^* G_i - \gamma I_r & * \\ C_i & L_i & -\gamma I_p \end{bmatrix} < 0, \tag{8.73a}$$

$$\begin{bmatrix} Y_iA_i^* + A_iY_i + \lambda_{ii}Y_i + \beta_iB_iB_i^* & * & * & * \\ J_i^* & -\gamma I_r & * & * \\ C_iY_i + \beta_iD_iB_i^* & L_i & \beta_iD_iD_i^* - \gamma I_p & * \\ \mathcal{R}_i(\mathbf{Y})^* & 0 & 0 & -\mathcal{D}_i(\mathbf{Y}) \end{bmatrix} < 0, \tag{8.73b}$$

$$\begin{bmatrix} X_i & I \\ I & Y_i \end{bmatrix} > 0, \qquad \alpha_i < 0, \qquad \beta_i < 0, \tag{8.73c}$$

are satisfied.

The second main result is related to the design of full-order compensators and reads as follows (the proof will be presented in Sect. 8.6.3):

Theorem 8.15 *Given $\gamma > 0$, suppose that suitable* $\mathbf{X}$, $\mathbf{Y}$, $\mathbf{U}$, $\mathbf{V}$, *and* $\mathbf{W}$ *satisfying the conditions of Theorem* 8.13 *may be found. Then, bearing in mind* (8.68), *the following full-order dynamic controller guarantees that internal MSS of the closed-loop system* (8.66) *is achieved along with $\|\mathcal{G}_w^{\mathcal{K}}\|_\infty < \gamma$:*

$$\begin{aligned} \widehat{A}_i = -\big(Y_i^{-1} - X_i\big)^{-1}\Bigg\{ & X_i(A_iY_i + B_iV_i) + (U_i - X_iB_iW_i)H_iY_i + \bar{A}_i^* \\ & + \sum_{j=1}^{N} \lambda_{ij}Y_j^{-1}Y_i + \gamma^{-1}\bar{C}_i^*(C_iY_i + D_iV_i) + \big[X_iJ_i + U_iG_i + \gamma^{-1}\bar{C}_i^*\bar{L}_i\big] \times \\ & \times \big(\gamma I - \gamma^{-1}\bar{L}_i^*\bar{L}_i\big)^{-1}\big[\bar{J}_i + \gamma^{-1}(C_iY_i + D_iV_i)^*\bar{L}_i\big]^*\Bigg\} Y_i^{-1}, \end{aligned} \tag{8.74}$$

$$\widehat{B}_i = \big(Y_i^{-1} - X_i\big)^{-1}(U_i - X_iB_iW_i), \tag{8.75}$$

$$\widehat{C}_i = (V_i - W_iH_iY_i)Y_i^{-1}, \tag{8.76}$$

$$\widehat{D}_i = W_i, \tag{8.77}$$

where, for all $i \in \mathcal{S}$,

$$\begin{bmatrix} \bar{A}_i & \bar{J}_i \\ \bar{C}_i & \bar{L}_i \end{bmatrix} := \begin{bmatrix} A_i & J_i \\ C_i & L_i \end{bmatrix} + \begin{bmatrix} B_i \\ D_i \end{bmatrix} W_i \begin{bmatrix} H_i & G_i \end{bmatrix}. \tag{8.78}$$

8.6.2 Analysis of Dynamic Controllers

Our objective in this subsection is to prove Theorem 8.13. In addition to (8.68)–(8.70), we use the following notation:

$$\mathcal{M}_i = \begin{bmatrix} \mathsf{A}_i^* P_i + P_i \mathsf{A}_i + \sum_{j=1}^N \lambda_{ij} P_j & P_i \mathsf{J}_i & \mathsf{C}_i^* \\ \mathsf{J}_i^* P_i & -\gamma I_r & L_i^* \\ \mathsf{C}_i & L_i & -\gamma I_p \end{bmatrix}, \qquad \mathcal{P}_i = \begin{bmatrix} P_i & 0 & 0 \\ 0 & I_r & 0 \\ 0 & 0 & I_p \end{bmatrix},$$

$$\mathcal{B}_i = \begin{bmatrix} \mathsf{B}_i^* & 0_{(n+m)\times r} & \mathsf{D}_i^* \end{bmatrix}, \qquad \mathcal{H}_i = \begin{bmatrix} \mathsf{H}_i & \mathsf{G}_i & 0_{(n+q)\times p} \end{bmatrix},$$

from which it is not difficult to check that (8.58) becomes

$$\mathcal{M}_i + (\mathcal{B}_i \mathcal{P}_i)^* \mathcal{K}_i \mathcal{H}_i + \mathcal{H}_i^* \mathcal{K}_i^* (\mathcal{B}_i \mathcal{P}_i) < 0. \tag{8.79}$$

In order to proceed further, the following auxiliary result, commonly known as the *projection* or *elimination* lemma, will be necessary (see, for instance, [163] or [49, Chap. 2]).

Lemma 8.16 *Given* $\mathbf{M} = (M_1, \dots, M_N) \in \mathbb{H}^{(n_q)*}$, $\boldsymbol{\Omega} = (\Omega_1, \dots, \Omega_N) \in \mathbb{H}^{n_q, n_m}$, *and* $\boldsymbol{\Xi} = (\Xi_1, \dots, \Xi_N) \in \mathbb{H}^{n_q, n_r}$, *there is* $\mathcal{K} = (\mathcal{K}_1, \dots, \mathcal{K}_N)$ *in* $\mathbb{H}^{n_r, n_m}$ *satisfying*

$$M_i + \Omega_i^* \mathcal{K}_i \Xi_i + \Xi_i^* \mathcal{K}_i^* \Omega_i < 0, \quad i \in \mathcal{S}, \tag{8.80}$$

if and only if

$$M_i < 0 \quad \text{on } \mathcal{N}(\Xi_i) \cup \mathcal{N}(\Omega_i) \tag{8.81}$$

is also satisfied for all $i \in \mathcal{S}$.

The importance of Lemma 8.16 is due to the fact that it allows us to remove the controller parameters from (8.79). As a consequence, we have the following result.

Proposition 8.17 *Given* $\gamma > 0$, *there exists a dynamic compensator* $\mathcal{K}$ *such as* (8.65)–(8.68) *for which system* (8.66) *is internally MSS with* $\|\mathcal{G}_w^{\mathcal{K}}\|_\infty < \gamma$ *if and only if there is* $\mathbf{P} = (P_1, \dots, P_N) > 0$ *in* $\mathbb{H}^n$ *such that*

$$\mathcal{M}_i < 0 \quad \text{on } \mathcal{N}(\mathcal{H}_i) \tag{8.82}$$

and

$$\mathcal{P}_i^{-1} \mathcal{M}_i \mathcal{P}_i^{-1} < 0 \quad \text{on } \mathcal{N}(\mathcal{B}_i) \tag{8.83}$$

are satisfied for all $i \in \mathcal{S}$.

Proof (*Necessity*) Assume the existence of such $\mathcal{K}$. Then, by Corollary 8.11, there must exist $\mathbf{P}$ such that (8.79) is satisfied. By Lemma 8.16, (8.82) is just a consequence of this fact, which also implies that

$$y_i^* \mathcal{M}_i y_i < 0 \quad \text{whenever } \mathcal{B}_i \mathcal{P}_i y_i = 0 \text{ for } y_i \in \mathbb{R}^{2n+r+p}. \tag{8.84}$$

Now notice that the linear map $\mathcal{N}(\mathcal{B}_i\mathcal{P}_i) \ni y_i \mapsto \mathcal{P}_i y_i = x_i \in \mathcal{N}(\mathcal{B}_i)$ is a bijection between $\mathcal{N}(\mathcal{B}_i\mathcal{P}_i)$ and $\mathcal{N}(\mathcal{B}_i)$. Thus, (8.84) holds if and only if, for all $i \in \mathcal{S}$,

$$\left(\mathcal{P}_i^{-1}x_i\right)^*\mathcal{M}_i\left(\mathcal{P}_i^{-1}x_i\right) < 0 \quad \text{whenever } \mathcal{B}_i x_i = 0, \tag{8.85}$$

which is just (8.83).

(*Sufficiency*) Since (8.84) is equivalent to (8.83) holding for all $i \in \mathcal{S}$, it follows that the existence of such $\mathbf{P} > 0$ in $\mathbb{H}^{2n}$ guarantees (again from Lemma 8.16) that there is $\mathcal{K}$ such that (8.79) is satisfied, from which the result follows. □

At this point, we are able to prove that the feasibility of (8.71a)–(8.71c) for all $i = 1, \dots, N$ guarantees the existence of a suitable H_∞ controller such as (8.65)–(8.68). In order to do so, let us consider the following parameterization of $\mathbf{P}$:

$$P_i = \begin{bmatrix} X_i & Y_i^{-1} - X_i \\ Y_i^{-1} - X_i & X_i - Y_i^{-1} \end{bmatrix}, \quad i \in \mathcal{S}. \tag{8.86}$$

Lemma 8.18 *Given $\gamma > 0$, there exists a dynamic compensator $\mathcal{K}$ such as* (8.65)–(8.68) *for which system* (8.66) *is internally MSS with $\|\mathcal{G}_w^{\mathcal{K}}\|_\infty < \gamma$ whenever there are $\mathbf{X} = (X_1, \dots, X_N)$ and $\mathbf{Y} = (Y_1, \dots, Y_N)$ in $\mathbb{H}^n$ such that* (8.71a)–(8.71c) *is satisfied for every $i \in \mathcal{S}$.*

Proof The idea behind the proof is that, if $\mathbf{P}$ is chosen as in (8.86), then (8.82) reduces to (8.71a), and (8.83) reduces to (8.71b). Additionally, from Schur complements (8.71c) implies $X - Y^{-1} > 0$, so it can be easily shown that $\mathbf{P} = (P_1, \dots, P_N)$ in (8.86) is positive definite.

In fact, notice that $\mathcal{N}(\mathcal{H}_i)$ may be written

$$\mathcal{N}\begin{bmatrix} \mathsf{H}_i & \mathsf{G}_i & 0_{(n+q)\times p} \end{bmatrix} = \mathcal{N}\begin{bmatrix} 0 & I_n & 0 & 0 \\ H_i & 0 & G_i & 0_{q\times p} \end{bmatrix} = \mathcal{R}\begin{bmatrix} \Delta_{1i} & 0 \\ 0 & 0_n \\ \Delta_{2i} & 0 \\ 0 & I_p \end{bmatrix}$$

whenever $\mathcal{N}[H_i \ G_i] = \mathcal{R}\left[\begin{smallmatrix} \Delta_{1i} \\ \Delta_{2i} \end{smallmatrix}\right]$. Thus, bearing in mind that $P_i \mathsf{A}_i$ is of the form $\left[\begin{smallmatrix} X_i A_i & \star \\ \star & \star \end{smallmatrix}\right]$, we have that (8.82) is equivalent to

$$0 > \begin{bmatrix} \Delta_{1i} & 0 \\ 0 & 0 \\ \Delta_{2i} & 0 \\ 0 & I \end{bmatrix}^* \begin{bmatrix} A_i^* X_i + X_i A_i + \sum_{j=1}^N \lambda_{ij} X_j & \star & X_i J_i & C_i^* \\ \star & \star & \star & \star \\ J_i^* X_i & \star & -\gamma I_r & L_i^* \\ C_i & \star & L_i & -\gamma I_p \end{bmatrix} \begin{bmatrix} \Delta_{1i} & 0 \\ 0 & 0 \\ \Delta_{2i} & 0 \\ 0 & I \end{bmatrix}$$

$$= \begin{bmatrix} \Delta_{1i} & 0 \\ \Delta_{2i} & 0 \\ 0 & I \end{bmatrix}^* \begin{bmatrix} A_i^* X_i + X_i A_i + \sum_{j=1}^N \lambda_{ij} X_j & X_i J_i & C_i^* \\ J_i^* X_i & -\gamma I_r & L_i^* \\ C_i & L_i & -\gamma I_p \end{bmatrix} \begin{bmatrix} \Delta_{1i} & 0 \\ \Delta_{2i} & 0 \\ 0 & I \end{bmatrix},$$

from which (8.71a) follows. As for (8.83), we have that, letting $S_i := P_i^{-1}$ and bearing in mind the notation at the beginning of this subsection,

$$\mathcal{P}_i^{-1}\mathcal{M}_i\mathcal{P}_i^{-1} = \begin{bmatrix} S_i\mathsf{A}_i^* + \mathsf{A}_i S_i + \sum_{j=1}^N \lambda_{ij} S_i S_j^{-1} S_i & \mathsf{J}_i & S_i\mathsf{C}_i^* \\ \mathsf{J}_i^* & -\gamma I & L_i^* \\ \mathsf{C}_i S_i & L_i & -\gamma I \end{bmatrix}.$$

Also, notice that $\mathcal{N}(\mathcal{B}_i)$ may be written

$$\mathcal{N}(\mathcal{B}_i) = \mathcal{N}\begin{bmatrix}\mathsf{B}_i^* & 0_{(n+m)\times r} & \mathsf{D}_i^*\end{bmatrix} = \mathcal{R}\begin{bmatrix}\hat{\Theta}_{1i} & 0 \\ 0 & I_r \\ \Theta_{2i} & 0\end{bmatrix}$$

with

$$\mathcal{R}\begin{bmatrix}\hat{\Theta}_{1i} \\ \Theta_{2i}\end{bmatrix} = \mathcal{N}\begin{bmatrix}\mathsf{B}_i^* & \mathsf{D}_i^*\end{bmatrix} = \mathcal{N}\begin{bmatrix}0 & I_n & 0 \\ B_i^* & 0 & D_i^*\end{bmatrix} = \mathcal{R}\begin{bmatrix}\Theta_{1i} & 0 \\ 0 & 0_n \\ \Theta_{2i} & 0\end{bmatrix},$$

in which $\mathcal{R}\left[\begin{smallmatrix}\Theta_{1i} \\ \Theta_{2i}\end{smallmatrix}\right] = \mathcal{N}[B_i^* \ D_i^*]$. Finally, due to the fact that (8.86) implies

$$S_i = \begin{bmatrix} Y_i & Y_i \\ Y_i & Y_i - (Y_i^{-1} - X_i)^{-1}\end{bmatrix},$$

it follows that (8.83) reduces to

$$0 > \begin{bmatrix}\Theta_{1i} & 0 \\ 0 & 0 \\ 0 & I_r \\ \Theta_{2i} & 0\end{bmatrix}^* \begin{bmatrix} A_iY_i + Y_iA_i^* + \sum_{j=1}^N \lambda_{ij} Y_i Y_j^{-1} Y_i & \star & J_i & Y_iC_i^* \\ \star & \star & \star & \star \\ J_i^* & \star & -\gamma I_r & L_i^* \\ C_iY_i & \star & L_i & -\gamma I_p \end{bmatrix}$$

$$\times \begin{bmatrix}\Theta_{1i} & 0 \\ 0 & 0 \\ 0 & I_r \\ \Theta_{2i} & 0\end{bmatrix}$$

$$= \begin{bmatrix}\Theta_{1i} & 0 \\ 0 & I_r \\ \Theta_{2i} & 0\end{bmatrix}^* \begin{bmatrix} A_iY_i + Y_iA_i^* + \sum_{j=1}^N \lambda_{ij} Y_i Y_j^{-1} Y_i & J_i & Y_iC_i^* \\ J_i^* & -\gamma I_r & L_i^* \\ C_iY_i & L_i & -\gamma I_p \end{bmatrix}$$

$$\times \begin{bmatrix}\Theta_{1i} & 0 \\ 0 & I_r \\ \Theta_{2i} & 0\end{bmatrix},$$

which, from Schur complements, is simply (8.71b). □

The following result states that, in order to exist a controller such as the one from the preceding lemma, it is necessary that (8.72a)–(8.72c) be feasible for each $i \in \mathcal{S}$.

Lemma 8.19 *Given $\gamma > 0$, suppose that there is a dynamic compensator $\mathcal{K}$ such as* (8.65)–(8.68) *for which system* (8.66) *is internally MSS with $\|\mathcal{G}_w^{\mathcal{K}}\|_\infty < \gamma$. Then there exist $\mathbf{X} = (X_1, \ldots, X_N)$ and $\mathbf{Y} = (Y_1, \ldots, Y_N)$ in $\mathbb{H}^n$, together with $\mathbf{U} = (U_1, \ldots, U_N) \in \mathbb{H}^{q,n}$, $\mathbf{V} = (V_1, \ldots, V_N) \in \mathbb{H}^{n,m}$, and $\mathbf{W} = (W_1, \ldots, W_N) \in \mathbb{H}^{q,m}$, such that* (8.72a)–(8.72c) *is satisfied for all $i \in \mathcal{S}$.*

Proof By the hypothesis there must exist $\mathbf{P} = (P_1, \ldots, P_N) > 0$ in $\mathbb{H}^{2n}$ satisfying (8.58) for every $i \in \mathcal{S}$. For later use, let us write such $\mathbf{P}$ and its inverse, $\mathbf{S} = (S_1, \ldots, S_N) > 0$, in the following compatible form:

$$P_i = \begin{bmatrix} X_i & \widehat{X}_i \\ \widehat{X}_i^* & \widetilde{X}_i \end{bmatrix}, \qquad S_i = \begin{bmatrix} Y_i & \widehat{Y}_i \\ \widehat{Y}_i^* & \widetilde{Y}_i \end{bmatrix}. \tag{8.87}$$

Introducing now $\mathcal{I} = [0_n \ I_n]$, let us make explicit the affine dependence of Γ_i on $\widehat{A}_i$:

$$\Gamma_i = \begin{bmatrix} A_i + B_i \widehat{D}_i H_i & B_i \widehat{C}_i \\ \widehat{B}_i H_i & 0 \end{bmatrix} + \begin{bmatrix} 0_n & 0 \\ 0 & \widehat{A}_i \end{bmatrix} =: \widetilde{A}_i + \mathcal{I}^* \widehat{A}_i \mathcal{I}. \tag{8.88}$$

Next, define $\mathfrak{I} = [I_n \ 0_n]^*$ and notice that $\mathcal{N}(\mathrm{diag}(\mathfrak{I}, I_r, I_p)) = \{0\}$, so uniform definiteness is preserved under application of the congruence transformation $\mathrm{diag}(\mathfrak{I}^*, I_r, I_p)(\cdot)\mathrm{diag}(\mathfrak{I}, I_r, I_p)$ on (8.58). Moreover, since $\mathcal{I}\mathfrak{I} = 0$, the hypothesis yields

$$\begin{bmatrix} \mathfrak{I}^*(\widetilde{A}_i^* P_i + P_i \widetilde{A}_i + \sum_{j=1}^N \lambda_{ij} P_j)\mathfrak{I} & \mathfrak{I}^* P_i \Psi_i & \mathfrak{I}^* \Lambda_i^* \\ \Psi_i^* P_i \mathfrak{I} & -\gamma I_r & \Phi_i^* \\ \Lambda_i \mathfrak{I} & \Phi_i & -\gamma I_p \end{bmatrix} < 0. \tag{8.89}$$

Hence, by defining $W_i := \widehat{D}_i$ and $U_i := X_i B_i W_i + \widehat{X}_i \widehat{B}_i$ and performing the indicated calculations, the equivalence between (8.89) and (8.72a) follows immediately.

Proceeding further, we have from (8.58) that

$$\begin{aligned} 0 &> \begin{bmatrix} S_i & 0 & 0 \\ 0 & I & 0 \\ 0 & 0 & I \end{bmatrix} \begin{bmatrix} \Gamma_i^* P_i + P_i \Gamma_i + \sum_{j=1}^N \lambda_{ij} P_j & P_i \Psi_i & \Lambda_i^* \\ \Psi_i^* P_i & -\gamma I_r & \Phi_i^* \\ \Lambda_i & \Phi_i & -\gamma I_p \end{bmatrix} \\ &\quad \times \begin{bmatrix} S_i & 0 & 0 \\ 0 & I & 0 \\ 0 & 0 & I \end{bmatrix} \\ &= \begin{bmatrix} \Gamma_i S_i + S_i \Gamma_i^* + \sum_{j=1}^N \lambda_{ij} S_i S_j^{-1} S_i & \Psi_i & S_i \Lambda_i^* \\ \Psi_i^* & -\gamma I_r & \Phi_i^* \\ \Lambda_i S_i & \Phi_i & -\gamma I_p \end{bmatrix}, \end{aligned}$$

so that from Schur complements (Lemma 2.26) we have that

$$\begin{bmatrix} \Gamma_i S_i + S_i \Gamma_i^* + \lambda_{ii} Y_i & \Psi_i & S_i \Lambda_i^* & \mathcal{R}_i(\mathbf{S})^* \\ \Psi_i^* & -\gamma I_r & \Phi_i^* & 0 \\ \Lambda_i S_i & \Phi_i & -\gamma I_p & 0 \\ \mathcal{R}_i(\mathbf{S}) & 0 & 0 & -\mathcal{D}_i(\mathbf{S}) \end{bmatrix} < 0$$

is always satisfied. Thus, since $\mathcal{N}(\mathfrak{I}) = \{0\}$, it follows that

$$\begin{bmatrix} \mathfrak{I}^* & 0 & 0 & 0 \\ 0 & I_r & 0 & 0 \\ 0 & 0 & I_p & 0 \\ 0 & 0 & 0 & \widetilde{\mathfrak{I}}^* \end{bmatrix} \begin{bmatrix} \Gamma_i S_i + S_i \Gamma_i^* + \lambda_{ii} Y_i & \Psi_i & S_i \Lambda_i^* & \mathcal{R}_i(\mathbf{S})^* \\ \Psi_i^* & -\gamma I_r & \Phi_i^* & 0 \\ \Lambda_i S_i & \Phi_i & -\gamma I_p & 0 \\ \mathcal{R}_i(\mathbf{S}) & 0 & 0 & -\mathcal{D}_i(\mathbf{S}) \end{bmatrix}$$

$$\times \begin{bmatrix} \mathfrak{I} & 0 & 0 & 0 \\ 0 & I_r & 0 & 0 \\ 0 & 0 & I_p & 0 \\ 0 & 0 & 0 & \widetilde{\mathfrak{I}} \end{bmatrix} < 0,$$

in which $\widetilde{\mathfrak{I}} = \operatorname{diag}(\mathfrak{I}, \dots, \mathfrak{I})$. Finally, letting $V_i := W_i H_i Y_i + \widehat{C}_i \widehat{Y}_i^*$ and bearing in mind that $\mathfrak{I}^* \left[\begin{smallmatrix} (\cdot) & \star \\ \star & \star \end{smallmatrix}\right] \mathfrak{I} \equiv (\cdot)$, (8.72b) follows immediately.

To complete the proof, notice that $\widehat{X}_i$, $i \in \mathcal{S}$, may be assumed invertible without loss of generality. In fact, given any admissible $\mathbf{P} > 0$ in $\mathbb{H}^{2n+}$, just perturb it to $\mathbf{Q} > 0$ in $\mathbb{H}^{2n+}$, with $Q_i = \left[\begin{smallmatrix} Z_i & \widehat{Z}_i \\ \widehat{Z}_i^* & \widetilde{Z}_i \end{smallmatrix}\right]$ for all $i \in \mathcal{S}$, in such a way that $\widehat{\mathbf{Z}} = (\widehat{Z}_1, \dots, \widehat{Z}_N)$ is invertible and (8.58) continues to hold with $\mathbf{Q}$ in place of $\mathbf{P}$. In light of this,[1] we have that

$$0 < \begin{bmatrix} I & 0 \\ Y_i & -Y_i \widehat{X}_i \widetilde{X}_i^{-1} \end{bmatrix} P_i \begin{bmatrix} I & Y_i \\ 0 & -\widetilde{X}_i^{-1} \widehat{X}_i^* Y_i \end{bmatrix} = \begin{bmatrix} X_i & I \\ I & Y_i \end{bmatrix}$$

for all $i \in \mathcal{S}$, from which the result follows. □

We can now present the proof of the main full-order analysis result, Theorem 8.13, as follows.

Proof of Theorem 8.13 First, assume that (ii) holds. Then Lemma 8.18 guarantees that (i) is satisfied. Next, notice that (i) yields (iii) according to Lemma 8.19 (the fact that such $\mathbf{X}$ and $\mathbf{Y}$ satisfying (iii) may be chosen by (ii) will be proved last). Thus, it remains only to prove (iii) ⇒ (ii).

Assume that (iii) is true. Since (8.71c) and (8.72c) are the same, it suffices to prove that such $\mathbf{X}, \mathbf{Y}$ satisfy (8.71a) and (8.71b), respectively. We have that (8.72a)

[1] If, for a given $\gamma > 0$, (8.58) has a solution, then its feasible set is open. Hence, there always exists $\mathbf{Q} > 0$ that is arbitrarily close to $\mathbf{P}$ in $\mathbb{H}^{2n+}$.

may be written

$$\begin{bmatrix} A_i^* X_i + X_i A_i + \sum_{j=1}^N \lambda_{ij} X_j & X_i J_i & C_i^* \\ J_i^* X_i & -\gamma I & L_i^* \\ C_i & L_i & -\gamma I \end{bmatrix}$$
$$+ \begin{bmatrix} I & 0 & 0 \\ 0 & I & 0 \\ 0 & 0 & D_i^* \end{bmatrix}^* \begin{bmatrix} U_i \\ 0 \\ W_i \end{bmatrix} \begin{bmatrix} H_i & G_i & 0 \end{bmatrix}$$
$$+ \begin{bmatrix} H_i & G_i & 0 \end{bmatrix}^* \begin{bmatrix} U_i \\ 0 \\ W_i \end{bmatrix}^* \begin{bmatrix} I & 0 & 0 \\ 0 & I & 0 \\ 0 & 0 & D_i^* \end{bmatrix} < 0.$$

Analogously, (8.72b) is equivalent to

$$\begin{bmatrix} A_i Y_i + Y_i A_i^* + \sum_{j=1}^N \lambda_{ij} Y_i Y_j^{-1} Y_i & J_i & Y_i C_i^* \\ J_i^* & -\gamma I & L_i^* \\ C_i Y_i & L_i & -\gamma I \end{bmatrix}$$
$$+ \begin{bmatrix} B_i^* & 0 & D_i^* \end{bmatrix}^* \begin{bmatrix} V_i & W_i & 0 \end{bmatrix} \begin{bmatrix} I & 0 & 0 \\ 0 & G_i & 0 \\ 0 & 0 & I \end{bmatrix}$$
$$+ \begin{bmatrix} I & 0 & 0 \\ 0 & G_i & 0 \\ 0 & 0 & I \end{bmatrix}^* \begin{bmatrix} V_i & W_i & 0 \end{bmatrix}^* \begin{bmatrix} B_i^* & 0 & D_i^* \end{bmatrix} < 0.$$

We further have that

$$\mathcal{N} \begin{bmatrix} I & 0 & 0 \\ 0 & I & 0 \\ 0 & 0 & D_i^* \end{bmatrix} = \mathcal{R} \begin{bmatrix} 0 \\ 0 \\ \Theta_{D_i^*} \end{bmatrix}, \qquad \mathcal{N} \begin{bmatrix} I & 0 & 0 \\ 0 & G_i & 0 \\ 0 & 0 & I \end{bmatrix} = \mathcal{R} \begin{bmatrix} 0 \\ \Theta_{G_i} \\ 0 \end{bmatrix}, \tag{8.90}$$

for some adequate $\Theta_{D_i^*}$ and Θ_{G_i}. These facts, along with Lemma 8.16, lead to the fulfillment of statement (ii).

Finally, let us prove that any **X** and **Y** satisfying (ii) may be fed into (iii) and suitable **U**, **V**, and **W** will always exist. In fact, just assume that (ii) holds. Then (i) follows, and Lemma 8.19 guarantees that $W_i = \widehat{D}_i$, $U_i = X_i B_i W_i + (Y_i^{-1} - X_i)\widehat{B}_i$, and $V_i = W_i H_i Y_i + \widehat{C}_i Y_i$ are such that (8.72a)–(8.72c) is altogether satisfied, completing the proof. □

8.6.3 *Synthesis of Dynamic Controllers*

The purpose of this subsection is to prove Theorem 8.15.

Proof of Theorem 8.15 From the hypothesis, (8.79) must be satisfied by some $\mathcal{K}$ as in (8.68) if $\mathbf{P} = (P_1, \dots, P_N)$ is chosen as

$$P_i = \begin{bmatrix} X_i & Y_i^{-1} - X_i \\ Y_i^{-1} - X_i & X_i - Y_i^{-1} \end{bmatrix} =: \begin{bmatrix} X_i & \widehat{X}_i \\ \widehat{X}_i^* & \widetilde{X}_i \end{bmatrix}, \quad i \in \mathcal{S}. \tag{8.91}$$

Moreover, from Lemma 8.19 we have that $\mathbf{U} = (U_1, \dots, U_N)$, $\mathbf{V} = (V_1, \dots, V_N)$, and $\mathbf{W} = (W_1, \dots, W_N)$ are related to the controller matrices, as well as to P_i, according to

$$U_i = X_i B_i W_i + \widehat{X}_i \widehat{B}_i, \qquad V_i = W_i H_i Y_i + \widehat{C}_i \widehat{Y}_i^*, \qquad W_i = \widehat{D}_i,$$

which, in accordance with (8.87) and (8.91), yields (8.75), (8.76), and (8.77). Thus, it remains only to explicitly present $\widehat{A}_i$ for each $i \in \mathcal{S}$ in order to solve the LMIs problem (8.58); for easiness of notation, let us write $Z_i \equiv \widehat{A}_i$. From Schur complements (Lemma 2.26) we have that this holds true whenever $\mathbf{Z} = (Z_1, \dots, Z_N) \in \mathbb{H}^n$ satisfies

$$\mathcal{Q}_i + P_i \mathcal{I}^* Z_i \mathcal{I} + \mathcal{I}^* Z_i^* \mathcal{I} P_i < 0, \quad i \in \mathcal{S}, \tag{8.92}$$

where $\mathcal{I} = [0_n \ I_n]$, and

$$\mathcal{Q}_i := \widetilde{A}_i^* P_i + P_i \widetilde{A}_i + \sum_{j=1}^{N} \lambda_{ij} P_j + \gamma^{-1} \widehat{C}_i^* \widehat{C}_i$$
$$+ \left(P_i \widehat{J}_i + \gamma^{-1} \widehat{C}_i^* \widehat{L}_i\right)\left(\gamma I - \gamma^{-1} \widehat{L}_i^* \widehat{L}_i\right)^{-1}\left(\widehat{J}_i^* P_i + \gamma^{-1} \widehat{L}_i^* \widehat{C}_i\right)$$

with $\widetilde{A}_i$ defined as in (8.88). Since such $\mathbf{Z}$ is guaranteed to exist, we have along the same lines of Proposition 8.17 that

$$\mathcal{Q}_i < 0 \quad \text{on } \mathcal{N}(\mathcal{I}) \tag{8.93}$$

and

$$P_i^{-1} \mathcal{Q}_i P_i^{-1} < 0 \quad \text{on } \mathcal{N}(\mathcal{I}) \tag{8.94}$$

must be satisfied for all $i \in \mathcal{S}$. Also notice that, due to (8.91), the inverse of P_i is such as

$$S_i := P_i^{-1} = \begin{bmatrix} Y_i & Y_i \\ Y_i & \star \end{bmatrix}, \quad i \in \mathcal{S}.$$

Thus, because of $\mathcal{N}(\mathcal{I}) = \mathcal{R}\left[\begin{smallmatrix} I_n \\ 0_n \end{smallmatrix}\right]$, we have, by writing down

$$\mathcal{Q}_i \equiv \begin{bmatrix} Q_i & \widehat{Q}_i \\ \widehat{Q}_i^* & \widetilde{Q}_i \end{bmatrix} \quad \text{with } Q_i, \widehat{Q}_i, \widetilde{Q}_i \in \mathbb{B}(\mathbb{R}^n)$$

for each $i \in \mathcal{S}$ that (8.93) and (8.94) are respectively equivalent to

$$W_i < 0, \quad i \in \mathcal{S}, \tag{8.95}$$

and

$$0 > \begin{bmatrix} I & 0 \end{bmatrix} \begin{bmatrix} Y_i & Y_i \\ Y_i & \star \end{bmatrix} \begin{bmatrix} Q_i & \widehat{Q}_i \\ \widehat{Q}_i^* & \widetilde{Q}_i \end{bmatrix} \begin{bmatrix} Y_i & Y_i \\ Y_i & \star \end{bmatrix} \begin{bmatrix} I \\ 0 \end{bmatrix} = Y_i \{ Q_i + \widehat{Q}_i + \widehat{Q}_i^* + \widetilde{Q}_i \} Y_i,$$

so that

$$Q_i + \widehat{Q}_i + \widehat{Q}_i^* + \widetilde{Q}_i < 0, \quad i \in \mathcal{S}. \tag{8.96}$$

Proceeding further, notice that

$$\begin{aligned} &P_i \mathcal{I}^* Z_i \mathcal{I} + \mathcal{I}^* Z_i^* \mathcal{I} P_i \\ &\quad = \begin{bmatrix} 0 & (Y_i^{-1} - X_i) Z_i \\ Z_i^* (Y_i^{-1} - X_i) & (X_i - Y_i^{-1}) Z_i + Z_i^* (X_i - Y_i^{-1}) \end{bmatrix} \\ &\quad = \begin{bmatrix} I & 0 \\ -I & I \end{bmatrix} \begin{bmatrix} 0 & (Y_i^{-1} - X_i) Z_i \\ Z_i^* (Y_i^{-1} - X_i) & 0 \end{bmatrix} \begin{bmatrix} I & -I \\ 0 & I \end{bmatrix}. \end{aligned}$$

Hence, due to the fact that

$$\begin{bmatrix} I & -I \\ 0 & I \end{bmatrix}^{-1} = \begin{bmatrix} I & I \\ 0 & I \end{bmatrix},$$

we have that Z_i may be entirely removed of the main diagonal blocks of (8.92) by applying the inverse transformation:

$$\begin{bmatrix} I & 0 \\ I & I \end{bmatrix} \left(P_i \mathcal{I}^* Z_i \mathcal{I} + \mathcal{I}^* Z_i^* \mathcal{I} P_i \right) \begin{bmatrix} I & I \\ 0 & I \end{bmatrix} = \begin{bmatrix} 0 & (Y_i^{-1} - X_i) Z_i \\ Z_i^* (Y_i^{-1} - X_i) & 0 \end{bmatrix}.$$

Also, we have

$$\begin{bmatrix} I & 0 \\ I & I \end{bmatrix} \mathcal{Q}_i \begin{bmatrix} I & I \\ 0 & I \end{bmatrix} = \begin{bmatrix} Q_i & Q_i + \widehat{Q}_i \\ Q_i + \widehat{Q}_i^* & Q_i + \widehat{Q}_i + \widehat{Q}_i^* + \widetilde{Q}_i \end{bmatrix},$$

so that (8.92) is equivalent to

$$\begin{bmatrix} Q_i & Q_i + \widehat{Q}_i + (Y_i^{-1} - X_i) Z_i \\ Q_i + \widehat{Q}_i^* + Z_i^* (Y_i^{-1} - X_i) & Q_i + \widehat{Q}_i + \widehat{Q}_i^* + \widetilde{Q}_i \end{bmatrix} < 0, \quad i \in \mathcal{S}.$$

Hence, bearing in mind (8.95)–(8.96), it follows that

$$\widehat{A}_i \equiv Z_i = -\left(Y_i^{-1} - X_i \right)^{-1} (Q_i + \widehat{Q}_i) \tag{8.97}$$

yields the desired result. In order to see that this is simply (8.74), first notice that

$$\begin{aligned}
\begin{bmatrix} I & 0 \end{bmatrix} P_i \widetilde{A}_i \begin{bmatrix} I \\ I \end{bmatrix} &= \begin{bmatrix} I & 0 \end{bmatrix} \begin{bmatrix} X_i & Y_i^{-1} - X_i \\ Y_i^{-1} - X_i & X_i - Y_i^{-1} \end{bmatrix} \begin{bmatrix} A_i + B_i \widehat{D}_i H_i & B_i \widehat{C}_i \\ \widehat{B}_i H_i & 0 \end{bmatrix} \begin{bmatrix} I \\ I \end{bmatrix} \\
&= X_i A_i + \left\{ X_i B_i \widehat{D}_i + \left(Y_i^{-1} - X_i \right)^{-1} \widehat{B}_i \right\} H_i + X_i B_i \widehat{C}_i \\
&= X_i A_i + U_i H_i + X_i B_i \widehat{C}_i \\
&= X_i A_i + (U_i - X_i B_i W_i) H_i + X_i B_i V_i Y_i^{-1},
\end{aligned} \tag{8.98}$$

$$\begin{bmatrix} I & 0 \end{bmatrix} \widetilde{A}_i^* P_i \begin{bmatrix} I \\ I \end{bmatrix} = \begin{bmatrix} (A_i + B_i \widehat{D}_i H_i)^* & (\widehat{B}_i H_i)^* \end{bmatrix} \begin{bmatrix} Y_i^{-1} \\ 0 \end{bmatrix} = \widetilde{A}_i^* Y_i^{-1}, \tag{8.99}$$

and

$$\begin{bmatrix} I & 0 \end{bmatrix} P_j \begin{bmatrix} I \\ I \end{bmatrix} = \begin{bmatrix} I & 0 \end{bmatrix} \begin{bmatrix} Y_j^{-1} \\ 0 \end{bmatrix} = Y_j^{-1}. \tag{8.100}$$

Similarly, bearing in mind that $\bar{C}_i := C_i + D_i W_i H_i = C_i + D_i \widehat{D}_i H_i$, we have

$$\begin{aligned}
\begin{bmatrix} I & 0 \end{bmatrix} \widehat{C}_i^* \widehat{C}_i \begin{bmatrix} I \\ I \end{bmatrix} &= \bar{C}_i^* (C_i + D_i \widehat{D}_i H_i + D_i \widehat{C}_i) \\
&= \bar{C}_i^* (C_i Y_i + D_i V_i) Y_i^{-1}.
\end{aligned} \tag{8.101}$$

Finally,

$$\begin{bmatrix} I & 0 \end{bmatrix} P_i \widehat{J}_i = X_i J_i + \left\{ X_i B_i \widehat{D}_i + \left(Y_i^{-1} - X_i \right) \widehat{B}_i \right\} G_i = X_i J_i + U_i G_i, \tag{8.102}$$

$$\begin{bmatrix} I & 0 \end{bmatrix} \widehat{C}_i^* \widehat{L}_i = \bar{C}_i^* \bar{L}_i, \tag{8.103}$$

and

$$\widehat{B}_i^* P_i \begin{bmatrix} I \\ I \end{bmatrix} = \begin{bmatrix} \bar{B}_i^* & G_i^* \widehat{B}_i^* \end{bmatrix} \begin{bmatrix} Y_i^{-1} \\ 0 \end{bmatrix} = \bar{B}_i^* Y_i^{-1}, \tag{8.104}$$

$$\widehat{L}_i^* \widehat{C}_i \begin{bmatrix} I \\ I \end{bmatrix} = \bar{L}_i^* (C_i + D_i \widehat{D}_i H_i + D_i \widehat{C}_i) = \bar{L}_i^* (C_i Y_i + D_i V_i) Y_i^{-1}. \tag{8.105}$$

Hence, (8.74) follows from (8.97) by noticing that $Q_i + \widehat{Q}_i = [I\ 0] \mathcal{Q}_i \left[\begin{smallmatrix} I \\ I \end{smallmatrix}\right]$. □

8.6.4 H_∞ Analysis and Synthesis Algorithms

As a consequence of Theorem 8.13, the existence of internally stabilizing dynamic controllers such as (8.65)–(8.68) that guarantee a desired level of H_∞ performance for system (8.66) can be characterized by means of the following algorithm.

Algorithm 8.20 Given $\gamma > 0$, the existence of a dynamic controller $\mathcal{K}$ such as (8.65)–(8.68) for which system (8.66) is internally MSS with $\|\mathcal{G}_w^{\mathcal{K}}\|_\infty < \gamma$ is guaranteed by solving any one of the following convex feasibility problems:

e$_1$: Find $\mathbf{X} = (X_1, \dots, X_N)$, $\mathbf{Y} = (Y_1, \dots, Y_N) \in \mathbb{H}^n$, $\mathbf{U} = (U_1, \dots, U_N) \in \mathbb{H}^{q,n}$, $\mathbf{V} = (V_1, \dots, V_N) \in \mathbb{H}^{n,m}$, and $\mathbf{W} = (W_1, \dots, W_N) \in \mathbb{H}^{q,m}$ such that (8.72a)–(8.72c) is satisfied for all $i \in \mathcal{S}$;

e$_2$: Find $\mathbf{X} = (X_1, \dots, X_N)$, $\mathbf{Y} = (Y_1, \dots, Y_N) \in \mathbb{H}^{n*}$ such that (8.71a)–(8.71c) is satisfied for every $i = 1, \dots, N$; alternatively, solve the LMIs in Corollary 8.14 for all $i \in \mathcal{S}$.

On the other hand, whenever it may be proved that either of these problems does not have a solution, then such a compensator does not exist at all.

The next algorithm provides one possible way of computing a full-order controller such as the one presented in Theorem 8.15.

Algorithm 8.21 (Two-step design procedure) Given $\gamma > 0$, a controller $\mathcal{K}$ such as (8.65)–(8.68) that internally stabilizes system (8.66) in the mean-square sense with $\|\mathcal{G}_w^{\mathcal{K}}\|_\infty < \gamma$ may be designed according to Theorem 8.15 by the following steps:

d$_1$: Solve the existence problem by means of e$_1$ in Algorithm 8.20;
↪ If such a solution cannot be found, then **stop**.

d$_2$: With **X**, **Y**, **U**, **V**, and **W** from the previous step, design a compensator by means of relations (8.74)–(8.75).

Looking back at Theorem 8.13, it is possible to propose the following alternative to Algorithm 8.21. The main advantage here is that the existence of solutions depends on the feasibility of a problem of relatively smaller dimension.

Algorithm 8.22 (Three-step design procedure) Given $\gamma > 0$, a controller $\mathcal{K}$ such as (8.65)–(8.68) which internally stabilizes system (8.66) in the mean-square sense with $\|\mathcal{G}_w^{\mathcal{K}}\|_\infty < \gamma$ may be designed according to Theorem 8.15 by the following steps:

D$_1$: Solve the existence problem by means of e$_2$ in Algorithm 8.20;
↪ If such a solution cannot be found, then **stop**.

D$_2$: With **X** and **Y** from the above step, find $\mathbf{U} = (U_1, \dots, U_N) \in \mathbb{H}^{q,n}$, $\mathbf{V} = (V_1, \dots, V_N) \in \mathbb{H}^{n,m}$, and $\mathbf{W} = (W_1, \dots, W_N) \in \mathbb{H}^{q,m}$ such that (8.72a) and (8.72b) are satisfied (due to Theorem 8.13, there is always a solution to this problem);

D$_3$: With **X**, **Y**, **U**, **V**, and **W** obtained from the previous steps, design a compensator by means of relations (8.74)–(8.75).

8.7 Historical Remarks

The historical foundations of H_∞ control were laid out in 1981 by G. Zames with the paper [314]. The fundamental concepts that emerged in this paper, however, can be traced back to his doctoral dissertation [313] in 1960, when the interconnection of two systems was considered. Part of the appeal of H_∞ control methods is certainly due to their connection with various topics in systems theory. Some of these include minimum entropy problems, risk sensitive control, differential games, dissipativity theory, model reduction, and many others. A research area strongly related to H_∞ control and of particular interest in the next chapter is the *robust control methods*.

As far as the authors are aware of, the study of H_∞ control of MJLS was initiated in [157] (see also [104]), for the continuous-time case and in [158] for the discrete-time case. A seminal reference for the discrete-time case is [73], which considers the Markov chain with countably infinite state space. Before the end of the 1990 decade, a fairly expressive number of papers had been published, including [47, 122, 236–238, 247, 256]. The literature which came up ever since has proven so rich that it is by now virtually impossible to list all the works dealing with the H_∞ control of MJLS. Some academic applications of H_∞ control methods in the MJLS context include but, of course, are not limited to: [261–263] in robotics; [101, 266] in aeronautics; [1, 2, 254] in networked systems.

The main sources for this chapter were, essentially, [278, 279].

Chapter 9
Design Techniques

9.1 Outline of the Chapter

In this chapter we present some design techniques expressed as linear matrix inequalities optimization problems for continuous-time MJLS. The LMIs paradigm offers a flexible and efficient framework to computational applications for which many powerful numerical packages exist. The interest here is in robust H_2 and mixed H_2/H_∞ control (Sect. 9.4) and in the stationary robust linear filtering problem (Sect. 9.5). In Chap. 10 the use of these techniques is illustrated through some problems in the fields of robotics, economic modeling, and stationary filtering.

9.2 Stability Radius

In applications of Markov jump linear systems, one is faced with the nontrivial task of ascribing adequate values to the parameters that define the system. Thus, it is of germane interest to investigate whether qualitative properties of the system of interest should be preserved if these values are not sufficiently accurate. In addition to this kind of modeling error, parametric uncertainty may also come up due to linearization, aging of components, or slow time dependence, for instance.

In this section we shall present a preliminary study on this problem that concerns the preservation of mean-square stability. We study the robustness of system (3.1) under the following assumption, which will be carried throughout.

Assumption 9.1 *System* (3.1) *is mean-square stable as in Definition* 3.2.

Our interest lies on the following perturbed version of system (3.1):

$$\dot{x}(t) = (A_{\theta(t)} + E_{\theta(t)}\Delta_{\theta(t)}F_{\theta(t)})x(t), \tag{9.1}$$

in which, for a prespecified $\alpha > 0$, the constraint

$$\mathbf{\Delta} = (\Delta_1, \ldots, \Delta_N), \quad \|\mathbf{\Delta}\|_{\max} < \alpha \tag{9.2}$$

O.L.V. Costa et al., *Continuous-Time Markov Jump Linear Systems*,
Probability and Its Applications, DOI 10.1007/978-3-642-34100-7_9,

must be satisfied for given $\mathbf{A} = (A_1, \dots, A_N) \in \mathbb{H}^n$, $\mathbf{E} = (E_1, \dots, E_N) \in \mathbb{H}^{n_e,n}$, and $\mathbf{F} = (F_1, \dots, F_N) \in \mathbb{H}^{n,n_f}$. It should be noted that (9.2) comprises parametric disturbances that are not necessarily small (because α is not so assumed), so that classical parametric sensitivity techniques are not an alternative.

The robustness measure we study in this section corresponds to the infimal α for which there is one $\boldsymbol{\Delta}$ that destabilizes system (9.1) in the MSS sense. According to whether or not $\boldsymbol{\Delta}$ is constrained to $\mathbb{H}^{n_f,n_e}$, we shall refer to *real* and *complex* stability radii, whose proper definition is as follows.

Definition 9.2 The stability radii of system (9.1) are defined as

$$\mathbf{r}_{\mathbb{F}} = \inf\{\alpha, \text{ s.t. (9.1) is not MSS for some } \boldsymbol{\Delta} \in \mathbb{H}_{\mathbb{F}} \text{ as in (9.2)}\}, \tag{9.3}$$

where either $\mathbb{F} = \mathbb{R}$ (for which $\mathbb{H}_{\mathbb{F}} = \mathbb{H}^{n_f,n_e}$) or $\mathbb{F} = \mathbb{C}$ (and $\mathbb{H}_{\mathbb{C}} = \mathbb{H}_{\mathbb{C}}^{n_f,n_e}$).

Remark 9.3 It should be noted that the complex stability radius is tacitly dependent on $\mathbf{A}$, $\mathbf{E}$, $\mathbf{F}$, and Π. For convenience, bearing in mind that such data are assumed known a priori, this dependence is omitted in Definition 9.2.

Remark 9.4 It is also important to mention that the real/complex terminology is entirely independent of whether the system data (i.e., $\mathbf{A}$, $\mathbf{E}$, $\mathbf{F}$) takes on real or complex values. In other words, a stable system whose a priori given parameters are real has both, real *and* complex, radii. The literature which came up in the last 25 years has shown, however, that computing the real stability radius is far more difficult than obtaining the complex radius. This is further discussed in Sect. 9.6.

The set of admissible disturbances in (9.2) can carry much more structure in applications. Block-diagonal real perturbations, for instance, are of great interest in many scenarios. However, dealing with structured disturbances is in general very difficult, and the advent of μ-values has not yet been proposed in the MJLS literature. Our objective throughout this section will consist on deriving easily computable bounds for stability radii, for norm-bounded but otherwise unconstrained $\boldsymbol{\Delta}$ as in (9.2).

As shown next, a lower bound for stability radii is given by the feasibility of the LMIs

$$\begin{bmatrix} \mathcal{T}_i(\mathbf{P}) + \alpha^2 s_i F_i^* F_i & P_i E_i \\ E_i^* P_i & -s_i I \end{bmatrix} < 0, \quad P_i > 0, \quad i \in \mathcal{S}, \tag{9.4}$$

or, alternatively,

$$\begin{bmatrix} \mathcal{L}_i(\mathbf{Q}) + \alpha^2 s_i E_i E_i^* & Q_i F_i^* \\ F_i Q_i & -s_i I \end{bmatrix} < 0, \quad Q_i > 0, \quad i \in \mathcal{S}. \tag{9.5}$$

The precise result is as follows.

Theorem 9.5 *The stability radii of system* (9.1) *satisfy*

$$\max\{\rho, \rho^\star\} \le \mathbf{r}_{\mathbb{C}} \le \mathbf{r}_{\mathbb{R}}, \tag{9.6}$$

where, by definition,

$$\rho = \sup\{\alpha;\ (9.4)\ \textit{is feasible}\}, \qquad \rho^\star = \sup\{\alpha;\ (9.5)\ \textit{is feasible}\}. \tag{9.7}$$

Proof Suppose that $0 \le \alpha < \rho$, so that (9.4) is feasible. Since

$$\bigl(s_i^{1/2}\Delta_i F_i - s_i^{-1/2}E_i^* P_i\bigr)^*\bigl(s_i^{1/2}\Delta_i F_i - s_i^{-1/2}E_i^* P_i\bigr) \ge 0,$$

we obtain from (9.2) that, for all $\boldsymbol{\Delta} \in \mathbb{H}_{\mathbb{C}}^{n_f,n_e}$,

$$\begin{aligned} F_i^*\Delta_i^*E_i^*P_i + P_iE_i\Delta_iF_i &\le s_iF_i^*\Delta_i^*\Delta_iF_i + s_i^{-1}P_iE_iE_i^*P_i \\ &\le \alpha^2 s_iF_i^*F_i + s_i^{-1}P_iE_iE_i^*P_i. \end{aligned}$$

Therefore, from Schur's complement (Lemma 2.26) (9.4) yields

$$\begin{aligned} 0 &> \mathcal{T}_i(\mathbf{P}) + \alpha^2 s_iF_i^*F_i + s_i^{-1}P_iE_iE_i^*P_i \\ &\ge (A_i + E_i\Delta_iF_i)^*P_i + P_i(A_i + E_i\Delta_iF_i) + \sum_{j\in\mathcal{S}}\lambda_{ij}P_j, \end{aligned}$$

and thus, by Theorem 3.25, system (9.1) is mean-square stable for all $\boldsymbol{\Delta}$ as in (9.2).

Conversely, suppose that $0 \le \alpha < \rho^\star$. Then the inequality $\alpha \le \mathbf{r}_{\mathbb{C}}$ is proven by analogous steps, bearing in mind that $0 \le (s_i^{1/2}E_i\Delta_i - s_i^{-1/2}Q_iF_i^*)(s_i^{1/2}E_i\Delta_i - s_i^{-1/2}Q_iF_i^*)^*$ and that (9.2) guarantees $\Delta_i\Delta_i^* \le \alpha^2 I$. Finally, the inequality $\mathbf{r}_{\mathbb{C}} \le \mathbf{r}_{\mathbb{R}}$ is an immediate consequence of the fact that $\boldsymbol{\Delta} \in \mathbb{H}^{n_f,n_e} \Rightarrow \boldsymbol{\Delta} \in \mathbb{H}_{\mathbb{C}}^{n_f,n_e}$. □

The following bisection algorithm can be used to obtain the robustness margins in (9.6).

Algorithm 9.6 The maximal robustness margins ρ and $\rho^\star$ in (9.7) can be computed with arbitrary precision $\varepsilon > 0$ as follows:

S_1: Find $0 \le \rho_{\min} < \rho_{\max}$ such that the LMIs problem (9.4) is feasible for $\alpha = \rho_{\min}$ and unfeasible for $\alpha = \rho_{\max}$, respectively.

↪ If such $\rho_{\max}$ cannot be found, then **stop**: the perturbed system is always MSS with (9.2), and $\rho = \infty$.

S_2: Let $\alpha \leftarrow (\rho_{\min} + \rho_{\max})/2$ and check whether (9.4) is feasible.

↪ If so, then let $\rho_{\min} \leftarrow \alpha$;
↪ otherwise, let $\rho_{\max} \leftarrow \alpha$;

S_3: Repeat S_2 until $(\rho_{\max} - \rho_{\min})/2 < \varepsilon$.

S$_4$: Return $\alpha \approx \rho$.
S$_5$: Repeat S$_1$ through S$_4$ with $(\rho^\star_{\min}, \rho^\star_{\max}, \rho^\star)$ in place of $(\rho_{\min}, \rho_{\max}, \rho)$ and (9.4) replaced by (9.5).

Determining the real stability radius is, even in the case without jumps, a difficult problem. We shall not directly address it here, but will rather obtain upper and lower bounds for it, in terms of the largest eigenvalues of specific matrices. For further simplicity, we focus on the so-called unstructured stability radii, which are defined next.

Definition 9.7 The unstructured stability radii of system (9.1) are

$$\widehat{\mathbf{r}}_{\mathbb{F}} = \inf\bigl\{\alpha, \text{ s.t. (9.1) with } E_i \equiv F_i \equiv I \\ \text{is not MSS for some } \boldsymbol{\Delta} \in \mathbb{H}_{\mathbb{F}} \text{ as in (9.2)}\bigr\}, \tag{9.8}$$

where $\mathbb{F}$ is either $\mathbb{R}$ or $\mathbb{C}$, with $\mathbb{H}_{\mathbb{R}} = \mathbb{H}^n$ and $\mathbb{H}_{\mathbb{C}} = \mathbb{H}^n_{\mathbb{C}}$.

Before proceeding further, the following result is in order. From now on, as in Corollary 3.23, $\lambda_{\max}(\cdot)$ denotes the largest eigenvalue of a given Hermitian matrix.

Proposition 9.8 *Given $\boldsymbol{\Delta} = (\Delta_1, \dots, \Delta_N) \in \mathbb{H}^n$, let $\mathbf{d}(\boldsymbol{\Delta}) := (d_1(\boldsymbol{\Delta}), \dots, d_N(\boldsymbol{\Delta}))$ with $d_i(\boldsymbol{\Delta}) := \frac{1}{2}\lambda_{\max}(\Delta_i + \Delta_i^*)$ for each $i \in \mathcal{S}$. Then we have $\|\mathbf{d}(\boldsymbol{\Delta})\|_{\max} \le \|\boldsymbol{\Delta}\|_{\max}$.*

Proof Indeed (with $\lambda_j(\cdot)$ denoting the jth eigenvalue, in no particular order, of a given matrix) we have

$$\begin{aligned}
\bigl\|\mathbf{d}(\boldsymbol{\Delta})\bigr\|_{\max} &= \max_{i\in\mathcal{S}}\bigl|d_i(\boldsymbol{\Delta})\bigr| \\
&= \max_{i\in\mathcal{S}}\left|\frac{1}{2}\lambda_{\max}\bigl(\Delta_i + \Delta_i^*\bigr)\right| = \max_{i\in\mathcal{S}}\left|\frac{1}{2}\max_{j=1,\dots,n}\lambda_j\bigl(\Delta_i + \Delta_i^*\bigr)\right| \\
&\le \max_{i\in\mathcal{S}}\left\{\frac{1}{2}\max_{j=1,\dots,n}\bigl|\lambda_j\bigl(\Delta_i + \Delta_i^*\bigr)\bigr|\right\} = \max_{i\in\mathcal{S}}\left\{\frac{1}{2}\bigl\|\Delta_i + \Delta_i^*\bigr\|\right\} \\
&\le \max_{i\in\mathcal{S}}\|\Delta_i\| = \|\boldsymbol{\Delta}\|_{\max},
\end{aligned}$$

where the last step follows from the triangle inequality. □

The aforementioned bounds on the unstructured stability radii are presented now.

Theorem 9.9 *The unstructured stability radii of system* (9.1) *satisfy* $\widehat{\mathbf{r}}_{\mathbb{C}} \le \widehat{\mathbf{r}}_{\mathbb{R}} \le -\frac{1}{2}\operatorname{Re}\{\lambda(\mathcal{A})\}$ *with $\mathcal{A}$ as in* (3.25). *If, in addition, the scalar MJLS*

$$\dot{\tilde{x}}(t) = \tilde{a}_{\theta(t)}\tilde{x}(t), \qquad \tilde{a}_i := \frac{1}{2}\lambda_{\max}\bigl(A_i + A_i^*\bigr), \quad i \in \mathcal{S}, \tag{9.9}$$

is mean-square stable, then we have

$$-\frac{1}{2}\operatorname{Re}\{\lambda(\tilde{\mathcal{A}})\} \le \widehat{\mathbf{r}}_{\mathbb{C}} \le \widehat{\mathbf{r}}_{\mathbb{R}} \le -\frac{1}{2}\operatorname{Re}\{\lambda(\mathcal{A})\} \tag{9.10}$$

with $\tilde{\mathcal{A}} := \Pi' + 2\operatorname{diag}(\tilde{a}_i)$. *That is,* $\tilde{\mathcal{A}} = [\tilde{\mathcal{A}}_{ij}] \in \mathbb{R}^{N\times N}$ *with*

$$\tilde{\mathcal{A}}_{ij} = \begin{cases} \lambda_{ji}, & i \ne j, \\ \lambda_{ii} + \lambda_{\max}(A_i + A_i^*), & i = j. \end{cases} \tag{9.11}$$

Proof If A_i is perturbed as $A_i \rightsquigarrow A_i + (\zeta/2)I$ for each $i \in \mathcal{S}$, then $\mathcal{A}$ is correspondingly perturbed as $\mathcal{A} \rightsquigarrow \mathcal{A} + \operatorname{Re}\{\zeta\}I$. Therefore, letting $\zeta := -\operatorname{Re}\{\lambda(\mathcal{A})\} > 0$ (bearing in mind Assumption 9.1), we have

$$A_i \rightsquigarrow A_i + (\zeta/2)I \quad \Longrightarrow \quad \operatorname{Re}\{\lambda(\mathcal{A}+\zeta)\} = \operatorname{Re}\{\lambda(\mathcal{A})\} + \zeta = 0,$$

so that this (real) perturbation renders the system unstable in the MSS sense. Since $\|(\zeta/2)I\| = -\frac{1}{2}\operatorname{Re}\{\lambda(\mathcal{A})\}$, we obtain $\widehat{\mathbf{r}}_{\mathbb{R}} \le -\frac{1}{2}\operatorname{Re}\{\lambda(\mathcal{A})\}$, and the fact that $\widehat{\mathbf{r}}_{\mathbb{C}} \le \widehat{\mathbf{r}}_{\mathbb{R}}$ is once again due to $\mathbb{H}^n \subset \mathbb{H}^n_{\mathbb{C}}$.

Proceeding further, let $d_i(\mathbf{\Delta}) := \frac{1}{2}\lambda_{\max}(\Delta_i + \Delta_i^*)$ as in Proposition 9.8 and assume that the system

$$\dot{\tilde{x}}(t) = \big(\tilde{a}_{\theta(t)} + d_{\theta(t)}(\mathbf{\Delta})\big)\tilde{x}(t) \tag{9.12}$$

is mean-square stable. Under this condition, there must be $\mathbf{p} = (p_1, \dots, p_N)$ in $\mathbb{H}^{1+}$ with $\mathbf{p} > 0$ such that, letting $P_i := p_i I_n > 0$,

$$\begin{aligned}
&(A_i + \Delta_i)^* P_i + P_i(A_i + \Delta_i) + \sum_{j\in\mathcal{S}} \lambda_{ij} P_j \\
&\quad = p_i\big(A_i + A_i^*\big) + p_i\big(\Delta_i + \Delta_i^*\big) + \sum_{j\in\mathcal{S}} \lambda_{ij} p_j I \\
&\quad \le \Big\{ p_i \lambda_{\max}\big(A_i + A_i^*\big) + p_i \lambda_{\max}\big(\Delta_i + \Delta_i^*\big) + \sum_{j\in\mathcal{S}} \lambda_{ij} p_j \Big\} I < 0,
\end{aligned}$$

where the last inequality follows from the MSS of system (9.9) in conjunction with Theorem 3.33. Thus, by contraposition we obtain that whenever the system

$$\dot{x}(t) = (A_{\theta(t)} + \Delta_{\theta(t)})x(t) \tag{9.13}$$

is not mean-square stable for a given $\mathbf{\Delta}$, so is not system (9.12). This fact, along with Proposition 9.8, is necessary to prove that the complex stability radius of system (9.13) is larger than the auxiliary radius

$$\mathbf{s}_{\mathbb{C}} := \inf\big\{ \|\boldsymbol{\delta}\|_{\max};\ \dot{\tilde{x}}(t) = (\tilde{a}_{\theta(t)} + \delta_{\theta(t)})\tilde{x}(t) \text{ is not MSS} \big\}, \tag{9.14}$$

which is done as follows:

$$
\begin{aligned}
\widehat{\mathbf{r}}_{\mathbb{C}} &= \inf\bigl\{\|\boldsymbol{\Delta}\|_{\max};\ (9.13)\text{ is not MSS}\bigr\} \geq \inf\bigl\{\|\boldsymbol{\Delta}\|_{\max};\ (9.12)\text{ is not MSS}\bigr\} \\
&\geq \inf\bigl\{\bigl\|\mathbf{d}(\boldsymbol{\Delta})\bigr\|_{\max};\ (9.12)\text{ is not MSS}\bigr\} \\
&\geq \inf\bigl\{\|\boldsymbol{\delta}\|_{\max};\ (9.14)\text{ is not MSS}\bigr\} = \mathbf{s}_{\mathbb{C}}.
\end{aligned}
\tag{9.15}
$$

To complete the proof, it remains only to check that $\mathbf{s}_{\mathbb{C}} \geq -\frac{1}{2}\operatorname{Re}\{\lambda(\tilde{\mathcal{A}})\}$. Indeed, from Theorem 9.5 we get

$$
\mathbf{s}_{\mathbb{C}} \geq \sup\left\{\alpha;\ \begin{bmatrix} \operatorname{Her}(p_i\tilde{a}_i) + \sum_{j\in\mathcal{S}}\lambda_{ij}p_j + \alpha^2 s_i & p_i \\ p_i & -s_i \end{bmatrix} < 0,\ p_i > 0 \text{ is feasible}\right\},
$$

so that, from an application of Schur's complement (Lemma 2.26) together with the choice $s_i = p_i/\alpha$ we have

$$
\mathbf{s}_{\mathbb{C}} \geq \sup\left\{\alpha;\ \operatorname{Her}\bigl(p_i(\tilde{a}_i + \alpha)\bigr) + \sum_{j\in\mathcal{S}}\lambda_{ij}p_j < 0,\ p_i > 0 \text{ is feasible}\right\}.
$$

Finally, bearing in mind Theorem 3.21, we obtain

$$
\begin{aligned}
\mathbf{s}_{\mathbb{C}} &\geq \sup\bigl\{\alpha;\ \operatorname{Re}\bigl\{\lambda\bigl((\Pi'\otimes I) + \operatorname{diag}\bigl((\tilde{a}_i+\alpha)\oplus(\tilde{a}_i+\alpha)\bigr)\bigr)\bigr\} < 0\bigr\} \\
&= \sup\bigl\{\alpha;\ \operatorname{Re}\bigl\{\lambda\bigl((\Pi'\otimes I) + \operatorname{diag}(\tilde{a}_i\oplus\tilde{a}_i)\bigr)\bigr\} + 2\alpha < 0\bigr\} \\
&= \sup\bigl\{\alpha;\ \operatorname{Re}\bigl\{\lambda(\widetilde{\mathcal{A}})\bigr\} < -2\alpha\bigr\} = -\frac{1}{2}\operatorname{Re}\bigl\{\lambda(\widetilde{\mathcal{A}})\bigr\},
\end{aligned}
$$

which, once plugged in (9.15), yields the lower bound in (9.10). □

The following consequence of the preceding theorem shows that, in the scalar case, (9.10) provides the exact characterization of the real and complex stability radii.

Corollary 9.10 *Suppose that system* (9.1) *is scalar* ($n = 1$), *with* $E_i = F_i = 1$ *for all* $i \in \mathcal{S}$. *Then*

$$
\widehat{\mathbf{r}}_{\mathbb{C}} = \widehat{\mathbf{r}}_{\mathbb{R}} = -\frac{1}{2}\operatorname{Re}\bigl\{\lambda(\mathcal{A})\bigr\}.
\tag{9.16}
$$

Proof Bearing in mind Assumption 9.1, this follows from the fact that in the scalar case the hypotheses of Theorem 9.9 are satisfied with $\mathcal{A} = \tilde{\mathcal{A}}$. □

Example 9.11 In the scalar case with two operation modes, i.e., $x(t) \in \mathbb{R}$ with $\mathcal{S} = \{1, 2\}$, we obtain from Corollary 9.10 the explicit formula

$$\widehat{\mathbf{r}}_{\mathbb{C}} = \widehat{\mathbf{r}}_{\mathbb{R}} = -\frac{1}{4}\operatorname{tr}(\mathcal{A}) - \frac{1}{2}\sqrt{\lambda_{11}\lambda_{22} + \left(\operatorname{Re}(A_1 - A_2) + \frac{\lambda_{11} - \lambda_{22}}{2}\right)^2} \tag{9.17}$$

with $\operatorname{tr}(\mathcal{A}) = 2\operatorname{Re}(A_1 + A_2) - (\lambda_{11} + \lambda_{22})$.

Example 9.12 Suppose that we wish to obtain estimates for the stability radii of the system

$$\dot{x}(t) = (A_{\theta(t)} + \Delta_{\theta(t)})x(t) \tag{9.18}$$

for the case $\mathcal{S} = \{1, 2\}$, with

$$A_1 = \begin{bmatrix} 0.5 & -1 \\ 0 & -2 \end{bmatrix}, \quad A_2 = \begin{bmatrix} -2 & -1 \\ 0 & 0.5 \end{bmatrix}, \quad \text{and} \quad \Pi = \begin{bmatrix} -3 & 3 \\ 4 & -4 \end{bmatrix}, \tag{9.19}$$

which is, apart from the transition rates, the MJLS analyzed in (1.3) and Sect. 3.4.1.

By employing Algorithm 9.6 with precision $\varepsilon = 10^{-6}$ we obtain, after 28 iterations, that $\rho = 0.1578$ and $\rho^\star = 0.1406$ in (9.6). Furthermore, since the real abscissa of $\mathcal{A}$ is $\operatorname{Re}\{\lambda(\mathcal{A})\} \simeq -0.4174$, the upper bound on (9.10) yields

$$0.1578 \le \widehat{\mathbf{r}}_{\mathbb{C}} \le \widehat{\mathbf{r}}_{\mathbb{R}} \le 0.2087. \tag{9.20}$$

Also, notice that a lower bound such as in (9.10) may not be constructed in this case, since, as one can easily verify, the corresponding scalar system (9.9) is not mean-square stable.

Example 9.13 Consider now the other system treated in Sect. 3.4.1, but with different transition rates. The system matrices in (9.18) are, in this case,

$$A_1 = \begin{bmatrix} 0.25 & -2 \\ 2 & 0.25 \end{bmatrix}, \quad A_2 = \begin{bmatrix} -2 & 10 \\ -10 & -2 \end{bmatrix}, \quad \text{and} \quad \Lambda = \begin{bmatrix} -2 & 2 \\ 1 & -1 \end{bmatrix}. \tag{9.21}$$

Then, the scalar system (9.9) is such that, in view of (9.10) and (9.11),

$$\frac{1}{2} = -\frac{1}{2}\operatorname{Re}\{\lambda(\tilde{\mathcal{A}})\} \le \widehat{\mathbf{r}}_{\mathbb{C}} \le \widehat{\mathbf{r}}_{\mathbb{R}} \le -\frac{1}{2}\operatorname{Re}\{\lambda(\mathcal{A})\} = \frac{1}{2}, \tag{9.22}$$

yielding an exact characterization for the stability radii.

9.3 A Robustness Margin for Π

We now proceed with an application of the robustness bounds derived in Sect. 9.2, which treated the stability radii of system (3.1) with respect to perturbations on $\mathbf{A}$. Here, on the other hand, we shall assume $\mathbf{A}$ fixed and will consider that the transition rates of the jump process are uncertain in (3.1).

We shall assume that such perturbations are of the form $\Pi \rightsquigarrow K\Pi$ in which $K \in \mathbb{B}(\mathbb{R}^N)$ is a diagonal matrix with positive entries. More explicitly, our interest is on disturbances such as

$$K\Pi = [\lambda_{ij}^k], \quad \lambda_{ij}^k := \begin{cases} k_i\lambda_{ij}, & i \in \mathcal{Z}, \\ \lambda_{ij} & \text{otherwise}, \end{cases} \tag{9.23}$$

for a given set $\mathcal{Z} \subseteq \mathcal{S}$ and $K = \operatorname{diag}(k_1, \dots, k_N)$ such that $k_i = 1$ if $i \notin \mathcal{Z}$.

Such perturbation can be interpreted as follows. For each $i \in \mathcal{Z}$, k_i is a tuning of the exponential rate (λ_{ii}) that governs the sojourn time spent on this operation mode. If $k_i > 1$, the mean time spent on the ith mode is decreased, whereas $k_i < 1$ implies a typically larger sojourn time. Otherwise, $k_i = 1$, and there is no change as far as this particular mode is concerned. Such perturbation is interesting in that only the waiting times are affected; the transition probabilities, on the other hand, are not. Thus, many qualitative properties of the jump process (such as its ordering or the presence/absence of absorbing modes) are not destroyed.

Under the perturbation $\Pi \rightsquigarrow K\Pi$, (3.1) becomes

$$\begin{cases} \dot{x}(t) = A_{\Theta(t)}x(t), & t \in \mathbb{R}^+, \\ \Theta(0) = \theta_0, \qquad x(0) = x_0, & P(\Theta(0) = i) = \nu_i, \end{cases} \tag{9.24}$$

where $\Theta = \{(\Theta(t), \mathcal{F}_t), t \geq 0\}$ is a Markov process analogous to θ, but with transition rates $[\lambda_{ij}^k]$. Bearing this in mind, we have the following auxiliary lemma.

Lemma 9.14 *System* (9.24) *is mean-square stable if and only if the system*

$$\dot{\mathsf{x}}(t) = \mathsf{A}_{\theta(t)}\mathsf{x}(t), \qquad \mathsf{x}(0) = x_0, \qquad \mathsf{A}_i := \begin{cases} k_i^{-1}A_i, & i \in \mathcal{Z}, \\ A_i & \text{otherwise}, \end{cases} \tag{9.25}$$

is also mean-square stable.

Proof From Theorem 3.33 we have that system (9.24) is mean-square stable if and only if there is $\mathbf{P} = (P_1, \dots, P_N) \in \mathbb{H}^{n+}$, $\mathbf{P} > 0$, such that $\mathcal{T}^k(\mathbf{P}) = (\mathcal{T}_1^k(\mathbf{P}), \dots, \mathcal{T}_N^k(\mathbf{P})) < 0$, where $\mathcal{T}_i^k(\mathbf{P}) := A_i^* P_i + P_i A_i + \sum_{j\in\mathcal{S}} \lambda_{ij}^k P_j$ for each $i \in \mathcal{S}$. To prove the result, simply notice that this is equivalent to $\tilde{\mathcal{T}}^k(\mathbf{P}) = (\tilde{\mathcal{T}}_1^k(\mathbf{P}), \dots, \tilde{\mathcal{T}}_N^k(\mathbf{P})) < 0$, where $\tilde{\mathcal{T}}_i^k(\mathbf{P}) := \mathsf{A}_i^* P_i + P_i \mathsf{A}_i + \sum_{j\in\mathcal{S}} \lambda_{ij} P_j$. □

In what follows we shall use the fact that A_i/k_i may always be decomposed in the form $A_i/k_i = A_i + E_i\Delta_i F_i$ with $E_i, F_i \in \mathbb{B}(\mathbb{R}^n)$, as long as

$$E_i F_i = A_i, \qquad \Delta_i = (k_i^{-1} - 1)I_n. \tag{9.26}$$

Remark 9.15 Notice that, besides (9.26), many other choices for decomposing A_i are valid. For example, $\operatorname{rank}(A_i) = m \leq n$ yields (9.26) satisfied with $E_i \in \mathbb{B}(\mathbb{C}^m, \mathbb{C}^n)$, $F_i \in \mathbb{B}(\mathbb{C}^n, \mathbb{C}^m)$, and $\Delta_i := (k_i^{-1} - 1)I_m$. In either case, however, notice that the disturbance Δ_i is *real*, which makes it difficult to obtain the exact robustness margin. Furthermore, $A_i = E_i F_i$ implies $\mathrm{rank}(A_i) \le \min\{\mathrm{rank}(E_i), \mathrm{rank}(F_i)\}$, which is a constraint on which pairs (E_i, F_i) are admissible for factorizing A_i.

The main result of this section is the following one. It is proven that a margin for the robust stochastic stability of system (3.1) with respect to perturbations such as $\Pi \rightsquigarrow K\Pi$ is given by the real stability radius $\mathbf{r}_{\mathbb{R}}$ of the system

$$\dot{x}(t) = (A_{\theta(t)} + E_{\theta(t)}\Delta_{\theta(t)}F_{\theta(t)})x(t), \qquad \begin{cases} E_{\theta(t)}F_{\theta(t)} = A_{\theta(t)}, & \theta(t) \in \mathcal{Z}, \\ E_{\theta(t)} = 0, \quad F_{\theta(t)} = 0 & \text{otherwise}, \end{cases}$$

or that, in a more refined way, it may be computed via an extension of Algorithm 9.6 to the LMIs problems

$$\begin{bmatrix} \mathcal{T}_i(\mathbf{P}) + \alpha^2 F_i^* S_i F_i & P_i E_i \\ E_i^* P_i & -S_i \end{bmatrix} < 0, \quad P_i, S_i > 0, \quad i = 1, \dots, N, \tag{9.27}$$

and

$$\begin{bmatrix} \mathcal{L}_i(\mathbf{Q}) + \alpha^2 E_i S_i E_i^* & Q_i F_i^* \\ F_i Q_i & -S_i \end{bmatrix} < 0, \quad Q_i, S_i > 0, \quad i = 1, \dots, N, \tag{9.28}$$

together with their respective robustness margins

$$\sigma := \sup\{\alpha;\ (9.27) \text{ is feasible}\}, \qquad \sigma^\star := \sup\{\alpha;\ (9.28) \text{ is feasible}\}. \tag{9.29}$$

Theorem 9.16 *System* (9.24) *is mean-square stable for all* $(k_1, \dots, k_N) \in \mathbb{H}^{1+}$ *such that*

$$\frac{1}{1 + \mathbf{r}_{\mathbb{R}}} < k_i < k_\infty \quad \textit{with } k_\infty := \begin{cases} \dfrac{1}{1 - \mathbf{r}_{\mathbb{R}}}, & \mathbf{r}_{\mathbb{R}} < 1, \\ \infty & \textit{otherwise}, \end{cases} \tag{9.30}$$

for each $i \in \mathcal{Z}$. *Moreover,* (9.24) *is MSS whenever* $k_i \in (k_{\min}, k_{\max})$ *for each* $i \in \mathcal{Z} \subset \mathcal{S}$, *in which, letting* $\varsigma := \max\{\sigma, \sigma^\star\}$,

$$k_{\min} := \frac{1}{1 + \varsigma}, \qquad k_{\max} := \begin{cases} \dfrac{1}{1 - \varsigma}, & \varsigma < 1, \\ \infty & \textit{otherwise}. \end{cases} \tag{9.31}$$

Proof The sufficiency of (9.30) is (bearing in mind Remark 9.15) due to mean-square stability being preserved whenever $\|\Delta_i\| = |1 - k_i^{-1}| < \mathbf{r}_{\mathbb{R}}$, from which (9.30) follows without much effort. Analogously, (9.31) is an immediate consequence of $\|\Delta_i\| = |1 - k_i^{-1}| < \max\{\sigma, \sigma^\star\}$, which, as we prove next, guarantees the mean-square stability of the perturbed system. In fact, similarly to Theorem 9.5, the feasibility of (9.27) should guarantee, for any such $\Delta_i = (k_i^{-1} - 1)I$,

$$
\begin{aligned}
0 &> \mathcal{T}_i(\mathbf{P}) + \alpha^2 F_i^* S_i F_i + P_i E_i S_i^{-1} E_i^* P_i \\
&\geq \mathcal{T}_i(\mathbf{P}) + \left(k_i^{-1} - 1\right)^2 F_i^* S_i F_i + P_i E_i S_i^{-1} E_i^* P_i \\
&= \mathcal{T}_i(\mathbf{P}) + \left(F_i^* \Delta_i^* S_i^{1/2} - P_i E_i S_i^{-1/2}\right)\left(S_i^{1/2} \Delta_i F_i - S_i^{-1/2} E_i^* P_i\right) \\
&\quad + \operatorname{Her}(P_i E_i \Delta_i F_i) \\
&\geq \operatorname{Her}\left\{P_i(A_i + E_i \Delta_i F_i)\right\} + \sum_{j \in \mathcal{S}} \lambda_{ij} P_j,
\end{aligned}
$$

yielding the mean-square stability of the perturbed system by Lemma 9.14. Alternatively, the feasibility of (9.28) implies, owing to the fact that Δ_i is a multiple of the identity,

$$
\begin{aligned}
0 &> \mathcal{L}_i(\mathbf{Q}) + \alpha^2 E_i S_i E_i^* + Q_i F_i^* S_i^{-1} F_i Q_i \\
&\geq \mathcal{L}_i(\mathbf{Q}) + \left(k_i^{-1} - 1\right)^2 E_i S_i E_i^* + Q_i F_i^* S_i^{-1} F_i Q_i \\
&= \mathcal{L}_i(\mathbf{Q}) + \left(E_i \Delta_i S_i^{1/2} - Q_i F_i^* S_i^{-1/2}\right)\left(S_i^{1/2} \Delta_i^* E_i^* - S_i^{-1/2} F_i Q_i\right) \\
&\quad + \operatorname{Her}(E_i \Delta_i F_i Q_i) \\
&\geq \operatorname{Her}\left\{(A_i + E_i \Delta_i F_i) Q_i\right\} + \sum_{j \in \mathcal{S}} \lambda_{ji} Q_j,
\end{aligned}
$$

once again ensuring the mean-square stability of system (9.24). □

Remark 9.17 Notice that, whenever the conditions of the preceding theorem are satisfied, we get the robust MSS of system (9.24) on a polytope such as (5.2), with $\ell = 2^N$ vertices of the form $\Pi_\varsigma^\kappa = K_\varsigma^\kappa \Pi$, where Π is the nominal transition rate matrix, and $\{K_\varsigma^\kappa\}$ are diagonal matrices whose nontrivial entries cover all possible 2^N-uples in $\{k_{\min}, k_{\max}\}$. For example:

$$
K_\varsigma^1 = \begin{bmatrix} k_{\min} & & 0 \\ & \ddots & \\ 0 & & k_{\min} \end{bmatrix},
$$

$$
K_\varsigma^2 = \begin{bmatrix} k_{\max} & & 0 \\ & \ddots & \\ 0 & & k_{\min} \end{bmatrix},
$$

$$
K_\varsigma^\ell = \begin{bmatrix} k_{\max} & & 0 \\ & \ddots & \\ 0 & & k_{\max} \end{bmatrix}.
$$

9.4 Robust Control

This section is concerned with the robust control of the system

$$\mathcal{G}_u = \begin{cases} \dot{x}(t) = A_{\theta(t)}x(t) + B_{\theta(t)}u(t) + J^{w}_{\theta(t)}w(t) + J^{\varpi}_{\theta(t)}\varpi(t), \\ z(t) = C_{\theta(t)}x(t) + D_{\theta(t)}u(t), \\ \zeta(t) = E_{\theta(t)}x(t) + F_{\theta(t)}u(t), \end{cases} \tag{9.32}$$

in which $w \mapsto z$ is a *performance* channel whose H_2 cost we desire to optimize, whereas $\varpi \mapsto \zeta$ is a *robustness* channel, in the sense that finite-gain uncertain perturbations of the form $\varpi(t) = \Delta_{\theta(t)}(\zeta(t))$ may affect (9.32). It is assumed that these finite-gain uncertain perturbations satisfy, for a prespecified $\alpha > 0$, the growth condition

$$\Delta_i(0) = 0, \qquad \big\|\Delta_i(\zeta) - \Delta_i(\hat{\zeta})\big\| \le \alpha\|\zeta - \hat{\zeta}\| \quad \forall\zeta, \hat{\zeta} \in \mathbb{R}^{p'},\ i \in \mathcal{S}. \tag{9.33}$$

In a more particular setup, the linear feedback

$$\varpi(t) = \Delta_{\theta(t)}\zeta(t), \quad \mathbf{\Delta} = (\Delta_1, \ldots, \Delta_N) \in \mathbb{H}^{p',r'}, \ \|\mathbf{\Delta}\|_{\max} < \alpha \tag{9.34}$$

will also be considered.

Our interest here is in the design of controllers of the form

$$u(t) = K_{\theta(t)}x(t) \tag{9.35}$$

such that, for given $\alpha > 0$ and $\beta > 0$, the closed-loop system

$$\mathcal{G}_K = \begin{cases} \dot{x}(t) = \widetilde{A}_{\theta(t)}x(t) + J^{w}_{\theta(t)}w(t) + J^{\varpi}_{\theta(t)}\varpi(t), \\ z(t) = \widetilde{C}_{\theta(t)}x(t), \\ \zeta(t) = \widetilde{E}_{\theta(t)}x(t), \end{cases} \tag{9.36}$$

with $\widetilde{A}_i := A_i + B_iK_i$, $\widetilde{C}_i := C_i + D_iK_i$, and $\widetilde{E}_i := E_i + F_iK_i$, is robustly mean-square stable in face of all disturbances of the form (9.34), while at the same time the performance $\|\mathcal{G}_K\|_2^2 < \beta$ is achieved. Finally, since Π may not be known a priori, later on we further assume that it belongs to the polytope (5.2), repeated below for convenience:

$$\mathbb{V} = \left\{\Pi;\ \Pi = \sum_{\kappa=1}^{\ell} \rho^{\kappa}\Pi^{\kappa};\ \rho^{\kappa} \ge 0,\ \sum_{\kappa=1}^{\ell}\rho^{\kappa} = 1\right\}. \tag{9.37}$$

9.4.1 Preliminary Results

Before proceeding further, we need to state some basic facts regarding the system

$$\mathcal{G} = \begin{cases} \dot{x}(t) = A_{\theta(t)}x(t) + J^{w}_{\theta(t)}w(t), \\ z(t) = C_{\theta(t)}x(t), \end{cases} \tag{9.38}$$

such as the following convex characterization of *upper bounds* on the H_2 norm, which is an easy consequence of the results stated in the generalized Gramians approach of Chap. 5.

Lemma 9.18 *Suppose that* (3.1) *is mean-square stable. Then, given* $\beta > 0$, *the inequality* $\|\mathcal{G}\|_2^2 < \beta$ *is satisfied whenever there is* $\mathbf{P} = (P_1, \dots, P_N) > 0$ *in* $\mathbb{H}^n$ *such that, for all* $i \in \mathcal{S}$,

$$\sum_{j \in \mathcal{S}} \nu_j \operatorname{tr}\big((J_j^w)^* P_j J_j^w\big) < \beta, \qquad \mathcal{T}_i(\mathbf{P}) + C_i^* C_i < 0, \tag{9.39}$$

or, equivalently, whenever there is $\mathbf{Q} = (Q_1, \dots, Q_N) > 0$ *in* $\mathbb{H}^n$ *such that, for all* $i \in \mathcal{S}$,

$$\sum_{j \in \mathcal{S}} \operatorname{tr}\big(C_j Q_j C_j^*\big) < \beta, \qquad \mathcal{L}_i(\mathbf{Q}) + \nu_i J_i^w \big(J_i^w\big)^* < 0. \tag{9.40}$$

Moreover, $\|\mathcal{G}\|_2^2 = \inf\{\beta,$ *subject to* (9.39)$\} = \inf\{\beta,$ *subject to* (9.40)$\}$.

Conversely, the robustness of the internal MSS of the system

$$\mathcal{G} = \begin{cases} \dot{x}(t) = A_{\theta(t)}x(t) + J^{\varpi}_{\theta(t)}\varpi(t), \\ \zeta(t) = E_{\theta(t)}x(t), \end{cases} \tag{9.41}$$

against feedback disturbances $\varpi(t) = \Delta_{\theta(t)}(\zeta(t))$ as in (9.33) or (9.34) may be guaranteed by the following lemma (its proof is presented at the end of this subsection).

Lemma 9.19 *Given* $\alpha > 0$, *the following assertions are true.*

(i) *System* (9.41) *is mean-square stable for any nonlinear disturbance of the form* (9.33) *whenever there are* $\mathbf{P} = (P_1, \dots, P_N) > 0$ *in* $\mathbb{H}^n$ *and* $\mathbf{s} = (s_1, \dots, s_N) > 0$ *in* $\mathbb{H}^1$ *such that*

$$\begin{bmatrix} \mathcal{T}_i(\mathbf{P}) + (1 \vee \alpha^2) s_i E_i^* E_i & P_i J_i^{\varpi} \\ (J_i^{\varpi})^* P_i & -(1 \vee \alpha^{-2}) s_i I_{r'} \end{bmatrix} < 0, \quad i \in \mathcal{S}, \tag{9.42}$$

with $(\cdot \vee \cdot)$ *standing for the maximum.*

(ii) *System* (9.41) *is mean-square stable for any linear disturbance of the form* (9.34) *whenever* (i) *is satisfied or, alternatively, whenever there are* $\mathbf{Q} = (Q_1, \dots, Q_N) > 0$ *in* $\mathbb{H}^n$ *and* $\mathbf{s} = (s_1, \dots, s_N) > 0$ *in* $\mathbb{H}^1$ *such that*

$$\begin{bmatrix} \mathcal{L}_i(\mathbf{Q}) + (1 \wedge \alpha^2) s_i J_i^{\varpi} (J_i^{\varpi})^* & Q_i E_i^* \\ E_i Q_i & -(1 \wedge \alpha^{-2}) s_i I_{p'} \end{bmatrix} < 0, \quad i \in \mathcal{S}, \tag{9.43}$$

with $(\cdot \wedge \cdot)$ *standing for the minimum.*

The proof of Lemma 9.19 relies on exchanging (9.41) by the *scaled* input–output system

$$\begin{cases} \dot{x}(t) = A_{\theta(t)} x(t) + \widehat{J}^{\varpi}_{\theta(t)} \varpi_s(t), \quad x(0) = 0, \\ \zeta_s(t) = \widehat{E}_{\theta(t)} x(t), \end{cases} \tag{9.44}$$

with $\widehat{J}_i^{\varpi} := s_i^{-1/2} J_i^{\varpi}$, $\widehat{E}_i := s_i^{1/2} E_i$, and $\widehat{\Delta}_i(\cdot) \equiv s_i^{1/2} \Delta_i(s_i^{-1/2}(\cdot))$, from which it is straightforward to check that (9.33) (and therefore (9.34) as well) also holds with $\boldsymbol{\Delta}$ replaced by $\widehat{\boldsymbol{\Delta}}$. Additionally, the following result will be necessary (the proof follows the same lines as that of Proposition 8.3 and thus will be omitted).

Proposition 9.20 *Given* $\mathbf{P} = (P_1, \dots, P_N) > 0$ *in* $\mathbb{H}^n$ *and* x *satisfying* (9.44), *let* $V(t) = E(\langle x(t); P_{\theta(t)} x(t)\rangle)$. *Then, for any* $T > 0$,

$$V(T) = V(0) + \int_0^T E\big(\langle \mathbf{x}(t); \mathfrak{Z}_{\theta(t)}(\mathbf{P}) \mathbf{x}(t)\rangle\big)\, dt$$

with

$$\mathbf{x}(t) := \begin{bmatrix} x(t) \\ \widehat{\Delta}_{\theta(t)}(s_{\theta(t)}^{1/2} E_{\theta(t)} x(t)) \end{bmatrix}, \qquad \mathfrak{Z}_{\theta(t)}(\mathbf{P}) := \begin{bmatrix} \mathcal{T}_{\theta(t)}(\mathbf{P}) & * \\ s_{\theta(t)}^{-1/2} (J^{\varpi}_{\theta(t)})^* P_{\theta(t)} & 0 \end{bmatrix}.$$

Proof of Lemma 9.19 For easiness of notation, let $\hat{\alpha} := (1 \vee \alpha^2)$ and $\check{\alpha} := (1 \vee \alpha^{-2})$. Then, since $\hat{\alpha} = \check{\alpha}\alpha^2$ for any $\alpha > 0$, we get

$$\begin{aligned} 0 &= (\hat{\alpha} - \check{\alpha}\alpha^2) \int_0^T E\big[\big\| s_{\theta(t)}^{1/2} E_{\theta(t)} x(t)\big\|^2\big]\, dt \\ &\le \int_0^T E\big[\hat{\alpha} \big\| s_{\theta(t)}^{1/2} E_{\theta(t)} x(t)\big\|^2 - \check{\alpha} \big\| \widehat{\Delta}_{\theta(t)}\big(s_{\theta(t)}^{1/2} E_{\theta(t)} x(t)\big)\big\|^2\big]\, dt \\ &= \int_0^T E\left(\left\langle \mathbf{x}(t); \begin{bmatrix} \hat{\alpha} s_{\theta(t)} E^*_{\theta(t)} E_{\theta(t)} & 0 \\ 0 & -\check{\alpha} I \end{bmatrix} \mathbf{x}(t) \right\rangle\right) dt, \end{aligned}$$

which, by Proposition 9.20, guarantees that there is $\delta > 0$ such that

$$
\begin{aligned}
0 \leq\ & E\big(\langle x(0); P_{\theta(0)} x(0)\rangle\big) - E\big(\langle x(T); P_{\theta(T)} x(T)\rangle\big) \\
& + \int_0^T E\left(\left\langle \mathbf{x}(t); \begin{bmatrix} \mathcal{T}_{\theta(t)}(\mathbf{P}) + \hat{\alpha} s_{\theta(t)} E^*_{\theta(t)} E_{\theta(t)} & s_{\theta(t)}^{-1/2} P_{\theta(t)} J^{\varpi}_{\theta(t)} \\ s_{\theta(t)}^{-1/2} (J^{\varpi}_{\theta(t)})^* P_{\theta(t)} & -\check{\alpha} I \end{bmatrix} \mathbf{x}(t) \right\rangle\right) dt \\
\leq\ & \|\mathbf{P}\|_{\max} E\big[\|x(0)\|^2\big] + 0 + \delta \int_0^T E\big[\|x(t)\|^2\big]\, dt
\end{aligned}
$$

for all $T > 0$, by the hypothesis in (i). Thus, stochastic stability of the uncertain system follows immediately:

$$
\lim_{T\to\infty} \int_0^T E\big[\|x(t)\|^2\big]\, dt \leq \frac{1}{\delta} \|\mathbf{P}\|_{\max} E\big[\|x(0)\|^2\big] < \infty, \tag{9.45}
$$

from which MSS is guaranteed as well (see Theorem 3.15 in page 44), yielding the validity of (i) and, in the case of linear disturbances, the first part of (ii) as well (since (9.34) is a particular case of (9.33)).

The second part of (ii) is proven as follows. Set

$$
\mathfrak{H}_i := (1 \vee \alpha^{-1}) s_i^{-1/2} \Delta_i E_i Q_i - (1 \wedge \alpha) s_i^{1/2} (J_i^{\varpi})^*,
$$

which yields

$$
\begin{aligned}
0 \leq \mathfrak{H}_i^* \mathfrak{H}_i &= (1 \vee \alpha^{-1})^2 s_i^{-1} Q_i E_i^* \Delta_i^* \Delta_i E_i Q_i + (1 \wedge \alpha)^2 s_i J_i^{\varpi} (J_i^{\varpi})^* \\
&\quad - \operatorname{Her}\big((1 \vee \alpha^{-1})(1 \wedge \alpha) J_i^{\varpi} \Delta_i E_i Q_i\big) \\
&\leq \alpha^2 (1 \vee \alpha^{-2}) s_i^{-1} Q_i E_i^* E_i Q_i + (1 \wedge \alpha^2) s_i J_i^{\varpi} (J_i^{\varpi})^* \\
&\quad - \operatorname{Her}\big(J_i^{\varpi} \Delta_i E_i Q_i\big) \\
&= (1 \vee \alpha^2) s_i^{-1} Q_i E_i^* E_i Q_i + (1 \wedge \alpha^2) s_i J_i^{\varpi} (J_i^{\varpi})^* - \operatorname{Her}\big(J_i^{\varpi} \Delta_i E_i Q_i\big),
\end{aligned}
$$

because $\Delta_i^* \Delta_i \leq \|\Delta_i\|^2 I < \alpha^2 I$. Thus, we easily obtain from Schur's complement (Lemma 2.26) and (9.43) that

$$
\begin{aligned}
0 &> \mathcal{L}_i(\mathbf{Q}) + (1 \wedge \alpha^2) s_i J_i^{\varpi} (J_i^{\varpi})^* + \big((1 \wedge \alpha^{-2}) s_i\big)^{-1} Q_i E_i^* E_i Q_i \\
&\geq \operatorname{Her}\big((A_i + J_i^{\varpi} \Delta_i E_i) Q_i\big) + \sum_{j\in\mathcal{S}} \lambda_{ji} Q_j,
\end{aligned}
$$

so that, bearing in mind Theorem 3.21 (page 48), internal MSS is guaranteed whenever $\|\boldsymbol{\Delta}\|_{\max} < \alpha$. □

9.4.2 Robust H_2 Control

Substituting (9.35) into (9.32), we obtain the closed-loop system (9.36) with $\widetilde{A}_i := A_i + B_i K_i$, $\widetilde{C}_i := C_i + D_i K_i$, and $\widetilde{E}_i := E_i + F_i K_i$. In this setting, the following notion of *robust stability* against perturbations such as (9.33) or (9.34), with uncertain transition rates, will be necessary.

Definition 9.21 Given $\alpha > 0$, system (9.36) is said to be robustly mean-square stable in the internal sense whenever

$$\lim_{t\to\infty} E\big[\|x(t)\|^2\big] = 0 \quad \text{for an arbitrary initial condition } \vartheta_0 \tag{9.46}$$

is satisfied, regardless of $\varpi(t) = \Delta_{\theta(t)}(\zeta(t))$ as in (9.33) (or (9.34), alternatively) and of $\Pi = [\lambda_{ij}] \in \mathbb{V}$ as in (9.37).

Introducing now the definition

$$\mathcal{R}_i^\kappa(\mathbf{Y}) := \Big[(\lambda_{i1}^\kappa)^{1/2} Y_i \quad \dots \quad (\lambda_{i(i-1)}^\kappa)^{1/2} Y_i \quad (\lambda_{i(i+1)}^\kappa)^{1/2} Y_i \quad \dots \quad (\lambda_{iN}^\kappa)^{1/2} Y_i\Big],$$

we are able to state the main result of this subsection, in terms of the feasibility of the following set of LMIs:

$$\sum_{j\in\mathcal{S}} \nu_j \operatorname{tr}(T_j) < \beta, \qquad \begin{bmatrix} T_i & (J_i^w)^* \\ J_i^w & Y_i \end{bmatrix} > 0, \tag{9.47a}$$

$$\begin{bmatrix} \Phi_i^\kappa(\mathbf{Y},\mathbf{V}) & (C_i Y_i + D_i V_i)^* & \mathcal{R}_i^\kappa(\mathbf{Y}) \\ C_i Y_i + D_i V_i & -I & 0 \\ \mathcal{R}_i^\kappa(\mathbf{Y})^* & 0 & -\mathcal{D}_i(\mathbf{Y}) \end{bmatrix} < 0, \tag{9.47b}$$

$$\begin{bmatrix} \Phi_i^\kappa(\mathbf{Y},\mathbf{V}) + (1\wedge\alpha^2) z_i^\kappa J_i^\varpi (J_i^\varpi)^* & (E_i Y_i + F_i V_i)^* & \mathcal{R}_i^\kappa(\mathbf{Y}) \\ E_i Y_i + F_i V_i & -(1\wedge\alpha^{-2}) z_i^\kappa I & 0 \\ \mathcal{R}_i^\kappa(\mathbf{Y})^* & 0 & -\mathcal{D}_i(\mathbf{Y}) \end{bmatrix} < 0, \tag{9.47c}$$

with

$$\Phi_i^\kappa(\mathbf{Y},\mathbf{V}) := \operatorname{Her}(A_i Y_i + B_i V_i) + \lambda_{ii}^\kappa Y_i, \tag{9.47d}$$

or, in the case of linear feedback disturbances, by the following ones:

$$\sum_{j\in\mathcal{S}} \operatorname{tr}(T_j) < \beta, \qquad \begin{bmatrix} T_i & C_i Y_i + D_i V_i \\ (C_i Y_i + D_i V_i)^* & Y_i \end{bmatrix} > 0, \tag{9.48a}$$

$$\Psi_i^\kappa(\mathbf{Y},\mathbf{V}) + \nu_i J_i^w (J_i^w)^* < 0, \tag{9.48b}$$

$$\begin{bmatrix} \Psi_i^\kappa(\mathbf{Y},\mathbf{V}) + (1\wedge\alpha^2) s_i^\kappa J_i^\varpi (J_i^\varpi)^* & (E_i Y_i + F_i V_i)^* \\ E_i Y_i + F_i V_i & -(1\wedge\alpha^{-2}) s_i^\kappa I \end{bmatrix} < 0, \tag{9.48c}$$

with

$$\Psi_i^\kappa(\mathbf{Y},\mathbf{V}) := \operatorname{Her}(A_i Y_i + B_i V_i) + \sum_{j\in\mathcal{S}} \lambda_{ji}^\kappa Y_j. \tag{9.48d}$$

Theorem 9.22 *Given $\alpha > 0$ and $\beta > 0$, there is a controller such as* (9.35) *which renders system* (9.36) *robustly mean-square stable in face of all nonlinear disturbances of the form* (9.33), *with $\|\mathcal{G}_K\|_2^2 < \beta$ for whatever transition rates $\Pi = [\lambda_{ij}]$ in $\mathbb{V}$, whenever there are $\mathbf{T} = (T_1, \dots, T_N) \in \mathbb{H}^r$, $\mathbf{Y} = (Y_1, \dots, Y_N) \in \mathbb{H}^n$, and $\mathbf{V} = (V_1, \dots, V_N) \in \mathbb{H}^{n,m}$, together with, for each $\kappa = 1, \dots, \ell$, scalings $\mathbf{s}^\kappa = (s_1^\kappa, \dots, s_N^\kappa)$ and $\mathbf{z}^\kappa = (z_1^\kappa, \dots, z_N^\kappa)$ in $\mathbb{H}^1$, such that the following is true*:

(i) *The LMIs* (9.47a)–(9.47c) *are satisfied for all $i \in \mathcal{S}$ and $\kappa \in \{1, \dots, \ell\}$.*

Furthermore, in the scenario of linear perturbations of the form (9.34), *the above also holds whenever the following assertion is true for some $\mathbf{T} \in \mathbb{H}^p$*:

(ii) *The LMIs* (9.48a)–(9.48c) *are satisfied for all $i \in \mathcal{S}$ and $\kappa \in \{1, \dots, \ell\}$.*

Moreover, in either case one such controller is given by

$$u(t) = V_{\theta(t)} Y_{\theta(t)}^{-1} x(t). \tag{9.49}$$

Proof If (i) is true, then, from Schur's complement (Lemma 2.26) we get from (9.47a) that $\beta > \sum_{j \in \mathcal{S}} \nu_j \operatorname{tr}(T_j) > \sum_{j \in \mathcal{S}} \nu_j \operatorname{tr}((J_j^w)^* Y_j^{-1} J_j^w)$, together with (after multiplying both sides of (9.47b) by ρ^κ and summing for $\kappa = 1, \dots, \ell$)

$$Y_i \left\{ \operatorname{Her}\big(Y_i^{-1} \widetilde{A}_i\big) + \sum_{j \in \mathcal{S}} \lambda_{ij} Y_j^{-1} + \widetilde{C}_i^* \widetilde{C}_i \right\} Y_i < 0$$

for $K_i := V_i Y_i^{-1}$, $\widetilde{A}_i = A_i + B_i K_i$, and $\widetilde{C}_i = C_i + D_i K_i$. Hence, after applying $Y_i^{-1}\{\cdot\}Y_i^{-1}$ to the preceding inequality, we get that the closed-loop data satisfies (9.39) with $\mathbf{P} = \mathbf{Y}^{-1}$.

Similarly, (9.47c) allows us to arrive at

$$\begin{aligned} Y_i \Bigg\{ & \operatorname{Her}\big(Y_i^{-1} \widetilde{A}_i\big) + \sum_{j \in \mathcal{S}} \lambda_{ij} Y_j^{-1} \\ & + \big(1 \vee \alpha^2\big) s_i \widetilde{E}_i^* \widetilde{E}_i + \big(\big(1 \vee \alpha^{-2}\big) s_i\big)^{-1} Y_i^{-1} J_i^\varpi \big(J_i^\varpi\big)^* Y_i^{-1} \Bigg\} Y_i < 0 \end{aligned}$$

with $s_i := (\sum_{\kappa=1}^{\ell} \rho^\kappa z_i^\kappa)^{-1}$, from which (9.42) is recovered.

Conversely, assume that (ii) is true, and let $K_i := V_i Y_i^{-1}$, $\widetilde{A}_i = A_i + B_i K_i$, $\widetilde{C}_i = C_i + D_i K_i$, $\widetilde{E}_i = E_i + F_i K_i$, and $s_i := \sum_{\kappa=1}^{\ell} \rho^\kappa s_i^\kappa$. Then the proof follows by rewriting (9.48a)–(9.48c) in the form

$$\beta > \sum_{j \in \mathcal{S}} \operatorname{tr}\big((C_i Y_i + D_i V_i) Y_i^{-1} (C_i Y_i + D_i V_i)^*\big) = \sum_{j \in \mathcal{S}} \operatorname{tr}\big(\widetilde{C}_i Y_i \widetilde{C}_i^*\big), \tag{9.50a}$$

$$\operatorname{Her}(\widetilde{A}_i Y_i) + \sum_{j \in \mathcal{S}} \lambda_{ji} Y_j + \nu_i J_i^w \big(J_i^w\big)^* < 0, \tag{9.50b}$$

$$\begin{bmatrix} \mathrm{Her}(\widetilde{A}_i Y_i) + \sum_{j\in\mathcal{S}} \lambda_{ji} Y_j + (1 \wedge \alpha^2) s_i J_i^{\varpi} (J_i^{\varpi})^* & (\widetilde{E}_i Y_i)^* \\ \widetilde{E}_i Y_i & -(1 \wedge \alpha^{-2}) s_i I \end{bmatrix} < 0, \tag{9.50c}$$

and then invoking Lemmas 9.18 and 9.19. □

9.4.3 The Equalized Case

In this section we shall assume that $J_i^w = J_i^{\varpi} := J_i$, along with $C_i = E_i$ and $D_i = F_i$ for all $i \in \mathcal{S}$, so that $z \equiv \zeta$. In this case, (9.36) reduces to

$$\mathcal{G}_K = \begin{cases} \dot{x}(t) = \widetilde{A}_{\theta(t)} x(t) + J_{\theta(t)} w(t), \\ z(t) = \widetilde{C}_{\theta(t)} x(t), \end{cases} \tag{9.51}$$

and thus we have the following refinement of Theorem 9.22.

Theorem 9.23 *Given $\alpha > 0$ and $\beta > 0$, there is a controller such as* (9.35) *which renders system* (9.51) *robustly mean-square stable in face of all nonlinear disturbances of the form* (9.33), *with $\|\mathcal{G}_K\|_2^2 < \beta$ for whatever transition rates $\Pi = [\lambda_{ij}]$ in $\mathbb{V}$, whenever there are $\mathbf{T} = (T_1, \ldots, T_N) \in \mathbb{H}^r$, $\mathbf{Y} = (Y_1, \ldots, Y_N) \in \mathbb{H}^n$, and $\mathbf{V} = (V_1, \ldots, V_N) \in \mathbb{H}^{n,m}$, together with, for each $\kappa = 1, \ldots, \ell$, scalings $\mathbf{s}^\kappa = (s_1^\kappa, \ldots, s_N^\kappa)$ and $\mathbf{z}^\kappa = (z_1^\kappa, \ldots, z_N^\kappa)$ in $\mathbb{H}^1$, such that at least one of the following assertions holds (with $C_i \equiv E_i$ and $D_i \equiv F_i$).*

(i) *The LMIs* (9.47a)–(9.47c) *are satisfied for all $i \in \mathcal{S}$ and $\kappa \in \{1, \ldots, \ell\}$.*
(ii) *The LMIs* (9.47a) *and* (9.47c), *together with $z_i^\kappa \le (1 \vee \alpha^2)$, are satisfied for all $i \in \mathcal{S}$ and $\kappa \in \{1, \ldots, \ell\}$.*

Furthermore, in the scenario of linear perturbations of the form (9.34), *the above is also true whenever at least one of the following assertions holds.*

(iii) *The LMIs* (9.48a)–(9.48c) *are satisfied for all $i \in \mathcal{S}$ and $\kappa \in \{1, \ldots, \ell\}$.*
(iv) *The LMIs* (9.48a) *and* (9.48c), *together with $(1 \wedge \alpha^2) s_i^\kappa \ge \nu_i$, are satisfied for all $i \in \mathcal{S}$ and $\kappa \in \{1, \ldots, \ell\}$.*

Moreover, in either case one such controller is given by (9.49).

Proof Since (i) and (iii) are merely a restatement of Theorem 9.22, it remains only to prove the sufficiency of (ii) and (iv). The former assertion is proven by noticing that, from Schur's complement (Lemma 2.26) (9.47c) implies (with $z_i := \sum_{\kappa=1}^{\ell} \rho^\kappa z_i^\kappa$ and $s_i := z_i^{-1}$)

$$
\begin{aligned}
0 &> \operatorname{Her}(A_i Y_i + B_i V_i) + \sum_{j\in\mathcal{S}} \lambda_{ij} Y_i Y_j^{-1} Y_i + \left(1 \wedge \alpha^2\right) z_i J_i J_i^* \\
&\quad + \left(1 \wedge \alpha^{-2}\right)^{-1} s_i (C_i Y_i + D_i V_i)^* (C_i Y_i + D_i V_i) \\
&= Y_i \left\{ \operatorname{Her}\left(Y_i^{-1} \widehat{A}_i\right) + \sum_{j\in\mathcal{S}} \lambda_{ij} Y_j^{-1} + \left(1 \vee \alpha^2\right) s_i \widetilde{C}_i^* \widetilde{C}_i \right. \\
&\quad \left. + \left(\left(1 \vee \alpha^{-2}\right) s_i\right)^{-1} Y_i^{-1} J_i J_i^* Y_i^{-1} \right\} Y_i \\
&\geq Y_i \left\{ \operatorname{Her}\left(Y_i^{-1} \widehat{A}_i\right) + \sum_{j\in\mathcal{S}} \lambda_{ij} Y_j^{-1} + \widetilde{C}_i^* \widetilde{C}_i \right\} Y_i,
\end{aligned}
$$

since $(1 \vee \alpha^2)s_i = (1 \vee \alpha^2)(\sum_{\kappa=1}^{\ell} \rho^\kappa z_i^\kappa)^{-1} \geq (1 \vee \alpha^2)(1 \wedge \alpha^{-2}) = 1$. Hence, letting $P_i := Y_i^{-1}$ and $\widetilde{T}_i(\mathbf{P}) := \widetilde{A}_i^* P_i + P_i \widetilde{A}_i + \sum_{j\in\mathcal{S}} \lambda_{ij} P_j$, an application of $P_i(\cdot)P_i$ to the preceding inequalities promptly yields the fulfillment of

$$
\sum_{j\in\mathcal{S}} \nu_j \operatorname{tr}\left(J_j^* P_j J_j\right) < \beta, \qquad \widetilde{T}_i(\mathbf{P}) + \widetilde{C}_i^* \widetilde{C}_i < 0, \tag{9.52}
$$

and, from Schur's complement,

$$
\begin{bmatrix} \widetilde{T}_i(\mathbf{P}) + (1 \vee \alpha^2) s_i \widetilde{C}_i^* \widetilde{C}_i & P_i J_i \\ J_i^* P_i & -(1 \vee \alpha^{-2}) s_i I \end{bmatrix} < 0, \tag{9.53}
$$

and therefore the result follows from Lemmas 9.18 and 9.19. Following similar lines, the sufficiency of (iv) is proven by noticing that (9.50b) is easily obtained from (9.50c) whenever $(1 \wedge \alpha^2) s_i^\kappa \geq \nu_i$. □

Remark 9.24 It is promptly noticed that conditions (ii) and (iv) in the preceding theorem are more restrictive (and hence more conservative) than (i) and (iii), respectively, because (ii) ⇒ (i) and (iv) ⇒ (iii). However, the advantage of (ii) and (iv) lies in the fact that less LMIs need to be solved, making them more appealing from the computational point of view. □

This section finishes with the following brief discussion of the mixed H_2/H_∞ control problem.

9.4.4 Robust Mixed H_2/H_∞ Control

A standard approach to the design of controllers with simultaneous optimality and robustness guarantees is well known to stem from the combination of H_2 and H_∞ design methods. Our objective in this last part of the section is to show that the usual convex approach to mixed H_2/H_∞ control may be recovered in a particular instance of Theorem 9.23 if scaling is not employed (i.e., $z_i^\kappa \equiv 1$ for all $i \in \mathcal{S}$ and $\kappa \in \{1, \dots, \ell\}$).

Corollary 9.25 *Suppose that $z_i^\kappa \equiv 1$ for all $i \in \mathcal{S}$ and $\kappa \in \{1, \dots, \ell\}$. Then, as long as either* (i) *or* (ii) *is satisfied in Theorem* 9.23, *the controller* (9.49) *renders system* (9.36) *internally MSS with $\|\mathcal{G}_K\|_2 < \beta^{1/2}$ and $\|\mathcal{G}_K\|_\infty < \alpha^{-1}$, regardless of $\Pi = [\lambda_{ij}]$ in $\mathbb{V}$.*

Proof Since internal MSS with $\|\mathcal{G}_K\|_2^2 < \beta$ follows from the very statement of Theorem 9.23, it remains only to notice that, with $K_i = V_i Y_i^{-1}$, $\widetilde{A}_i = A_i + B_i K_i$, and $\widetilde{C}_i = C_i + D_i K_i$, the feasibility of (9.47c) yields, from Schur's complement (Lemma 2.26),

$$Y_i \left\{ \operatorname{Her}\bigl(Y_i^{-1}\widetilde{A}_i\bigr) + \sum_{j\in\mathcal{S}} \lambda_{ij} Y_j^{-1} + \bigl(1 \wedge \alpha^2\bigr) Y_i^{-1} J_i J_i^* Y_i^{-1} + \bigl(1 \wedge \alpha^{-2}\bigr)^{-1} \widetilde{C}_i^* \widetilde{C}_i \right\} Y_i < 0$$

or, equivalently,

$$\widetilde{\Upsilon}_i := \begin{bmatrix} \operatorname{Her}(Y_i^{-1}\widetilde{A}_i) + \sum_{j\in\mathcal{S}} \lambda_{ij} Y_j^{-1} + (1 \wedge \alpha^{-2})^{-1} \widetilde{C}_i^* \widetilde{C}_i & Y_i^{-1} J_i \\ J_i^* Y_i^{-1} & -(1 \wedge \alpha^2)^{-1} I \end{bmatrix} < 0,$$

so that

$$0 > \bigl(1 \wedge \alpha^{-2}\bigr)\widetilde{\Upsilon}_i = \begin{bmatrix} \operatorname{Her}(X_i \widetilde{A}_i) + \sum_{j\in\mathcal{S}} \lambda_{ij} X_j + \widetilde{C}_i^* \widetilde{C}_i & X_i J_i \\ J_i^* X_i & -\gamma^2 I \end{bmatrix}$$

with $X_i := (1 \wedge \alpha^{-2}) Y_i^{-1}$ and $\gamma := \alpha^{-1}$. Hence, the result follows from the bounded real lemma (Lemma 8.2, p. 154). □

9.5 Robust Linear Filtering Problem via an LMIs Formulation

In this section we formulate the filter problem defined in Sect. 7.6.1 as an *LMIs optimization problem*. In addition, we present the *robust version* of the filter based on this LMIs formulation. Furthermore, it is shown in Sect. 9.5.3 how approximations of a certain algebraic Riccati equation can yield a solution of the LMIs problem. We illustrate these results with some examples in Sect. 10.5.

In this section we suppose that Assumption 7.22 holds, and we recall from Definition 7.23 that the stationary filtering problem consists of finding (A_f, B_f, L_f) such that A_f is stable and minimizes $\operatorname{tr}(\widetilde{L} P \widetilde{L}^*)$, where P is a unique solution of (7.72), with Υ_i defined as in (7.60) and $\widetilde{F}$, $\widetilde{L}$, $\widetilde{J}$ defined as in (7.53), (7.54). Notice that $\widetilde{F}$, $\widetilde{L}$, $\widetilde{J}$ depend on (A_f, B_f, L_f). We set $\mathcal{J}^* := \operatorname{tr}(\widetilde{L} P \widetilde{L}^*)$.

9.5.1 The LMIs Formulation

We start by making the following definition.

Definition 9.26 We say that $(P, W, A_f, B_f, L_f) \in \mathcal{R}_1$ if it satisfies

$$P = \begin{bmatrix} Z & U \\ U^* & \widehat{Z} \end{bmatrix} > 0, \quad Z = \operatorname{diag}(Q_i), \tag{9.54}$$

$$\begin{bmatrix} P & P\widetilde{L}^* \\ \widetilde{L}P & W \end{bmatrix} > 0, \tag{9.55}$$

$$\begin{bmatrix} \widetilde{F}P + P\widetilde{F}^* & \Upsilon_1 \operatorname{dg}(Q_1) & \dots & \Upsilon_N \operatorname{dg}(Q_N) & \widetilde{J} \\ \operatorname{dg}(Q_1)\Upsilon_1^* & -\operatorname{dg}(Q_1) & \dots & 0 & 0 \\ \vdots & \vdots & \ddots & \vdots & \vdots \\ \operatorname{dg}(Q_N)\Upsilon_N^* & 0 & \dots & -\operatorname{dg}(Q_N) & 0 \\ \widetilde{J}^* & 0 & \dots & 0 & -I \end{bmatrix} < 0. \tag{9.56}$$

From the results of Sect. 7.6 we have the following result.

Proposition 9.27 *We have that*

$$\inf\{\operatorname{tr}(W); (P, W, A_f, B_f, L_f) \in \mathcal{R}_1\} = \mathcal{J}^*. \tag{9.57}$$

Proof As seen in Sect. 7.6.2, there exists an optimal solution $(A_{f,\text{op}}, B_{f,\text{op}}, L_{f,\text{op}})$ for the stationary filtering problem posed in Definition 7.23, and P_{op} is as in (7.76) satisfying (7.72). Let us show first that any feasible solution $(\bar{P}, \bar{W}, A_f, B_f, L_f)$ satisfying (9.54), (9.55), and (9.56) will be such that $\operatorname{tr}(\bar{W}) > \operatorname{tr}(L\widetilde{Z}L^*)$, where $\widetilde{Z}$ is a unique positive semi-definite solution of the algebraic Riccati equation (7.38). From Schur's complement (Lemma 2.26) we have that (9.55) and (9.56) are equivalent to $\bar{W} > \widetilde{L}\bar{P}\widetilde{L}^*$ and

$$\widetilde{F}\bar{P} + \bar{P}\widetilde{F}^* + \sum_{\ell\in\mathcal{S}} \Upsilon_\ell \operatorname{dg}(\bar{Q}_\ell)\Upsilon_\ell^* + \widetilde{J}\widetilde{J}^* < 0.$$

From Proposition 7.20 we have that system (7.51) is MSS, and, for $P(t)$ as in (7.52), we have from Proposition 7.21 that $P(t) \to P$ as $t \to \infty$, where P is a unique solution of (7.72), and (7.70) holds. Furthermore from Proposition 7.21 we have that $\bar{P} > P$, and from (7.75) we get that

$$\operatorname{tr}(\bar{W}) > \operatorname{tr}(\widetilde{L}\bar{P}\widetilde{L}^*) \ge \operatorname{tr}(\widetilde{L}P\widetilde{L}^*) \ge \operatorname{tr}(L\widetilde{Z}L^*).$$

Let us show now that we have a sequence $(P_\epsilon, W_\epsilon, A_f, B_f, L_f)$ satisfying (9.54), (9.55), and (9.56) and converging to the optimal solution of the minimization problem posed in (9.57) as $\epsilon \downarrow 0$. Fix $A_f = A_{f,\text{op}}$, $B_f = B_{f,\text{op}}$, $L_f = L$, and

$$P_\epsilon = \begin{bmatrix} Z_\epsilon & U_\epsilon \\ U_\epsilon^* & \widehat{Z}_\epsilon \end{bmatrix}, \quad Z_\epsilon = \operatorname{diag}(Q_{\epsilon,i}),$$

the unique solution of $\widetilde{F}P_\epsilon + P_\epsilon\widetilde{F}^* + \sum_{\ell\in\mathcal{S}}\Upsilon_\ell \operatorname{dg}(Q_{\epsilon,\ell})\Upsilon_\ell^* + \widetilde{J}\widetilde{J}^* + \epsilon I = 0$, and $W_\epsilon = \widetilde{L}P_\epsilon\widetilde{L}^* + \epsilon I$. Then, by Proposition 7.21, $P_\epsilon \downarrow P_{\text{op}}$ as $\epsilon \downarrow 0$, $P_\epsilon > 0$ for each $\epsilon > 0$, and clearly $W_\epsilon \downarrow \widetilde{L}P_{\text{op}}\widetilde{L}^*$ as $\epsilon \downarrow 0$. Since

$$0 > \widetilde{F}P_\epsilon + P_\epsilon\widetilde{F}^* + \sum_{\ell\in\mathcal{S}}\Upsilon_\ell \operatorname{dg}(\bar{Q}_{\epsilon,\ell})\Upsilon_\ell^* + \widetilde{J}\widetilde{J}^*,$$

we have from Schur's complement that (9.54), (9.55), and (9.56) are satisfied. Taking the limit as $\epsilon \downarrow 0$, we obtain that (9.57) holds. □

We consider from now on that $n_f = Nn$. The next theorem rewrites the above problem as an LMIs optimization problem. First, we need the following definition (in what follows, we recall the following definitions: L as in (7.45), F as in (7.5), H as in (7.6), Ψ_i as in (7.59), J^p as in (7.39), and G^p as in (7.40)).

Definition 9.28 We say that $(X_i, i = 1, \dots, N, Y, W, R, S, J) \in \mathcal{R}_2$ if it satisfies the following LMIs:

$$X = \operatorname{diag}(X_i), \tag{9.58}$$

$$\begin{bmatrix} X & X & L^* - J^* \\ X & Y & L^* \\ L - J & L & W \end{bmatrix} > 0, \tag{9.59}$$

$$\begin{bmatrix} \mathrm{P}(X, Y, S, R) & \mathrm{S}(X, Y) & \mathrm{T}(X, Y) \\ \mathrm{S}(X, Y)^* & -\mathrm{D}(X) & 0 \\ \mathrm{T}(X, Y)^* & 0 & -I \end{bmatrix} < 0, \tag{9.60}$$

where

$$\mathrm{P}(X, Y, S, R) = \begin{bmatrix} XF + F^*X & XF + F^*Y + H^*S^* + R^* \\ F^*X + YF + SH + R & F^*Y + YF + SH + H^*S^* \end{bmatrix},$$

$$\mathrm{S}(X, Y) = \begin{bmatrix} X\Psi_1 & \dots & X\Psi_N \\ Y\Psi_1 & \dots & Y\Psi_N \end{bmatrix},$$

$$\mathrm{T}(X, Y) = \begin{bmatrix} XJ^p \\ YJ^p + SG^p \end{bmatrix},$$

$$\mathrm{D}(X) = \begin{bmatrix} \operatorname{dg}(X_1) & \dots & 0 \\ \vdots & \ddots & \vdots \\ 0 & \dots & \operatorname{dg}(X_N) \end{bmatrix}.$$

Theorem 9.29 *There exists $(P, W, A_f, B_f, L_f) \in \mathcal{R}_1$ if and only if there exists $(X_i, i = 1, \dots, N, Y, W, R, S, J) \in \mathcal{R}_2$. Moreover, if $(X_i, i = 1, \dots, N, Y, W, R, S, J) \in \mathcal{R}_2$, then by choosing an arbitrary nonsingular $(Nn \times Nn)$ matrix U, setting P as in* (9.54) *with*

$$Z = X^{-1} = \operatorname{diag}(X_i^{-1}) = \operatorname{diag}(Q_i), \tag{9.61}$$

$$\widehat{Z} = U^*\big(X^{-1} - Y^{-1}\big)^{-1}U \tag{9.62}$$

(*notice that, by* (9.59), $X > XY^{-1}X \Rightarrow X^{-1} > Y^{-1}$ *and therefore* $X^{-1} - Y^{-1} > 0$), *and by taking*

$$V = \big(I - YX^{-1}\big)\big(U^*\big)^{-1}, \tag{9.63}$$

$$A_f = V^{-1}R\big(U^*X\big)^{-1}, \tag{9.64}$$

$$B_f = V^{-1}S, \tag{9.65}$$

$$L_f = J\big(U^*X\big)^{-1}, \tag{9.66}$$

we have $(P, W, A_f, B_f, L_f) \in \mathcal{R}_1$. *Furthermore,*

$$\inf\big\{\mathrm{tr}(W); (X_i, i = 1, \dots, N, Y, W, R, S, J) \in \mathcal{R}_2\big\} = \mathcal{J}^*. \tag{9.67}$$

Proof For (A_f, B_f, L_f) fixed, consider P, W satisfying (9.54)–(9.56). Without loss of generality, suppose further that U is nonsingular (if not, redefine U as $U + \epsilon I$ so that it is nonsingular and $\epsilon > 0$ is small enough so that (9.54)–(9.56) still hold). As in [166], define

$$P^{-1} = \begin{bmatrix} Y & V \\ V^* & \widehat{Y} \end{bmatrix} > 0,$$

where $Y > 0$ and $\widehat{Y} > 0$ are $Nn \times Nn$. We have that

$$ZY + UV^* = I, \tag{9.68}$$

$$U^*Y + \widehat{Z}V^* = 0, \tag{9.69}$$

and thus $Y^{-1} = Z + UV^*Y^{-1} = Z - U\widehat{Z}^{-1}U^* < Z$, $V^* = U^{-1} - U^{-1}ZY = U^{-1}(Y^{-1} - Z)Y$, implying that V is nonsingular. Define the nonsingular $2Nn \times 2Nn$ matrix

$$T = \begin{bmatrix} Z^{-1} & Y \\ 0 & V^* \end{bmatrix}$$

and the non-singular matrices T_1, T_2 as follows:

$$T_1 = \begin{bmatrix} T & 0 \\ 0 & I \end{bmatrix}, \qquad T_2 = \begin{bmatrix} T & 0 & 0 & 0 & 0 \\ 0 & \mathrm{dg}(Q_1^{-1}) & 0 & 0 & 0 \\ 0 & 0 & \ddots & 0 & 0 \\ 0 & 0 & 0 & \mathrm{dg}(Q_N^{-1}) & 0 \\ 0 & 0 & 0 & 0 & I \end{bmatrix}.$$

Set $X = Z^{-1} = \mathrm{diag}(X_i)$, $X_i = Q_i^{-1}$, $i \in \mathcal{S}$, $J = L_f U^* Z^{-1} = L_f U^* X$, $S = V B_f$, and $R = V A_f U^* Z^{-1} = V A_f U^* X$. With these definitions, we have that

$$T_1^* \begin{bmatrix} P & P\widetilde{L}^* \\ \widetilde{L}P & W \end{bmatrix} T_1 = \begin{bmatrix} X & X & L^* - J^* \\ X & Y & L^* \\ L - J & L & W \end{bmatrix} \tag{9.70}$$

and

$$T_2^* \begin{bmatrix} \widetilde{F}P + P\widetilde{F}^* & \Upsilon_1 \,\mathrm{dg}(Q_1) & \dots & \Upsilon_N \,\mathrm{dg}(Q_N) & \widetilde{J} \\ \mathrm{dg}(Q_1)\Upsilon_1^* & -\mathrm{dg}(Q_1) & \dots & 0 & 0 \\ \vdots & \vdots & \ddots & \vdots & \vdots \\ \mathrm{dg}(Q_N)\Upsilon_N^* & 0 & \dots & -\mathrm{dg}(Q_N) & 0 \\ \widetilde{J}^* & 0 & \dots & 0 & -I \end{bmatrix} T_2$$

$$= \begin{bmatrix} \mathrm{P}(X,Y,S,R) & \mathrm{S}(X,Y) & \mathrm{T}(X,Y) \\ \mathrm{S}(X,Y)^* & -\mathrm{D}(X) & 0 \\ \mathrm{T}(X,Y)^* & 0 & -I \end{bmatrix}, \tag{9.71}$$

and thus from (9.55) and (9.56) we have that (9.59) and (9.60) are satisfied. Consider now X, Y, W, R, S, J satisfying (9.58), (9.59), and (9.60). From (9.59) it follows that

$$\begin{bmatrix} X & X \\ X & Y \end{bmatrix} > 0,$$

and from Schur's complement (Lemma 2.26), $Y - X > 0$, that is,

$$\begin{bmatrix} Z & I \\ I & Y \end{bmatrix} > 0,$$

where $Z = X^{-1}$. According to Lemma 2.28, after choosing an arbitrary nonsingular $(Nn \times Nn)$ matrix U, we can find $\widehat{Z}, \widehat{Y}, V$ such that

$$\begin{bmatrix} Z & U \\ U^* & \widehat{Z} \end{bmatrix}^{-1} = \begin{bmatrix} Y & V \\ V^* & \widehat{Y} \end{bmatrix} > 0,$$

and $\widehat{Z}$, V satisfy respectively (9.62) and (9.63). We set P as

$$P = \begin{bmatrix} Z & U \\ U^* & \widehat{Z} \end{bmatrix} > 0$$

and A_f, B_f, L_f as in (9.64)–(9.66), so that (9.70) and (9.71) are satisfied. Pre and pos multiplying (9.70) by $(T_1^{-1})^*$ and (T_1^{-1}), respectively, and pre and pos multiplying (9.71) by $(T_2^{-1})^*$ and (T_2^{-1}), respectively, we get that (9.59) and (9.60) imply (9.55) and (9.56), showing that P, W, A_f, B_f, L_f satisfy (9.54)–(9.56). Finally, from (9.57) we obtain (9.67). □

9.5.2 Robust Filter

Assume now that $\mathbf{A} = (A_1, \ldots, A_N) \in \mathbb{H}^n$, $\mathbf{J} = (J_1, \ldots, J_N) \in \mathbb{H}^{p,n}$, $\mathbf{H} = (H_1, \ldots, H_N) \in \mathbb{H}^{n,m}$, $\mathbf{G} = (G_1, \ldots, G_N) \in \mathbb{H}^{p,m}$, and Π are not exactly known, but instead there are known matrices $\mathbf{A}^\kappa = (A_1^\kappa, \ldots, A_N^\kappa) \in \mathbb{H}^n$, $\mathbf{J}^\kappa = (J_1^\kappa, \ldots, J_N^\kappa) \in \mathbb{H}^{p,n}$, $\mathbf{H}^\kappa = (H_1^\kappa, \ldots, H_N^\kappa) \in \mathbb{H}^{n,m}$, $\mathbf{G}^\kappa = (G_1^\kappa, \ldots, G_N^\kappa) \in \mathbb{H}^{p,m}$, and irreducible stationary transition rate matrices Π^κ such that for $0 \le \rho^\kappa \le 1$, $\kappa = 1, \ldots, \ell$, $\sum_{\kappa=1}^{\ell} \rho^\kappa = 1$, we have that

$$\begin{aligned} \mathbf{A} &= \sum_{\kappa=1}^{\ell} \rho^\kappa \mathbf{A}^\kappa, \qquad \mathbf{J} = \sum_{\kappa=1}^{\ell} \rho^\kappa \mathbf{J}^\kappa, \qquad \mathbf{H} = \sum_{\kappa=1}^{\ell} \rho^\kappa \mathbf{H}^\kappa, \\ \mathbf{G} &= \sum_{\kappa=1}^{\ell} \rho^\kappa \mathbf{G}^\kappa, \qquad \Pi = \sum_{\kappa=1}^{\ell} \rho^\kappa \Pi^\kappa. \end{aligned} \tag{9.72}$$

We denote by $\pi_i^\kappa > 0$, $i = 1, \ldots, N$, the stationary distribution associated to Π^κ. Define F^κ, H^κ, $J^{p\kappa}$, $G^{p\kappa}$, Ψ_i^κ, Υ_i^κ as respectively in (7.5), (7.6), (7.39), (7.40), (7.59), (7.60) replacing $A_i, H_i, J_i, G_i, \lambda_{ij}, \pi_i$ by A_i^κ, H_i^κ, J_i^κ, G_i^κ, λ_{ij}^κ, π_i^κ. Define also

$$J^{r\kappa} := \operatorname{diag}(J_i^\kappa), \tag{9.73}$$

$$G^{r\kappa} := [G_1^\kappa \ldots G_N^\kappa]. \tag{9.74}$$

Our next result presents the robust linear filter for system (7.41)–(7.43).

Theorem 9.30 *Suppose that the following LMIs optimization problem has an* (ϵ-) *optimal solution* $\bar{X}, \bar{Y}, \bar{W}, \bar{R}, \bar{S}, \bar{J}$:

$$\inf \operatorname{tr}(W)$$

subject to (9.58), (9.59), *and for* $\kappa = 1, \ldots, \ell$,

$$\begin{bmatrix} \mathrm{P}^\kappa(X,Y,S,R) & \mathrm{S}^\kappa(X,Y) & \mathrm{T}^\kappa(X,Y) \\ \mathrm{S}(X,Y)^* & -\mathrm{D}(X) & 0 \\ \mathrm{T}^\kappa(X,Y)^* & 0 & -I \end{bmatrix} < 0, \tag{9.75}$$

where

$$\mathrm{P}^\kappa(X,Y,S,R) = \begin{bmatrix} XF^\kappa + (F^\kappa)^*X & XF^\kappa + (F^\kappa)^*Y + (H^\kappa)^*S^* + R^* \\ (F^\kappa)^*X + YF^\kappa + SH^\kappa + R & (F^\kappa)^*Y + YF^\kappa + SH^\kappa + (H^\kappa)^*S^* \end{bmatrix},$$

$$\mathrm{T}^\kappa(X,Y) = \begin{bmatrix} XJ^{r\kappa} \\ YJ^{r\kappa} + SG^{r\kappa} \end{bmatrix},$$

$$S^\kappa(X,Y) = \begin{bmatrix} X\Psi_1^\kappa & \dots & X\Psi_N^\kappa \\ Y\Psi_1^\kappa & \dots & Y\Psi_N^\kappa \end{bmatrix}.$$

Then for the filter given as in (9.64)–(9.66), *we have that system* (7.51) *is MSS and* $\lim_{t\to\infty} E(\|e(t)\|^2) \le \mathrm{tr}(\bar{W})$.

Proof Consider **A**, **H**, **J**, **G**, Π as in (9.72) for some $0 \le \rho^\kappa \le 1$, $\sum_{\kappa=1}^{\ell} \rho^\kappa = 1$, and set F, H, J^p, G^p as respectively in (7.5), (7.6), (7.39), (7.40), with $\pi_i > 0$, $i = 1, \dots, N$, the stationary distribution associated to Π. Notice that, by Proposition 2.12, Π is irreducible, which guarantees the existence of the stationary distribution $\pi_i > 0$. Since (9.75) holds for each $\kappa = 1, \dots, \ell$, we have from Theorem 9.29 that there exists $(P, \bar{W}, A_f, B_f, L_f)$ (which, according to (9.61), (9.62), (9.64), (9.65), and (9.66) depend only on $\bar{X}, \bar{Y}, \bar{R}, \bar{F}, \bar{J}$) such that $\mathrm{tr}(\bar{W}) > \mathrm{tr}(\widetilde{L}P\widetilde{L}^*)$ and for each $\kappa = 1, \dots, \ell$,

$$\widetilde{F}^\kappa P + P(\widetilde{F}^\kappa)^* + \sum_{i\in\mathcal{S}} \Upsilon_i^\kappa \,\mathrm{dg}(Q_i)(\Upsilon_i^\kappa)^* + \widetilde{J}^{r\kappa}(\widetilde{J}^{r\kappa})^* < 0, \tag{9.76}$$

where $\widetilde{F}^\kappa$ is as in (7.53), and

$$\widetilde{J}^{r\kappa} := \begin{bmatrix} J^{r\kappa} \\ B_f G^{r\kappa} \end{bmatrix}.$$

From (7.37) and (7.58) we have that (9.76) can be rewritten as

$$\widetilde{F}^\kappa P + P(\widetilde{F}^\kappa)^* + \begin{bmatrix} \mathcal{V}^\kappa(\mathbf{Q}) & 0 \\ 0 & 0 \end{bmatrix} + \widetilde{J}^{r\kappa}(\widetilde{J}^{r\kappa})^* < 0, \tag{9.77}$$

where $\mathcal{V}^\kappa$ is as in (7.37) replacing λ_{ij} and Π by respectively λ_{ij}^κ and Π^κ. Noticing that $\sum_{\kappa=1}^{\ell} \rho^\kappa \widetilde{F}^\kappa = \widetilde{F}$ and $\sum_{\kappa=1}^{\ell} \rho^\kappa \mathcal{V}^\kappa(\mathbf{Q}) = \mathcal{V}(\mathbf{Q})$, we get, after taking the sum of (9.77) multiplied by ρ^κ, over κ from 1 to ℓ, that

$$\widetilde{F}P + P(\widetilde{F})^* + \begin{bmatrix} \mathcal{V}(\mathbf{Q}) & 0 \\ 0 & 0 \end{bmatrix} + \sum_{\kappa=1}^{\ell} \rho^\kappa \widetilde{J}^{r\kappa}(\widetilde{J}^{r\kappa})^* < 0. \tag{9.78}$$

Set $\widetilde{J}^r = \sum_{\kappa=1}^{\ell} \rho^\kappa \widetilde{J}^{r\kappa}$. Since

$$0 \le \sum_{\kappa=1}^{\ell} \rho^\kappa (\widetilde{J}^{r\kappa} - \widetilde{J}^r)(\widetilde{J}^{r\kappa} - \widetilde{J}^r)^* = \sum_{\kappa=1}^{\ell} \rho^\kappa \widetilde{J}^{r\kappa}(\widetilde{J}^{r\kappa})^* - \widetilde{J}^r \widetilde{J}^{r*},$$

we conclude that $\widetilde{J}^r \widetilde{J}^{r*} \le \sum_{\kappa=1}^{\ell} \rho^\kappa \widetilde{J}^{r\kappa}(\widetilde{J}^{r\kappa})^*$ and, by (9.78),

$$\widetilde{F}P + P(\widetilde{F})^* + \begin{bmatrix} \mathcal{V}(\mathbf{Q}) & 0 \\ 0 & 0 \end{bmatrix} + \widetilde{J}^r(\widetilde{J}^r)^* < 0. \tag{9.79}$$

Let us show now that $\widetilde{J}\widetilde{J}^* \leq \widetilde{J}^r(\widetilde{J}^r)^*$, where $\widetilde{J}$ is as in (7.54). For any vector $x^* = [x_1^* \ \dots \ x_N^*]$ and y of appropriate dimensions, we have that

$$\begin{aligned}
\begin{bmatrix} x^* & y^* \end{bmatrix} \widetilde{J}\widetilde{J}^* \begin{bmatrix} x \\ y \end{bmatrix} &= \sum_{i=1}^{N} \pi_i \begin{bmatrix} x_i^* & y^* \end{bmatrix} \begin{bmatrix} J_i \\ B_f G_i \end{bmatrix} \begin{bmatrix} J_i^* & G_i^* B_f^* \end{bmatrix} \begin{bmatrix} x_i \\ y \end{bmatrix} \\
&\leq \sum_{i=1}^{N} \begin{bmatrix} x_i^* & y^* \end{bmatrix} \begin{bmatrix} J_i \\ B_f G_i \end{bmatrix} \begin{bmatrix} J_i^* & G_i^* B_f^* \end{bmatrix} \begin{bmatrix} x_i \\ y \end{bmatrix} \\
&= \begin{bmatrix} x^* & y^* \end{bmatrix} \widetilde{J}^r (\widetilde{J}^r)^* \begin{bmatrix} x \\ y \end{bmatrix},
\end{aligned}$$

showing that indeed $\widetilde{J}\widetilde{J}^* \leq \widetilde{J}^r(\widetilde{J}^r)^*$. Combining this with (9.77) and (9.79) and identity (7.58), we get that

$$\widetilde{F}P + P\widetilde{F}^* + \sum_{\ell=1}^{N} \Upsilon_\ell \operatorname{dg}(Q_\ell)(\Upsilon_\ell)^* + \widetilde{J}\widetilde{J}^* < 0. \tag{9.80}$$

From (9.80) and Propositions 7.20 and 7.21 we get the result. □

9.5.3 ARE Approximations for the LMIs Problem

In this subsection we show how approximations of the filtering ARE derived in Sect. 7.5 can yield a solution of the LMIs optimization problem presented in Sect. 9.5.1, that is, how they can yield a sequence of feasible solutions for the LMIs (9.58), (9.59), (9.60) which converges to the optimal solution. Consider $\epsilon > 0$. We have the following theorem (recall the definition of T in (7.73)).

Theorem 9.31 *Consider the unique positive definite solution $\widetilde{Z}_\epsilon$ of the following ARE:*

$$0 = F\widetilde{Z}_\epsilon + \widetilde{Z}_\epsilon F^* - \widetilde{Z}_\epsilon H^* \left(G^p G^{p*}\right)^{-1} H \widetilde{Z}_\epsilon + J^p J^{p*} + \mathcal{V}(\mathbf{Q}_\epsilon) + \epsilon I, \tag{9.81}$$

where $\mathbf{Q}_\epsilon = (Q_{\epsilon,1}, \dots, Q_{\epsilon,N})$, and $Z_\epsilon = \operatorname{diag}(Q_{\epsilon,i}) > 0$ is the unique solution of

$$0 = FZ_\epsilon + Z_\epsilon F^* + J^p J^{p*} + \mathcal{V}(\mathbf{Q}_\epsilon) + 2\epsilon I. \tag{9.82}$$

Set $X_i = Q_{\epsilon,i}^{-1}$, $i \in \mathcal{S}$, $X = \operatorname{diag}(X_i)$, $Y = \widetilde{Z}_\epsilon^{-1}$, $S = -\widetilde{Z}_\epsilon^{-1} T(\widetilde{Z}_\epsilon)$, $R = -(YF + SH)(I - Y^{-1}X)$, $J = L(I - Y^{-1}X)$, and $W > L\widetilde{Z}_\epsilon L^*$. Then with this choice of X, Y, W, R, S, J, we have that (9.58), (9.59), and (9.60) are satisfied, and moreover, by taking $U = (X^{-1} - Y^{-1})$ we have that $A_f = F - T(\widetilde{Z}_\epsilon)H$, $B_f = T(\widetilde{Z}_\epsilon)$, $L_f = L$, P_ϵ as in (9.54) is given by

$$P_\epsilon = \begin{bmatrix} Z_\epsilon & Z_\epsilon - \widetilde{Z}_\epsilon \\ Z_\epsilon - \widetilde{Z}_\epsilon & Z_\epsilon - \widetilde{Z}_\epsilon \end{bmatrix}, \tag{9.83}$$

and $\operatorname{tr}(\widetilde{L}P_\epsilon\widetilde{L}^*) = \operatorname{tr}(L\widetilde{Z}_\epsilon L^*)$.

Proof First, notice that, by Schur's complement, (9.59) is equivalent to

$$\begin{bmatrix} X - XY^{-1}X & (I - XY^{-1})L^* - J^* \\ L(I - Y^{-1}X) - J & W - LY^{-1}L^* \end{bmatrix} > 0. \tag{9.84}$$

By choosing $J = L(I - Y^{-1}X)$ we have that (9.84) is equivalent to $X^{-1} > Y^{-1}$ and $W > LY^{-1}L^*$. Consider now the following matrices:

$$T_3 = \begin{bmatrix} X^{-1} & X^{-1} \\ 0 & -Y^{-1} \end{bmatrix}, \qquad T_4 = \begin{bmatrix} \mathrm{dg}(X_1^{-1}) & \dots & 0 \\ \vdots & \ddots & \vdots \\ 0 & \dots & \mathrm{dg}(X_N^{-1}) \end{bmatrix},$$

$$T_5 = \begin{bmatrix} T_3 & 0 & 0 \\ 0 & T_4 & 0 \\ 0 & 0 & I \end{bmatrix}.$$

Pre and pos multiplying (9.60) by T_5^* and T_5 and considering $R = -(YF + SH)(I - Y^{-1}X)$ yields

$$\begin{bmatrix} \Gamma_{11} & \Gamma_{12} & \Gamma_{13} \\ \Gamma_{12}^* & \Gamma_{22} & 0 \\ \Gamma_{13}^* & 0 & -I \end{bmatrix} < 0, \tag{9.85}$$

where

$$\Gamma_{\mathrm{aux},1,1} = FX^{-1} + X^{-1}F^*,$$

$$\Gamma_{\mathrm{aux},1,2} = F\big(X^{-1} - Y^{-1}\big) + \big(X^{-1} - Y^{-1}\big)F^* - Y^{-1}SHY^{-1},$$

$$\Gamma_{\mathrm{aux},2,2} = F\big(X^{-1} - Y^{-1}\big) + \big(X^{-1} - Y^{-1}\big)F^*,$$

$$\Gamma_{11} = \begin{bmatrix} \Gamma_{\mathrm{aux},1,1} & \Gamma_{\mathrm{aux},1,2}^* \\ \Gamma_{\mathrm{aux},1,2} & \Gamma_{\mathrm{aux},2,2} \end{bmatrix},$$

$$\Gamma_{12} = \begin{bmatrix} \Psi_1\,\mathrm{dg}(X_1^{-1}) & \dots & \Psi_N\,\mathrm{dg}(X_N^{-1}) \\ 0 & \dots & 0 \end{bmatrix},$$

$$\Gamma_{13} = \begin{bmatrix} J^p \\ -Y^{-1}SG^p \end{bmatrix},$$

$$\Gamma_{22} = -\begin{bmatrix} \mathrm{dg}(X_1^{-1}) & \dots & 0 \\ \vdots & \ddots & \vdots \\ 0 & \dots & \mathrm{dg}(X_N^{-1}) \end{bmatrix}.$$

From Schur's complement, (9.85) is equivalent to

$$\begin{bmatrix} \widetilde{\Gamma}_{11} & \widetilde{\Gamma}_{21}^* \\ \widetilde{\Gamma}_{21} & \widetilde{\Gamma}_{22} \end{bmatrix} = \Gamma_{11} - \begin{bmatrix} \Gamma_{12} & \Gamma_{13} \end{bmatrix} \begin{bmatrix} -\Gamma_{22}^{-1} & 0 \\ 0 & I \end{bmatrix} \begin{bmatrix} \Gamma_{12}^* \\ \Gamma_{13}^* \end{bmatrix} > 0. \tag{9.86}$$

After some algebraic manipulations we get that

$$\widetilde{\Gamma}_{11} = -\Big(FX^{-1} + X^{-1}F^* + \sum_{i\in\mathcal{S}} \Psi_i \operatorname{dg}(X_i^{-1})\Psi_i^* + J^p J^{p*}\Big), \tag{9.87}$$

$$\widetilde{\Gamma}_{21} = -(F(X^{-1} - Y^{-1}) + (X^{-1} - Y^{-1})F^* - Y^{-1}SHY^{-1}), \tag{9.88}$$

$$\widetilde{\Gamma}_{22} = -(F(X^{-1} - Y^{-1}) + (X^{-1} - Y^{-1})F^* + Y^{-1}SG^pG^{p*}S^*Y^{-1}). \tag{9.89}$$

From (9.81) and (9.82) we have that

$$0 = F(Z_\epsilon - \widetilde{Z}_\epsilon) + (Z_\epsilon - \widetilde{Z}_\epsilon)F^* + \widetilde{Z}_\epsilon H^*\big(G^pG^{p*}\big)^{-1}H\widetilde{Z}_\epsilon + \epsilon I. \tag{9.90}$$

From the assumption that $\operatorname{Re}\{\lambda(\mathcal{L})\} < 0$ and Proposition 3.13 it follows that F is stable, and from (9.90) we have that $Z_\epsilon > \widetilde{Z}_\epsilon$. Thus, with the choice $X = Z_\epsilon^{-1}$, $Y = \widetilde{Z}_\epsilon^{-1}$, $S = -\widetilde{Z}_\epsilon^{-1}T(\widetilde{Z}_\epsilon)$, we have that $X^{-1} > Y^{-1}$ and, from (9.89) and (9.90),

$$\widetilde{\Gamma}_{22} = -(F(X^{-1} - Y^{-1}) + (X^{-1} - Y^{-1})F^* + Y^{-1}SG^pG^{p*}S^*Y^{-1}) = \epsilon I.$$

From (7.58), (9.82), and (9.87), we have

$$\widetilde{\Gamma}_{11} = -(F(X^{-1} - Y^{-1}) + (X^{-1} - Y^{-1})F^* + Y^{-1}SG^pG^{p*}S^*Y^{-1}) = 2\epsilon I.$$

Finally, noticing that $Y^{-1}SHY^{-1} = -\widetilde{Z}_\epsilon H^*(G^pG^p)^{-1}H\widetilde{Z}_\epsilon$, we get from (9.88) and (9.90) that

$$\begin{aligned}\widetilde{\Gamma}_{21} &= -(F(X^{-1} - Y^{-1}) + (X^{-1} - Y^{-1})F^* - Y^{-1}SHY^{-1}) \\ &= -\big(F(Z_\epsilon - \widetilde{Z}_\epsilon) + (Z_\epsilon - \widetilde{Z}_\epsilon)F^* + \widetilde{Z}_\epsilon H^*\big(G^pG^{p*}\big)^{-1}H\widetilde{Z}_\epsilon\big) = \epsilon I.\end{aligned}$$

Thus, we have that (9.86) is indeed verified. It remains to show that $A_f = F - T(\widetilde{Z}_\epsilon)H$, $B_f = T(\widetilde{Z}_\epsilon)$, $L_f = L$, and P_ϵ is as in (9.83). By taking $R = -(YF + SH)(I - Y^{-1}X)$, $J = L(I - Y^{-1}X)$, $U = (X^{-1} - Y^{-1})$ we have that $V = -Y$ and from (9.64), (9.65), (9.66) that $A_f = F + Y^{-1}SH = F - T(\widetilde{Z}_\epsilon)H$, $B_f = -Y^{-1}S = T(\widetilde{Z}_\epsilon)$, $L_f = L$, and P_ϵ is as in (9.83). Finally it is easy to see that $\widetilde{L}P_\epsilon\widetilde{L} = L\widetilde{Z}_\epsilon L$, completing the proof. □

Remark 9.32 As $\epsilon \downarrow 0$, it is easy to see that $\mathbf{Q}_\epsilon \downarrow \mathbf{Q}$ and $\widetilde{Z}_\epsilon \downarrow \widetilde{Z}$, where $\mathbf{Q}$ and $\widetilde{Z}$ are as in Theorem 7.15. Therefore, the approximating ARE solutions in Theorem 9.31 indeed lead to an optimal solution for the LMIs optimization problem posed in Theorem 9.29.

9.6 Historical Remarks

A great deal of the early criticism to "modern" control theory (which historically refers to the time period which followed R.E. Kalman's work, at the beginning of the 1960 decade) was due to the fact that the obtained results were of a purely theoretical fashion, and practical design methods did not exist. The design tools used in practice at that time were restricted to frequency-domain graphical criteria such as the ones derived by H. Nyquist, H.W. Bode, N.B. Nichols, and W.R. Evans. In addition, it became clear that robustness matters could not be tackled in the time-domain framework in the same way as "classical" design methods did (an exposition on this can be found in [27]).

The robustness issues that were observed in the early days of state-space analysis defined a mainstream of research in the intervening decades and have led to what is known today as state-space *robust* control methods. It is fair to say that the main reason why such methods turned out to be as popular as they did is that, dissimilarly to frequency-domain methods, the time-domain approach can be extended to many different scenarios, such as stochastic systems, MJLS, time-varying systems, nonlinear systems, hybrid systems, and so on.

Another major boost for the popularization of state-space methods in robust control was the fact that many analysis and control problems admitted elegant solutions, with explicit formulae for offline computation. Some notorious examples include the H_2 and H_∞ control solutions presented in [117] and the study of *stability radii* by D. Hinrichsen and A.J. Pritchard. This latter problem, proposed in [180, 181] (see also [294]), has proven to be of a particularly intriguing nature because of the differences between the complex and real setups (see [182, Chap. 5]). Stability radii lie at the heart of robustness analysis problems and thus constitute an area of central interest. In the MJLS context, two seminal references are [118, 128] (see [121] for a unified account on the subject).

One of the main venues of research that has gained wide popularity in systems theory in the latter part of the twentieth century was the development of efficient methods for analysis and synthesis of controllers. The main interest here was in a computational framework that should be simple enough as to be tractable by numerical analysis and yet able to cope with robustness issues. A cornerstone of the research in this direction was the observation that many control design problems could be described in terms of *convex optimization*. As a consequence, many linear controller design problems could be seen as particular problems of a much broader setup.

Another decisive step in the spread of convex optimization methods in control was the development of *interior-point methods* for their solution. According to [49], the inception of efficient interior-point methods for solving linear programming problems was possible only after the paper [195], which introduced a polynomial-time algorithm for solving them. This motivated a great deal of study on interior-point methods for more general problems and led to the development of an efficient algorithm to solve LMIs problems, which was proposed by Y. Nesterov and A. Nemirovskii in 1988. The standard textbooks on LMIs methods in control are [235], which duly documents this algorithm, and [49], whose comprehensiveness and accessible language had a definitive role in popularizing this approach. The last cornerstone that consolidated LMIs problems as a centerpiece in the control literature was certainly the popularization of free computer programs for tackling them, such as SeDuMi, LMILAB, LMISol [102], SDPT3 [285, 289], or SDPA-M [162], as well as friendly interfaces such as YALMIP [192] and Matlab's LMITOOLS.

As far as the authors are aware of, the inception of LMIs methods to the control of MJLS was made in [170, 245–247]. A huge amount of literature spurred from this point, and nowadays the LMIs paradigm is ubiquitous in the study of MJLS. More recently, much emphasis is being placed on the application of LMIs to approximate, e.g., by iterative methods, the solution of *bilinear* matrix inequality problems, which are NP-hard. The key reference which has been brought to the MJLS literature in this context is [169].

The main sources for this chapter were, essentially, [82, 280–284].

Chapter 10
Some Numerical Examples

10.1 Outline of the Chapter

In this chapter some numerical applications of continuous-time MJLS are treated by means of the theoretical results introduced earlier, in particular the design techniques presented for H_2 control, robust H_2 guaranteed cost control, mixed H_2/H_∞ control, and stationary filtering in Chaps. 5, 7, and 9. Examples in the fields of economic modeling, electrical systems, and robotics are presented in Sects. 10.2, 10.3, and 10.4, respectively.

10.2 An Example on Economics

In this section the problem treated in [32] is solved by means of the different algorithms presented in the previous chapters. The H_2 standpoint from Chap. 5, together with alternative results from the literature and the design methods from Theorem 9.22, is studied.

The example, which was previously described in Sect. 1.2, considers a simple economic system based on Samuelson's multiplier–accelerator model (see [32]). Three modes of operation for an economic system are considered: 1 = "normal", 2 = "boom", and 3 = "slump". The parameters for each of these operation modes are:

$$A_1 = \begin{bmatrix} 0 & 0 & 0 \\ 0 & -0.545 & 0.626 \\ 0 & -1.570 & 1.465 \end{bmatrix}, \qquad B_1 = \begin{bmatrix} 0 \\ -0.283 \\ 0.333 \end{bmatrix},$$

$$A_2 = \begin{bmatrix} 0 & 0 & 0 \\ 0 & -0.106 & 0.087 \\ 0 & -3.810 & 3.861 \end{bmatrix}, \qquad B_2 = \begin{bmatrix} 0 \\ 0 \\ 0.087 \end{bmatrix},$$

O.L.V. Costa et al., *Continuous-Time Markov Jump Linear Systems*,
Probability and Its Applications, DOI 10.1007/978-3-642-34100-7_10,

Table 10.1 Results for Situation 1 (Π exactly known, Theorem 5.7)

H_2-norm	$\sqrt{\mu} = 48.53$
Controller	$K_1 = -[2.0343\ 14.5181\ -23.5917]$
	$K_2 = -[1.0187\ 73.0961\ -78.7596]$
	$K_3 = -[93.6651\ -11.4921\ 11.6875]$

$$A_3 = \begin{bmatrix} 1.80 & -0.3925 & 4.52 \\ 3.14 & 0.100 & -0.28 \\ -19.06 & -0.148 & 1.56 \end{bmatrix}, \qquad B_3 = \begin{bmatrix} -0.064 \\ 0.195 \\ -0.080 \end{bmatrix},$$

$$C_1 = C_2 = C_3 = \begin{bmatrix} I_3 \\ 0_{1\times 3} \end{bmatrix}, \qquad D_1 = D_2 = D_3 = \begin{bmatrix} 0_{3\times 1} \\ 1 \end{bmatrix},$$

where we recall that I_r represents the $r \times r$ identity matrix, and $0_{r\times s}$ the null matrix of corresponding dimension.

Adopting $\nu_i = 1/3$ and $J_i = J_i^w := I_3$, $i \in \mathcal{S}$, three different situations are treated in this section. First, under the assumption that the transition rate matrix is entirely known, the method devised in Chap. 5 is employed to obtain the optimal H_2 control solution. In the second case it is assumed that the transition rate matrix is not entirely known but belongs to a set $\mathbb{V}$ as defined in (9.37). The third case is such that the transition rate matrix is not entirely known but belongs to a set $\mathbb{V}$ as defined in (9.37), and a static feedback disturbance affects the system. In this case the guaranteed cost H_2 control design from Chap. 5 is compared to the robust controllers introduced in Chap. 9.

Situation 1: Known Transition Rates

Just as in [32], let us consider the transition rates given by

$$\Pi = \begin{bmatrix} -0.53 & 0.32 & 0.21 \\ 0.50 & -0.88 & 0.38 \\ 0.40 & 0.13 & -0.53 \end{bmatrix}.$$

In this case the obtained optimal solution is given in Table 10.1. The control renders the closed-loop system stable in the mean-square sense, with $\mathrm{Re}\{\lambda(\mathcal{T})\} = -0.2503$.

Situation 2: Uncertain Transition Rates

Let us now assume that Π is not exactly known, and the set $\mathbb{V}$ in (9.37) is defined by the following extreme values:

$$\Pi^1 = \begin{bmatrix} -0.53 & 0.32 & 0.21 \\ 0.50 & -0.88 & 0.38 \\ 0.40 & 0.13 & -0.53 \end{bmatrix}, \qquad \Pi^2 = \begin{bmatrix} -0.7 & 0.38 & 0.32 \\ 0.6 & -1 & 0.4 \\ 0.5 & 0.2 & -0.7 \end{bmatrix}.$$

In this case, the H_2-guaranteed cost and associated control are given in Table 10.2. Notice that Π, as defined in Situation 1 above, belongs to $\mathbb{V}$ ($\Pi = \Pi^1$) and, as expected, $\beta > \mu$.

Table 10.2 Results for Situation 2 (Π not exactly known, Theorem 5.7)

H_2-guaranteed cost	$\sqrt{\beta} = 53.55$
Controller	$K_1 = -[2.2889\ 15.1983\ -25.4635]$
	$K_2 = -[1.1018\ 70.8596\ -77.9725]$
	$K_3 = -[100.8453\ -13.4936\ 10.5960]$

Table 10.3 Results for Situation 3 ($\beta = H_2$-guaranteed cost) via Theorem 9.22

Small-gain approach, (i)	Adjoint operator approach, (ii)
$\sqrt{\beta} = 97.61$	$\sqrt{\beta} = 58.22$
$K_1 = [0.0660\ 1.8413\ -11.5466]$	$K_1 = [3.1765\ 18.4408\ -36.0874]$
$K_2 = [1.0856\ 71.8829\ -72.2111]$	$K_2 = [2.3591\ 85.9429\ -87.6313]$
$K_3 = [174.6847\ -34.7378\ -14.0185]$	$K_3 = [135.4769\ -23.6030\ 0.1213]$

Situation 3: Uncertain transition rates with feedback disturbance

Let us further assume that a linear disturbance of the form (9.34), with shaping matrices

$$J_i^{\varpi} \equiv J_i, \qquad E_i \equiv 0_{1\times 3}, \qquad F_i \equiv 1/10,$$

affects the system robustness channel in (9.32). Notice that $\mathbf{\Delta}$ can be regarded as an unstructured perturbation of the input operator in this case, in the sense that $B_i \rightsquigarrow B_i + \Delta_i$ in (9.32). Letting $\alpha = 0.125$, an example of such a feedback gain is given by

$$\Delta_1 = \begin{bmatrix} 0.0971 \\ 0.0713 \\ -0.0333 \end{bmatrix}, \qquad \Delta_2 = \begin{bmatrix} 0.1060 \\ 0.0310 \\ -0.0584 \end{bmatrix}, \qquad \Delta_3 = \begin{bmatrix} 0.0938 \\ 0.0701 \\ 0.0436 \end{bmatrix},$$

which will be adopted from now on. It is easy to check that, in this case, $\|\mathbf{\Delta}\|_{\max} < 0.12499 < \alpha$.

The designs obtained via (i) and (ii) in Theorem 9.22 are displayed in Table 10.3. It should be noted that, although the H_2-guaranteed costs in this case are larger than the one in Table 10.2, they are uniformly guaranteed over all disturbances of the form (9.33).

It is worth mentioning that, due to the fact that Π is not entirely known and due to uncertainty of the form $\|\mathbf{\Delta}\|_{\max} < 0.125$, the guaranteed costs in Table 10.3 are inherently conservative. For the sake of illustration, an estimate of actual running costs of the form (with e_i standing for the ith vector of the standard basis in $\mathbb{R}^r$, and δ for the Dirac delta)

$$T \mapsto \left(\sum_{i=1}^{r} \sum_{j=1}^{N} \nu_j \int_0^T E\left[\|z(t)\|^2 \mid x_0 = 0,\ \theta_0 = j,\ w(t) = e_i \delta(t) \right] dt \right)^{1/2} \tag{10.1}$$

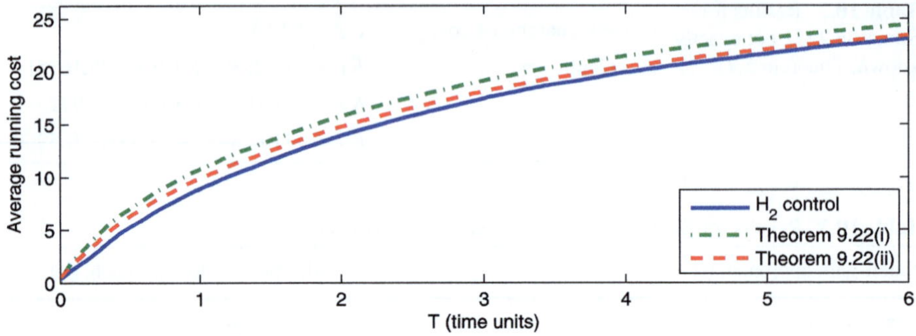

Fig. 10.1 Average running cost in (10.1), under a static feedback disturbance ($\alpha = 0.125$)

for the closed-loop system

$$\begin{cases} \dot{x}(t) = \big(A_{\theta(t)} + B_{\theta(t)}K_{\theta(t)} + J^{\varpi}_{\theta(t)}\Delta_{\theta(t)}(E_{\theta(t)} + F_{\theta(t)}K_{\theta(t)})\big)x(t) + J^{w}_{\theta(t)}w(t), \\ z(t) = (C_{\theta(t)} + D_{\theta(t)}K_{\theta(t)})x(t) \end{cases} \tag{10.2}$$

was also computed in this example by means of Monte Carlo simulation, as depicted in Fig. 10.1. The numerical results correspond to 2000 different trajectories of θ, with $\mathbf{K}$ given by each one of the controllers in Table 10.1 (with $\mathbf{K} \rightsquigarrow -\mathbf{K}$) and Table 10.3, in which, for simplicity, the fixed transition rate matrix

$$\widetilde{\Pi} = \frac{95}{100}\Pi^1 + \frac{5}{100}\Pi^2 = \begin{bmatrix} -0.5385 & 0.3230 & 0.2155 \\ 0.5050 & -0.8860 & 0.3810 \\ 0.4050 & 0.1335 & -0.5385 \end{bmatrix},$$

which belongs to $\mathbb{V}$, was adopted.

The terminal costs associated with $T = 6$ time units in (10.1) are 24.4512 and 23.4102 for the controllers (i) and (ii) in Table 10.3, respectively. The actual terminal cost offered by the controller in Table 10.1 is 23.1019, which, albeit close to that of the adjoint approach, is not guaranteed to be robust in the face of uncertainty as in (10.2). Finally, the Monte Carlo estimates of the mean return $t \mapsto E[\|x(t)\|^2]$ are depicted in Fig. 10.2. The results for all three controllers are quite similar in this case.

Robust Stability Margin for Π

Finally, let us assume that the transition rates Π are subject to uncertainty of the form discussed in Sect. 9.3. The closed-loop robustness margins in (9.29), along with the corresponding $k_{\min}$ and $k_{\max}$ as in Theorem 9.16, are displayed in Table 10.4.

The biggest margins are obtained via the design from the small-gain approach, Theorem 9.22(i). The robust mean-square stability is guaranteed on the polytope $\mathbb{V}$

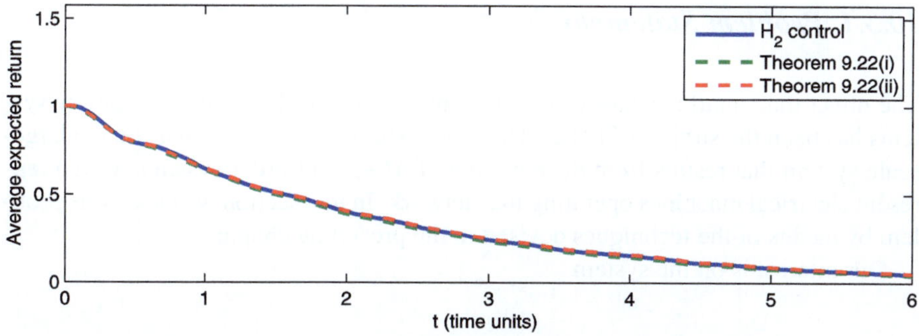

Fig. 10.2 Average expected return, $E[\|x(t)\|^2]$, under a static feedback disturbance ($\alpha = 0.125$)

Table 10.4 Closed-loop robust stability margins for $\Pi \rightsquigarrow K\Pi$ in Theorem 9.16

Robustness margins	σ	$\sigma^\star$	$k_{\min}$	$k_{\max}$
Situation 1	0.2174	0.2172	0.8214	1.2778
Situation 2	0.2375	0.2373	0.8081	1.3114
Situation 3, (i)	0.3654	0.3658	0.7322	1.5768
Situation 3, (ii)	0.3065	0.3064	0.7654	1.4420

as in (5.2), with the vertices (in this example, $2^N = 8$):

$$\Pi^\kappa = \begin{bmatrix} k_1^{(\kappa)} & 0 & 0 \\ 0 & k_2^{(\kappa)} & 0 \\ 0 & 0 & k_3^{(\kappa)} \end{bmatrix} \begin{bmatrix} -0.53 & 0.32 & 0.21 \\ 0.50 & -0.88 & 0.38 \\ 0.40 & 0.13 & -0.53 \end{bmatrix}, \quad \kappa = 1, \ldots, 2^N,$$

where $(k_1^{(\kappa)}, k_2^{(\kappa)}, k_3^{(\kappa)}) = (k_{\min}, k_{\min}, k_{\min})$ for $\kappa = 1$, $(k_1^{(\kappa)}, k_2^{(\kappa)}, k_3^{(\kappa)}) = (k_{\min}, k_{\min}, k_{\max})$ for $\kappa = 2$, and so on, until $(k_1^{(\kappa)}, k_2^{(\kappa)}, k_3^{(\kappa)}) = (k_{\max}, k_{\max}, k_{\max})$ for $\kappa = 8$. It should be emphasized, however, that Theorem 9.16 does not provide a guarantee that the H_2 cost should be that of Table 10.3 uniformly on the polytope. Instead, it only ensures that the mean-square stability should not be lost, were the transition rates so perturbed. The obtained robust stability margin should be understood as an additional feature, one which accounts for unmodeled uncertainty on Π.

10.3 Coupled Electrical Machines

In this section we address the control of coupled electrical machines, following the approach initiated by [212].[1]

[1]The convex programming problems in this section were tackled in Matlab 7.8.0 (release 2009a), with the aid of the Yalmip interface. The adopted solver was SeDuMi 1.3, with a perturbation (shift) of 10^{-5} to ensure strict feasibility.

10.3.1 Problem Statement

The effect that small stochastic couplings may cause in the stability of power systems has been the subject of [212]. The model studied therein consists on the large-scale system that results from the coupling of M second-order systems, which represent electrical machines operating in a network. In this section we tackle this problem by means of the techniques devised in the preceding chapters.

Our interest is on the system

$$\ddot{\varphi}_j(t) + 2\alpha_j\dot{\varphi}_j(t) + \omega_j^2\varphi_j(t) = \sum_{k=1}^{M} \epsilon\mu_{jk}(t)\varphi_k(t), \quad j = 1, \dots, M, \tag{10.3}$$

where M is the number of machines considered in the network, and φ_j is the deviation of the rotor angle of the jth machine with respect to its nominal value. The parameters α_j and ω_j stand respectively for the damping coefficient and natural frequency of the jth machine. Throughout this section we shall further borrow from [212] the following assumption:

$$0 = \alpha_1 < \alpha_2 \leq \alpha_3 \leq \alpha_M, \tag{10.4}$$

which means that the first machine has a zero net damping coefficient, whereas the other machines have strictly positive dampings.

As mentioned in [212], in this case it is expected that energy should be "pumped" from the stable, positively damped portion of the system ($j \geq 2$) into the marginally stable portion ($j = 1$), thereby driving the overall system to instability. This adverse coupling mechanism, whose amplitude is tuned by the small parameter $\epsilon > 0$, is represented by the term which appears at the right-hand side of (10.3). The processes μ_{jk} model the random fluctuations which affect the system. The results in [212] require only that μ_{jk} satisfy a mixing condition, but in order to arrive at more explicit results, we shall henceforth assume that they are homogeneous telegraph processes taking values in $\{-1, 1\}$.

For the sake of simplicity, for the remainder of this section, we shall restrict ourselves to the case of *two* machines, $M = 2$. In addition, let us assume that (for convenience, the time dependence is omitted) $\mu_{11} = \mu_{22} = \theta_1$ and $\mu_{12} = \mu_{21} = \theta_2$, where θ_1 and θ_2 are independent and homogeneous Markov chains taking values in $\{-1, 1\}$ with transitions governed by $\Pi_j = \left[\begin{smallmatrix} -\eta_j & \eta_j \\ \eta_j & -\eta_j \end{smallmatrix}\right]$, $j = 1, 2$.

In order to put the above model in the MJLS form (3.1), we may proceed as follows. Let

$$x(t) = \big(\varphi_1(t), \dot{\varphi}_1(t), \varphi_2(t), \dot{\varphi}_2(t)\big) \in \mathbb{R}^4 \tag{10.5}$$

be the state vector with $\theta(t) = f(\theta_1(t), \theta_2(t)) \in \mathcal{S} = \{1, 2, 3, 4\}$ mapping all the possible combinations of θ_1 and θ_2. For instance, we adopt

$$f(-1, -1) = 1, \qquad f(-1, 1) = 2, \qquad f(1, -1) = 3, \qquad f(1, 1) = 4, \tag{10.6}$$

meaning that $\theta = 1$ for $\theta_1 = \theta_2 = -1$, $\theta = 2$ for $\theta_1 = -1$, $\theta_2 = 1$, $\theta = 3$ for $\theta_1 = 1$, $\theta_2 = -1$, and $\theta = 4$ for $\theta_1 = \theta_2 = 1$. It is easy to check that, in this case, the system parameters are:

$$A_1 = \begin{bmatrix} 0 & 1 & 0 & 0 \\ -\omega_1^2 - \epsilon & 0 & -\epsilon & 0 \\ 0 & 0 & 0 & 1 \\ -\epsilon & 0 & -\omega_2^2 - \epsilon & -2\alpha_2 \end{bmatrix},$$

$$A_2 = \begin{bmatrix} 0 & 1 & 0 & 0 \\ -\omega_1^2 - \epsilon & 0 & \epsilon & 0 \\ 0 & 0 & 0 & 1 \\ \epsilon & 0 & -\omega_2^2 - \epsilon & -2\alpha_2 \end{bmatrix},$$

$$A_3 = \begin{bmatrix} 0 & 1 & 0 & 0 \\ -\omega_1^2 + \epsilon & 0 & -\epsilon & 0 \\ 0 & 0 & 0 & 1 \\ -\epsilon & 0 & -\omega_2^2 + \epsilon & -2\alpha_2 \end{bmatrix},$$

$$A_4 = \begin{bmatrix} 0 & 1 & 0 & 0 \\ -\omega_1^2 + \epsilon & 0 & \epsilon & 0 \\ 0 & 0 & 0 & 1 \\ \epsilon & 0 & -\omega_2^2 + \epsilon & -2\alpha_2 \end{bmatrix},$$

and the transition rate matrix of θ is $\Pi = \Pi_1 \otimes I + I \otimes \Pi_2$, that is,

$$\Pi = \begin{bmatrix} -(\eta_1 + \eta_2) & \eta_2 & \eta_1 & 0 \\ \eta_2 & -(\eta_1 + \eta_2) & 0 & \eta_1 \\ \eta_1 & 0 & -(\eta_1 + \eta_2) & \eta_2 \\ 0 & \eta_1 & \eta_2 & -(\eta_1 + \eta_2) \end{bmatrix}. \tag{10.7}$$

Let us further adopt the system data as in the numerical example of [212]. In this case,

$$\epsilon = 0.1, \qquad \omega_1 = 1, \qquad \omega_2 = 10, \qquad \alpha_2 = 0.05, \qquad \eta_1 = 2.5, \qquad \eta_2 = 0.5,$$

which renders the nominal system *unstable* in the mean-square sense, because $\mathrm{Re}\{\lambda(\mathcal{A})\} \simeq 1.723 \times 10^{-3} > 0$. In the next subsections we shall treat the control of the above system by means of some of the design techniques studied in the preceding chapters. Our interest lies on state feedback controls of the form $u(t) = K_{\theta(t)} x(t)$, for which the corresponding closed-loop system is

$$\dot{x}(t) = (A_{\theta(t)} + B_{\theta(t)} K_{\theta(t)}) x(t), \tag{10.8}$$

Table 10.5 Mean-square stabilizing controller (Lemma 3.37(SS3))

Stability check	$\mathrm{Re}\{\lambda(\mathcal{A})\} \simeq -0.9824 < 0$
Controller	$K_1 = -[1.8699\ 8.4810\ 74.5778\ 9.3089]$
	$K_2 = -[1.6835\ 8.5017\ 75.0598\ 9.3293]$
	$K_3 = -[2.3594\ 8.3003\ 72.6871\ 9.2169]$
	$K_4 = -[2.1783\ 8.2932\ 72.8735\ 9.2228]$

where we assume that

$$B_i = \begin{bmatrix} 0 \\ 1 \\ 0 \\ 1 \end{bmatrix}, \quad i \in \mathcal{S}. \tag{10.9}$$

10.3.2 Mean-Square Stabilization

The fundamental requirement we shall impose on the control (10.8) is that it must ensure the MSS of the closed-loop system. To this end, the LMIs design in Lemma 3.37(SS3) yields the controller in Table 10.5. It should be noted that even though the controller stabilizes the system, no performance guarantees are obtained from this design a priori; this issue is tackled next.

10.3.3 H_2 Control

The controller obtained in the preceding subsection ensures that the system mean-square stability is maintained even in the face of adverse coupling. We shall see next, however, that the design of a controller for system (10.3) may be done in such a way as to improve the *performance* of the closed-loop system. More specifically, we now turn our attention to mean-square stabilizing controllers that are *optimal* in the H_2 sense, following Chap. 5.

In order to design an H_2 controller, we need to define with which proportion states and controls should be penalized. A possible choice is

$$C_i \equiv \begin{bmatrix} I_4 \\ 0_{1\times 4} \end{bmatrix}, \quad D_i \equiv \begin{bmatrix} 0_{4\times 1} \\ 1 \end{bmatrix}, \quad i \in \mathcal{S}, \tag{10.10}$$

for which the corresponding quadratic cost, as studied in Chap. 4, is $\int_0^\infty E(\|x(t)\|^2 + |u(t)|^2)\,dt = \int_0^\infty E(\|x(t)\|^2 + |K_{\theta(t)}x(t)|^2)\,dt$. In comparison with the design treated in the preceding section, controllers which minimize this quadratic cost are expected to reduce the magnitude of swings in the closed-loop response, with smaller controller effort.

Table 10.6 H_2 controller for Π exactly known (Theorem 5.7)

Stability check	$\mathrm{Re}\{\lambda(\mathcal{A})\} \simeq -1.0048 < 0$
H_2 norm	$\sqrt{\beta} = 9.8362$
Controller	$K_1 = -[0.3821\ 1.3520\ 1.2296\ 0.9000]$
	$K_2 = -[0.3823\ 1.3509\ 1.2314\ 0.9017]$
	$K_3 = -[0.4243\ 1.3628\ 1.2403\ 0.9003]$
	$K_4 = -[0.4244\ 1.3617\ 1.2422\ 0.9021]$

Table 10.7 H_2 controller for uncertain transition rates (Corollary 5.9)

H_2 cost	$\sqrt{\mu} = 9.8527$
Controller	$K_1 = -[0.3415\ 1.3684\ 1.1435\ 0.9013]$
	$K_2 = -[0.3448\ 1.3586\ 1.3030\ 0.9049]$
	$K_3 = -[0.5139\ 1.3942\ 1.2153\ 0.9002]$
	$K_4 = -[0.5165\ 1.3900\ 1.3822\ 0.9036]$

We also choose $J_i \equiv I$ and assume that $\nu_i \equiv 1/4$ without loss of generality (see Remark 5.12). In this case, by solving the LMIs problem in Theorem 5.7 we obtain the closed-loop H_2 norm and corresponding controller indicated in Table 10.6. For the sake of comparison, notice that the closed-loop H_2 cost ensured via the controller in the preceding section (computed from the LMIs in Lemma 9.18) is higher, about 20.4211, and that the controller entries in Table 10.5 are larger than those in Table 10.6.

A situation of further interest comes up if we assume that the transition rates η_1 and η_2 are uncertain, with a variation of 100 % around the nominal values (that is, $0 \le \eta_1 \le 5$ and $0 \le \eta_2 \le 1$). In this case the polytope $\mathbb{V}$ has four vertices, and the corresponding controller obtained from Corollary 5.9 yields the results indicated in Table 10.7. Although the controller is guaranteed to stabilize the system for all $\Pi \in \mathbb{V}$, notice that the computation of the continuum of values assumed by $\mathrm{Re}\{\lambda(\mathcal{A})\}$ is entirely impractical in this case.

10.3.4 Stability Radius Analysis

Suppose now that the coupling term at the right-hand side of (10.3) is perturbed, giving rise to the uncertain system

$$\ddot{\varphi}_j(t) + 2\alpha_j\dot{\varphi}_j(t) + \omega_j^2\varphi_j(t) = \sum_{k=1}^{M} \epsilon\big(1 + \delta_{\theta(t)}^{jk}\big)\mu_j(t)\mu_k(t)\varphi_k(t), \tag{10.11}$$

with $j = 1, \dots, M$ as before. In the case of two machines, it is easy to check that this corresponds to a perturbation on system (10.8), of the form

$$\dot{x}(t) = \big(\widetilde{A}_{\theta(t)} + E_{\theta(t)}\Delta_{\theta(t)}F_{\theta(t)}\big)x(t), \tag{10.12}$$

Table 10.8 Closed-loop robust stability margins computed via Algorithm 9.6

	$\max\{\rho, \rho^\star\}$	H_2 cost
Stabilizing control (Table 10.5)	1.9921	20.6625
H_2 control, known rates (Table 10.6)	1.3063	9.8362
H_2 control, uncertain rates (Table 10.7)	1.3357	9.8527

where $\widetilde{A}_{\theta(t)} := A_{\theta(t)} + B_{\theta(t)}K_{\theta(t)}$, and

$$\Delta_{\theta(t)} = \begin{bmatrix} \delta^{11}_{\theta(t)} & \delta^{12}_{\theta(t)} \\ \delta^{21}_{\theta(t)} & \delta^{22}_{\theta(t)} \end{bmatrix}, \qquad E_1 = E_4 = \begin{bmatrix} 0 & 0 \\ 1 & 0 \\ 0 & 0 \\ 0 & 1 \end{bmatrix}, \qquad F_1 = F_4 = \begin{bmatrix} 1 & 0 & 0 & 0 \\ 0 & 0 & 1 & 0 \end{bmatrix},$$

$$E_2 = -E_3 = \begin{bmatrix} 0 & 0 \\ -1 & 0 \\ 0 & 0 \\ 0 & 1 \end{bmatrix}, \qquad F_2 = F_3 = \begin{bmatrix} 1 & 0 & 0 & 0 \\ 0 & 0 & -1 & 0 \end{bmatrix}.$$

System (10.8) can thus be seen as a particular instance of (10.12) if $\Delta_{\theta(t)} \equiv 0$. In general, (10.12) describes a situation where the coupling gain, ϵ, is subject to jump disturbances around its nominal value, which are typically not infinitesimal. In fact we may and will assume, as in Sect. 9.2, that $\mathbf{\Delta} = (\Delta_1, \dots, \Delta_4)$ merely satisfies, for a prespecified $\alpha > 0$, the constraint

$$\|\mathbf{\Delta}\|_{\max} < \alpha. \tag{10.13}$$

We already know that, for $\alpha = 0$, the mean-square stability of system (10.12) is guaranteed by each of the controllers in Tables 10.5, 10.6, and 10.7. In addition, a measure of robustness to these controllers is provided by the *stability radii* of (10.11), which, as seen in Definition 9.2, correspond to the smallest $\alpha > 0$ for which there is $\mathbf{\Delta}$ as in (10.13) that destabilizes (10.12).

For simplicity, we shall restrict ourselves to analyzing the *complex* stability radius. As shown in Theorem 9.5, an estimate to the complex radius is obtained by maximizing α in the LMIs problems (9.4) and (9.5). By running Algorithm 9.6 with precision $\varepsilon = 10^{-6}$ and bisection bounds $\rho_{\min} = \rho^\star_{\min} = 0$ and $\rho_{\max} = \rho^\star_{\max} = 100$, we obtain, after at most 30 iterations, the robustness margins shown in Table 10.8. The larger robustness is attained by the stabilizing controller of Table 10.5, which as seen before is the one with larger H_2 cost (shown in the last column for convenience). The H_2 optimal controller is greedy for performance and, not surprisingly, offers a relatively poor robustness. The H_2 controller of Table 10.7 achieves a trade-off between performance and robustness and has the great advantage of not relying on the precise knowledge of the transition rates.

Table 10.9 Robust H_2 controller for uncertain transition rates, with $\alpha = 15$ (Theorem 9.22(i))

H_2 cost	$\sqrt{\mu} = 21.4112$
Controller	$K_1 = -[23.7524\ 10.9740\ 30.5194\ 1.3973]$
	$K_2 = -[23.7099\ 10.9661\ 30.6861\ 1.4038]$
	$K_3 = -[23.9518\ 10.9883\ 30.5970\ 1.3874]$
	$K_4 = -[23.8786\ 10.9550\ 30.6849\ 1.3926]$

Table 10.10 Robust H_2 controller for uncertain transition rates, with $\alpha = 15$ (Theorem 9.22(ii))

H_2 cost	$\sqrt{\mu} = 19.3338$
Controller	$K_1 = -[17.7998\ 7.1251\ 19.1098\ 1.5738]$
	$K_2 = -[17.7381\ 7.1311\ 19.4306\ 1.5732]$
	$K_3 = -[18.0501\ 7.1319\ 18.9506\ 1.5864]$
	$K_4 = -[17.9843\ 7.1302\ 19.2597\ 1.5837]$

10.3.5 Synthesis of Robust Controllers

The applicability of the analysis carried out in the preceding subsection is somewhat limited, because it is assumed that a stabilizing controller is known a priori. In this final subsection we circumvent this issue and treat the design of a controller that provides a satisfactory trade-off between robustness and H_2 performance.

Let $\alpha = 15$ and consider the design of controllers for which, regardless of $\mathbf{\Delta}$ as in (10.13) or of $\Pi \in \mathbb{V}$ as in the preceding subsection, the closed-loop system (10.12) is mean-square stable, with guaranteed H_2 cost as small as possible. In this case the designs from Theorem 9.22(i) and Theorem 9.22(ii) yield the results in Tables 10.9 and 10.10.

10.4 Robust Control of an Underactuated Robotic Arm

Controlling the dynamics of an underactuated manipulator robot, which is nonlinear and depends on the arm's geometry, is not an easy task. Furthermore, in applications it will always be desirable for the robotic arm to undergo operation under prespecified time and energy constraints. Therefore, whenever actuator faults are likely to occur, the controller's response has to be sufficiently fast as to effectively capture sudden changes on the system's point of operation. This clearly represents a critical issue to the design of gain-scheduled controllers, because the success of such techniques strongly depends upon having a sufficiently fine gridding phase and, at the same time, being able to perform all the necessary computations online.

The gist of the application of MJLS theory to the control of underactuated manipulators through a given reference trajectory relies on (i) identifying a representative set of operation points, around which the original model is linearized, (ii) choosing the transition rates on the basis of an estimate of the waiting times that are necessary

Table 10.11 Linearization points of the underactuated manipulator

Domain	q_1	q_2	q_3	$\dot{q}_1$	$\dot{q}_2$	$\dot{q}_3$
ζ_1	0	0	0	0	0	0
ζ_2	4.9988	4.9974	4.9987	6.3611	6.3533	6.3608
ζ_3	9.9972	9.9592	9.9962	7.5048	7.3544	7.5011
ζ_4	15.0081	14.6371	15.0014	6.3770	5.4425	6.3652
ζ_5	17.9733	20.0718	21.9564	22.2354	0	−16.8097
ζ_6	20	20	20	0	0	0

for the system to move across the reference trajectory, and (iii) actually implementing, in the original system, the controller computed via MJLS design algorithms. Better results are obtained if one is able enough to properly tune a proportional-derivative controller, for the sake of precompensation of model imprecisions. Practical experiments were conducted by Siqueira and Terra et al., in the UARM II (Underactuated arm II) robotic manipulator, which is a device especially suited to the academic benchmark of control methods for underactuated manipulators. The UARM II is a planar horizontal manipulator comprised of three joints, which can be made either active or passive, by the activation of direct current motors and pneumatic brakes, as to simulate various faulty conditions. Due to smaller control effort and faster response after the occurrence of an actuator failure, the experimental results reported in [261] indicate that MJLS control techniques may sometimes outperform both gain-scheduling and standard nonlinear H_∞ controllers. Furthermore, the MJLS control strategy devised in [261] does not rely on completely stopping the arm (which requires the abrupt activation of breaks) after a fault is detected, so that the stress on mechanical parts is reduced.

In this section we tackle the control of an underactuated robotic system by means of the LMIs approach of Theorem 9.22. We partially follow the techniques reported in [283]. Our approach to the control of a planar three-link robotic arm whose second joint is failure prone follows closely the one in [261], except for some aspects that are explicitly discussed throughout this section. Unlike the work of Siqueira, Terra, and their coworkers, we do not consider the physical implementation of the robot, but only its numerical simulation.

Differently from all previous work in the MJLS literature, in this example we consider only *seven* possible configurations in the workspace (instead of 24 as in [261]), corresponding to the six different operation points indicated (in deg and deg /s) in Table 10.11. This choice is in accordance to an evenly spaced spatial sampling of the system configuration under a hypothetical faultless motion along the set-point trajectories given (in degrees) by $o(t) = \min\{t^3(1.6 - 0.48t + 0.0384t^2), 20\}$ for the first and third joints, and $\tilde{o}(t) = o(t) + 1.8\,e^{-2(t-4.5)^2}$ for the second joint, which is the task we would like the robot to perform. Moving across this trajectory causes the system to temporarily dwell within different configuration domains as time goes by. Table 10.12 indicates how long it takes for the system to

Table 10.12 Fault dynamics of the underactuated manipulator

i	Phase	Operation point	Exit trigger	Expected waiting time (s)
1	AAA	ζ_1	average position crosses 5 deg	1.7970
2	AAA	ζ_2	average position crosses 10 deg	0.7025
3	AAA	ζ_3	average position crosses 15 deg	0.7033
4	AAA	ζ_4	fast transition	0.0100
5	APAu	ζ_4	2nd link reaches desired final target	1.1313
6	APAl	ζ_5	all links reach desired final target	1.0000
7	APAl	ζ_6	end of simulation	1.0000

move across the corresponding configurations which were chosen for this example. As the table shows, we assume that the fault is expected to occur whenever *average position*, which stands for the averaged setpoint $\{o(t)+\tilde{o}(t)+o(t)\}/3 = o(t)+0.6e^{-2(t-4.5)^2}$, crosses 15 deg. Accordingly, in our simulations the system forcedly suffers an actuator failure whenever the actual average position reaches 15 degrees (that is, $(q_1+q_2+q_3)/3 \geq \pi/12$), driving the system to the control phase APAu (whose role has been thoroughly discussed in [261]). Afterwards, once the passive joint reaches a vicinity of the final position (up to 4 decimals around 20 degrees), phase APAl is activated. Our experiments were conducted under the observation that $E(T_{k+1}-T_k \mid \theta(T_k)=i) = -1/\lambda_{ii}$ for the jump process defined in (2.13), in which $\{T_k\}_{k\geq 0}$ stands for the random time instants at which the process jumps.[2] Bearing in mind this fact, the transition rates were chosen as the reciprocal of the measured waiting times on each mode of operation of the setpoint trajectories—with the exception of the instantaneous state transition from mode $i=4$ to $i=5$ (represented by a very large rate) and of an artificial reset from $i=7$ to $i=1$, which is introduced in order to ensure the positive recurrence of the related Markov process (so that θ is free of absorbing modes, who should compromise the applicability of infinite-horizon techniques). To be explicit, we chose

$$\Pi = \begin{bmatrix} -0.5565 & 0.5565 & 0 & 0 & 0 & 0 & 0 \\ 0 & -1.4235 & 1.4235 & 0 & 0 & 0 & 0 \\ 0 & 0 & -1.4219 & 1.4219 & 0 & 0 & 0 \\ 0 & 0 & 0 & -100 & 100 & 0 & 0 \\ 0 & 0 & 0 & 0 & -0.8839 & 0.8839 & 0 \\ 0 & 0 & 0 & 0 & 0 & -1 & 1 \\ 1 & 0 & 0 & 0 & 0 & 0 & -1 \end{bmatrix}$$

as the transition-rate matrix of θ. The bidiagonal structure of this choice naturally reflects the assumption that the robot's setpoint trajectory is of the "slope-like" shape

[2] This easily follows from the waiting times being exponentially distributed with rate λ_{ii} whenever $\theta = i$.

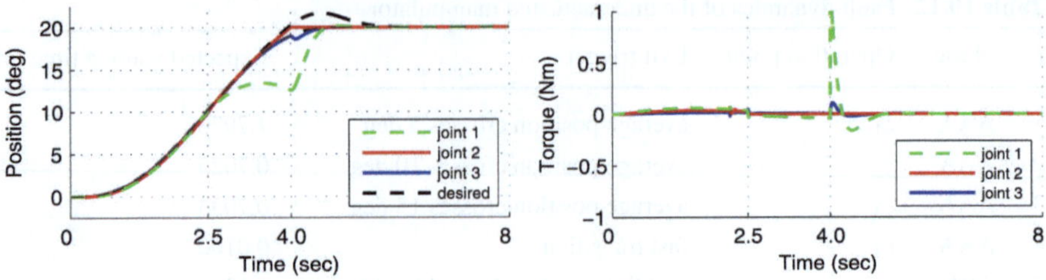

Fig. 10.3 Theorem 9.22(ii): Joint positions (*left*) and applied torques (*right*)

of $o(t)$ and $\tilde{o}(t)$ (see Fig. 10.3). The arrays $\mathbf{A}$, $\mathbf{B}$ in (9.51) were chosen exactly as in [262], with $\mathbf{C}$ and $\mathbf{D}$ given by

$$C_i \equiv 10 \times \begin{bmatrix} I_3 & 0 \\ 0 & 0 \end{bmatrix}, \quad D_i \equiv \begin{bmatrix} 0_3 \\ I_3 \end{bmatrix}, \quad i \in \mathcal{S},$$

and

$$J_i^{\varpi} = \begin{bmatrix} 0 & 0 & 0 \\ 0 & 0 & 0 \\ 0 & 0 & 0 \\ 1 & 0 & 0 \\ 0 & 1_{\{i \geq 5\}} & 0 \\ 0 & 0 & 1 \end{bmatrix}, \quad E_i \equiv \begin{bmatrix} I_3 \\ 0_3 \end{bmatrix}, \quad F_i \equiv 0_3, \quad i \in \mathcal{S},$$

whereas $J_i^w \equiv I$. Assume also that the initial distribution of θ equals the invariant distribution of $\Pi = [\lambda_{ij}]$, i.e.,

$$\nu_i = \frac{\lambda_{ii}^{-1}}{\sum_{j \in \mathcal{S}} \lambda_{jj}^{-1}}, \quad i \in \mathcal{S},$$

so that $\sum_{i \in \mathcal{S}} \nu_i \lambda_{ij} = 0$ for all $j \in \mathcal{S}$.

In this scenario, we solved both LMIs designs in Theorem 9.22 with $\alpha = 0.1$. The results reveal that the controllers obtained from (i) and (ii) (say, $\mathbf{K}^{(\mathrm{i})}$ and $\mathbf{K}^{(\mathrm{ii})}$) are essentially the same since $\|\mathbf{K}^{(\mathrm{i})} - \mathbf{K}^{(\mathrm{ii})}\|_{\max} < 10^{-1}$ in this example. However, the controller (ii) is obtained three and a half times faster than the one corresponding to (i) (about 4 seconds versus 14 seconds, on average, in an Intel Core i7 CPU with 2.8 GHz and 6 GB RAM). Just for illustration, the system trajectories yielded by the adjoint controller are depicted in Figs. 10.3 and 10.4.

10.5 An Example of a Stationary Filter

In this section we carry out a numerical example in order to illustrate the performance of the filters derived in Sect. 9.5. We confine our discussion performing the

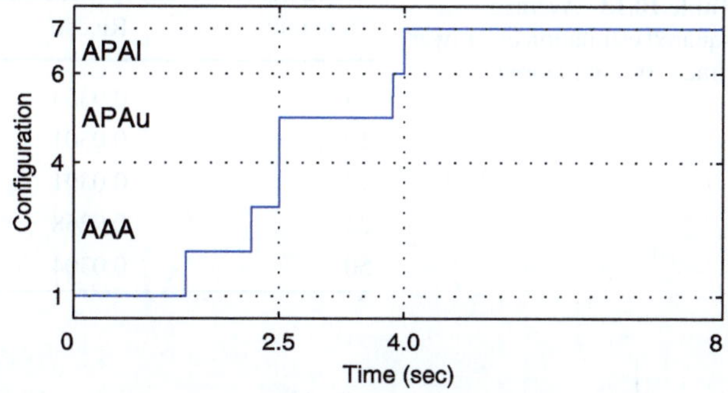

Fig. 10.4 Configuration process for the robust controller from (ii) in Theorem 9.22

comparison vis-à-vis the optimal filter for the case with complete observation of the regime (from now on denoted by COF), which can be found in [223]. In order to give just a glimpse of the performance, we focus our attention on the Riccati and LMIs versions of the stationary filter. Consider the two-mode scalar system given by

$$\left[\begin{array}{c|c} A_1 & J_1 \\ \hline H_1 & G_1 \\ \hline L_1 & - \end{array}\right] = \left[\begin{array}{c|cc} -1 & 0.1 & 0 \\ \hline 0 & 0 & 0.2 \\ \hline 1 & - & - \end{array}\right],$$

$$\left[\begin{array}{c|c} A_2 & J_2 \\ \hline H_2 & G_2 \\ \hline L_2 & - \end{array}\right] = \left[\begin{array}{c|cc} 0.1 & 0.15 & 0 \\ \hline 1 & 0 & 0.3 \\ \hline 1 & - & - \end{array}\right],$$

with the transition rate matrix

$$\Pi = \begin{bmatrix} -4 & 4 \\ 2 & -2 \end{bmatrix}.$$

Suppose also that $x(0) = x_0$ is independent of θ_0, so that

$$E\big(z_i(0)\big) = E(x_0 1_{\{\theta_0 = i\}}) = E(x_0)P(\theta_0 = i), \quad i = 1, 2,$$

where x_0 is Gaussian with

$$\hat{x}_0 := E(x_0) = 0.2 \quad \text{and} \quad E\big[(x_0 - \hat{x}_0)^2\big] = 0.01.$$

Finally, let us assume that the distribution of θ_0 is identical to the invariant distribution of θ:

$$\pi_1 := P(\theta_0 = 1) = 1/3, \qquad \pi_2 := P(\theta_0 = 2) = 2/3.$$

A solution obtained for the stationary filtering problem of Theorem 7.15 (Riccati equation formulation) is given by

$$\left[\begin{array}{c|c} A_{f,op} & B_{f,op} \\ \hline L_{f,op} & - \end{array}\right] = \left[\begin{array}{cc|c} -5.0000 & 2.0489 & -0.0489 \\ 4.0000 & -2.2652 & 0.3652 \\ \hline 1.0000 & 1.0000 & - \end{array}\right]$$

Table 10.13 Average squared estimation error over time (100 simulations)

Time units	Riccati	LMIs	COF
5	0.0353	0.0357	0.0348
10	0.0301	0.0302	0.0263
15	0.0301	0.0301	0.0265
25	0.0268	0.0268	0.0252
50	0.0304	0.0303	0.0265

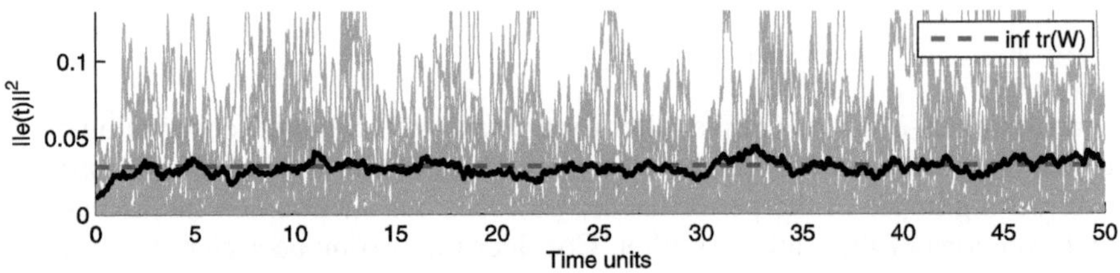

Fig. 10.5 Squared estimation errors for the Riccati filter

with $\hat{z}_{op}(0) = (0.0667, 0.1333)$. As for the filter corresponding to the LMIs formulation, we found that

$$\left[\begin{array}{c|c} A_f & B_f \\ \hline L_f & - \end{array}\right] = \left[\begin{array}{cc|c} 3.2638 & 61.7401 & -1.7448 \times 10^3 \\ -1.3719 & -23.2150 & 0.6665 \times 10^3 \\ \hline 0.0052 & 0.0142 & - \end{array}\right]$$

is a feasible solution to (9.58), (9.59), (9.60), with $U = I$,

$$\hat{z}_{op}(0)^* = (0.0667\ 0.1333),$$

and

$$\inf \operatorname{tr}(W) = 0.0311 \approx \lim_{t\to\infty} E\big(\|e(t)\|^2\big). \tag{10.14}$$

For this example, we performed a Monte Carlo simulation of 100 randomly generated trajectories of the system and compared the obtained results for each stationary filter (Riccati and LMIs versions) with those of the COF (see Sect. 5.3 of [223]) over 50 time units. The mean values for the corresponding quadratic errors (that is, the mean of $\|e(t)\|^2$ for 100 simulations) on different time instants is shown in Table 10.13. As one can see, all values are close to $\inf \operatorname{tr}(W)$ in (10.14).

A plot of the quadratic estimation errors for each simulation is depicted in Figs. 10.5, 10.6, and 10.7, where thick lines correspond to the average trajectories, and the dashed lines represent $\inf \operatorname{tr}(W)$ in (10.14). Also, a comparison of the average estimation errors between each of the filters is presented in Fig. 10.8. Finally, a sample trajectory of the estimation problem is shown in Fig. 10.9.

From the simulations we can see that, as expected, the COF performs better, although the estimation errors are quite similar. Perhaps this similarity has to do with

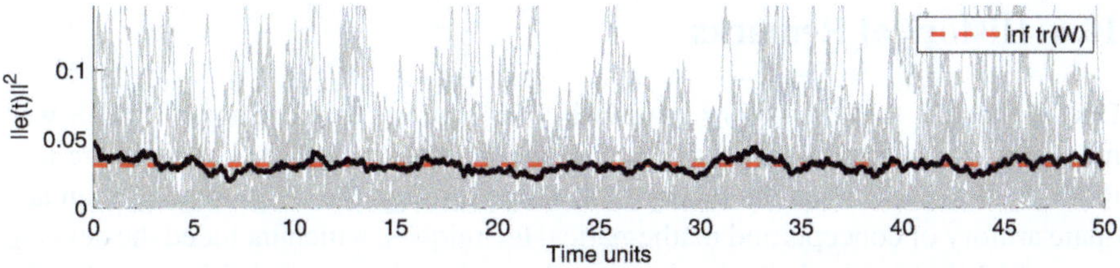

Fig. 10.6 Squared estimation errors for the LMIs filter

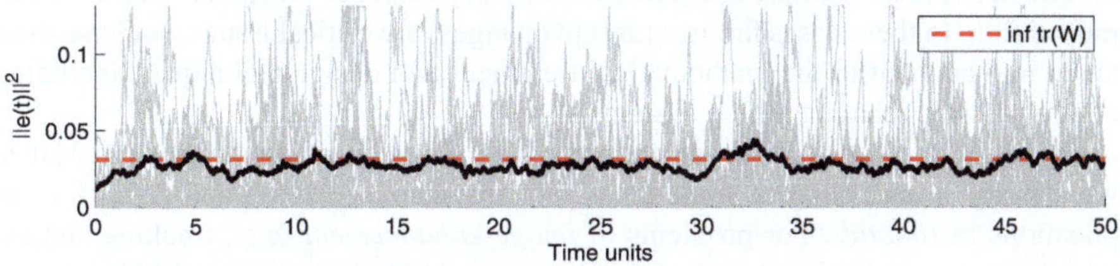

Fig. 10.7 Squared estimation errors for the COF

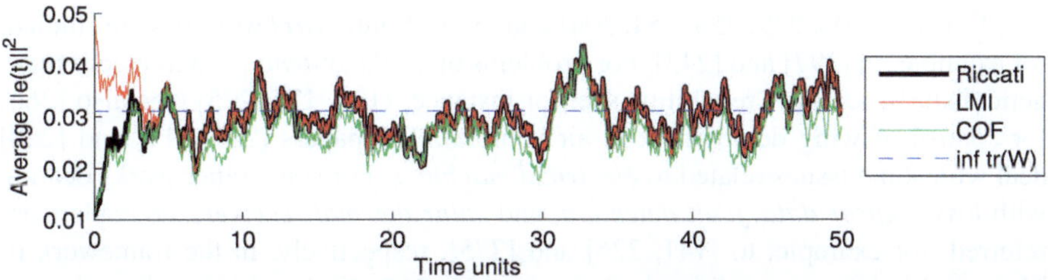

Fig. 10.8 Comparison of average squared errors via Monte Carlo (100 simulations)

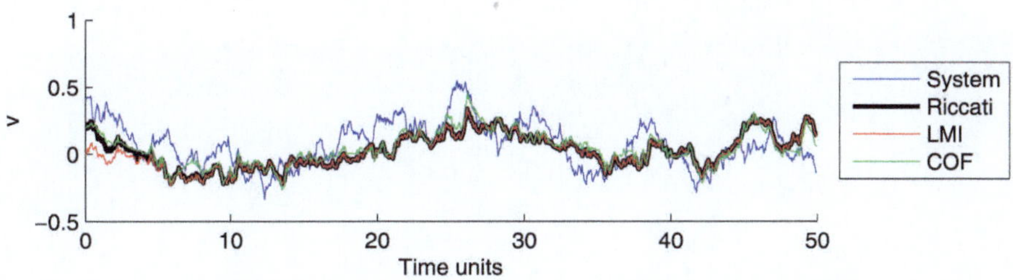

Fig. 10.9 Sample trajectory of the estimation problem

the required stability for the system, although we have not considered an exhaustive number of examples. The simulations seem, prima facie, strongly support the theoretical results.

10.6 Historical Remarks

Due to the fact that the theory took awhile to mature, applications of MJLS were meager at the beginning. The rare exceptions were [32] and [33], which date back to 1975. However, when the theory took shape through the development of an adequate armory of concepts and mathematical techniques, which included the development of LMIs-based results, applications began to emerge at a brisk pace. Another favorable point regarding applications is the fact that reliability is a main issue in many modern society complex systems, such as in aircrafts, nuclear power stations, mobile networks, robotics, integrated communication networks, and wireless communication. In these cases failures (abrupt changes) are critical issues that have to be taken into account in the mathematical treatment and design techniques, including control design, related to these systems.

In order to facilitate the organization of the references on applications of MJLS, we classify it according to some areas. For instance, see [261] and [262] for applications in *robotics*. For problems of *image-enhancement* (e.g., tracking and estimation), the reader is referred, for instance, to [124, 133, 272] and [276]. Some problems in mathematical *finance* are treated, for example, in [34, 68, 88, 108] and [315]. In the scenario of communication *networks*, issues such as packet loss, fading channels, and chaotic communication are considered, for instance, in [57, 173, 186, 208, 249, 253, 254, 304] and [323]. Some *wireless* issues are studied, for example, in [197] and [243]. For problems on *flight systems* such as electromagnetic disturbances and reliability, see, for instance, [174, 175, 225] (see also [266] for control of wing deployment in air vehicle). The papers [212, 219] and [220] deal with some issues related to *electrical machines*. For some other works dealing with *lossy sensor data*, *fault detection*, and *solar thermal receiver*, the readers are referred, for example, to [141, 226] and [275], respectively. In the framework of Multiple-Model, we mention, for instance, [35, 213, 296] and [184]. See also the books [22, 271] and references therein.

Appendix A
Coupled Differential and Algebraic Riccati Equations

A.1 Outline of the Appendix

This appendix is mainly concerned with the coupled differential and algebraic Riccati equations (CDRE and CARE, respectively) that are used throughout this book. Initially, we consider the problem of uniqueness, existence, positive definiteness, and continuity of the solution of the CDRE. After that we study the CARE. We deal, essentially, with conditions for the existence of solutions and asymptotic convergence, based on the concepts of mean-square stabilizability and detectability seen in Sect. 3.6. Regarding the existence of a solution, we are particularly interested in maximal and stabilizing solutions. The appeal of the maximal solution has to do with the fact that it can be obtained numerically via a certain LMIs optimization problem (see Lemma A.6). Although in control and filtering applications the interest lies essentially in the stabilizing solution, it is shown that the two concepts of solution coincide whenever the stabilizing solution exists (see Remark A.10). In Sect. A.7 we presented the proof of Theorem 7.15, related to the differential and algebraic Riccati equations for the filtering problem when $\theta(t)$ is unknown. This proof follows essentially the same ideas as for the CARE case.

A.2 Coupled Differential Riccati Equations

We consider in this section only the equations related to the control problem, and for this reason, we borrow the notation and definitions from Chap. 4. The results related to the filtering problem are obtained in a similar way, and, for this reason, we only state the theorems in Sect. A.5, with the details omitted. For the CDRE, we will consider a time-varying model as in [53], defined by the following jump controlled system $\mathcal{G}$:

$$\mathcal{G} = \begin{cases} \dot{x}(t) = A_{\theta(t)}(t)x(t) + B_{\theta(t)}(t)u(t), \\ z(t) = C_{\theta(t)}(t)x(t) + D_{\theta(t)}(t)u(t), \end{cases} \tag{A.1}$$

O.L.V. Costa et al., *Continuous-Time Markov Jump Linear Systems*, Probability and Its Applications, DOI 10.1007/978-3-642-34100-7,

where $A_i(t)$, $B_i(t)$, $C_i(t)$, $D_i(t)$ are real matrices of class **PC** (see Definition 2.3). We also assume that $D_i^*(t)D_i(t) > 0$ and $C_i^*(t)D_i(t) = 0$ for all $i \in \mathcal{S}$ and $t \in \mathbb{R}^+$ and that the matrices $(D_i^*(t)D_i(t))^{-1}$ are of class **PC**. We write $\mathbb{D} \subset \mathbb{R}^+$ as the union of the discontinuity points of A_i, B_i, C_i, D_i, $(D_i^* D_i)^{-1}$, $i \in \mathcal{S}$. As pointed out in [53], p. 7, for any $T \in \mathbb{R}^+$, $\mathbb{D} \cap [0, T]$ contains at most a finite number of points.

We recall from (4.27) the definition of the nonlinear operator $\mathcal{R}(t, \cdot) : \mathbb{H}^n \to \mathbb{H}^n$ and introduce the definition of the linear operator $\mathcal{K}(t, \cdot) \in \mathbb{B}(\mathbb{H}^n)$ in the following way: for any $\mathbf{X} = (X_1, \dots, X_N) \in \mathbb{H}^n$, $\mathcal{K}(\mathbf{X}, t) = (\mathcal{K}_1(\mathbf{X}, t), \dots, \mathcal{K}_N(\mathbf{X}, t))$ and $\mathcal{R}(\mathbf{X}, t) = (\mathcal{R}_1(\mathbf{X}, t), \dots, \mathcal{R}_N(\mathbf{X}, t))$ are defined as

$$\mathcal{K}_i(\mathbf{X}, t) = \left[D_i^*(t)D_i(t)\right]^{-1} B_i^*(t)X_i, \quad i \in \mathcal{S}, \tag{A.2}$$

and

$$\begin{aligned}\mathcal{R}_i(\mathbf{X}, t) = {} & A_i^*(t)X_i + X_i A_i(t) - X_i B_i(t)\big(D_i^*(t)D_i(t)\big)^{-1} B_i^*(t)X_i \\ & + \sum_{j=1}^{N} \lambda_{ij} X_j + C_i^*(t)C_i(t), \quad i \in \mathcal{S}.\end{aligned} \tag{A.3}$$

As in (4.17), for finite $T \in \mathbb{R}^+$ and $\mathbf{L} \in \mathbb{H}^{n+}$ arbitrarily fixed, the set of CDRE is defined as

$$\begin{cases} \dot{\mathbf{X}}^T(t) + \mathcal{R}(\mathbf{X}^T(t), t) = 0, & t \in (0, T), \\ \mathbf{X}^T(T) = \mathbf{L}, \end{cases} \tag{A.4}$$

where $\mathbf{X}^T(t) = (X_1^T(t), \dots, X_N^T(t))$. Equation (A.4) may be written as the following set of coupled differential Riccati equations:

$$\begin{aligned}& \dot{X}_i^T(t) + A_i^*(t)X_i^T(t) + X_i^T(t)A_i(t) - X_i^T(t)B_i(t)\left[D_i^*(t)D_i(t)\right]^{-1} B_i^*(t)X_i^T(t) \\ & \quad + \sum_{j=1}^{N} \lambda_{ij} X_j^T(t) + C_i^*(t)C_i(t) = 0, \quad i \in \mathcal{S},\end{aligned}$$

with boundary condition $\mathbf{X}^T(T) = \mathbf{L}$. The main result of this section is the following.

Theorem A.1 *There exists a unique set of N positive semi-definite and continuous $n \times n$ matrices $\mathbf{X}^T(t) = (X_1^T(t), \dots, X_N^T(t)) \in \mathbb{H}^{n+}$, $0 \le t \le T$, satisfying* (A.4) *for each $t \in [0, T] \setminus \mathbb{D}$.*

For notational simplicity, we will suppress the superscript T that indicates the final time T. The proof of Theorem A.1 follows the same steps as in Sect. 3 of [302], the difference being that we consider the matrices in the class **PC**, and, as a consequence, the solution of the coupled Riccati equations will be continuous, with derivative for all t except on the discontinuity points $\mathbb{D}$. We need first to introduce the following definitions and auxiliary result. For $\mathbf{\Psi}(t) = (\Psi_1(t), \dots, \Psi_N(t))$, $\Psi_i(t)$ in the class **PC** with discontinuity points $t \in \mathbb{D}$, define for $\widetilde{\mathbf{X}} = (\widetilde{X}_1, \dots, \widetilde{X}_N) \in \mathbb{H}^n$, $\mathcal{T}(\mathbf{\Psi}(t), \widetilde{\mathbf{X}}, t) = (\mathcal{T}_1(\mathbf{\Psi}(t), \widetilde{\mathbf{X}}, t), \dots, \mathcal{T}_N(\mathbf{\Psi}(t), \widetilde{\mathbf{X}}, t))$ as

$$\mathcal{T}_i\big(\mathbf{\Psi}(t), \widetilde{\mathbf{X}}, t\big) = \big(A_i(t) - B_i(t)\Psi_i(t)\big)^* \widetilde{X}_i + \widetilde{X}_i^*\big(A_i(t) - B_i(t)\Psi_i(t)\big)$$
$$+ \sum_{j=1}^{N} \lambda_{ij} \widetilde{X}_j + C_i^*(t)C_i(t) + \Psi_i^*(t)D_i^*(t)D_i(t)\Psi_i(t), \quad i \in \mathcal{S}. \tag{A.5}$$

Lemma A.2 *Consider $\widetilde{\mathbf{X}}_f = (\widetilde{X}_{f1}, \dots, \widetilde{X}_{fN}) \in \mathbb{H}^{n+}$. There exists a unique continuous $\widetilde{\mathbf{X}}(t) = (\widetilde{X}_1(t), \dots, \widetilde{X}_N(t)) \in \mathbb{H}^n$, $0 \le t \le T$, such that for all $t \in [0, T] \setminus \mathbb{D}$,*

$$\dot{\widetilde{\mathbf{X}}}(t) + \mathcal{T}\big(\mathbf{\Psi}(t), \widetilde{\mathbf{X}}(t), t\big) = 0, \quad \widetilde{\mathbf{X}}(T) = \widetilde{\mathbf{X}}_f. \tag{A.6}$$

Moreover, $\widetilde{\mathbf{X}}(t) \in \mathbb{H}^{n+}$, and for some $a \ge 0$, $\|\widetilde{X}_i(t)\| \le a$ for all $i = 1, \dots, N$ and $t \in [0, T]$.

Proof The existence, uniqueness, and continuity follow from Theorem 2.4 (see also the Fundamental Theorem B1.2-6 presented in [53], p. 470). To show that $\widetilde{\mathbf{X}}(t) \ge 0$, we adopt the same approach as in Sect. 3 of [302] by considering the fundamental matrix $\Phi_i(t, s)$ associated to the matrix $A_i(t) + \frac{\lambda_{ii}}{2} I - B_i(t)\Psi_i(t)$ and the sequence $\mathbf{Y}^k(t) = (Y_1^k(t), \dots, Y_N^k(t))$ for $t \in [0, T]$ obtained recursively as

$$Y_i^{k+1}(t) = \Phi_i^*(T, t) X_{fi} \Phi_i(T, t) + \int_t^T \Phi_i^*(s, t) \Bigg[\sum_{j \ne i} \lambda_{ij} Y_j^k(s) + C_i^*(s)C_i(s)$$
$$+ \Psi_i^*(s)D_i^*(s)D_i(s)\Psi_i(s)\Bigg] \Phi_i(s, t)\, ds \tag{A.7}$$

and starting with $\mathbf{Y}^0(t) = 0$. From the fact that $\lambda_{ij} \ge 0$ for $j \ne i$ it is easy to see that $\mathbf{Y}^{k+1}(t) \ge \mathbf{Y}^k(t) \ge 0$ for all $t \in [0, T]$. Since all matrices are in the class **PC**, we can find a constant $c > 0$ such that for all $t \in [0, T]$ and $i = 1, \dots, N$, $\|A_i(t) + \frac{\lambda_{ii}}{2} I - B_i(t)\Psi_i(t)\| \le c$, $\|C_i^*(t)C_i(t)\| \le c$, and $\|\Psi_i^*(t)D_i^*(t)D_i(t)\Psi_i(t)\| \le c$. Thus, defining the N-dimensional vector $\ell^*(t) = [\ell_1(t) \ \dots \ \ell_N(t)]$ as the solution of the linear differential equation given, for $i = 1, \dots, N$, by

$$\dot{\ell}_i(t) = 2c\ell_i(t) + \sum_{j \ne i} \lambda_{ij} \ell_j(t) + 2c, \quad \ell_i(T) = \|X_{fi}\|, \tag{A.8}$$

we have, for some $a \ge 0$, that

$$\big\|Y_i^k(t)\big\| \le \ell_i(t) \le a \tag{A.9}$$

for all $i = 1, \dots, N$ and all $t \in [0, T]$. To show (A.9) by induction on k, we first notice from (A.8) that $\ell(t) = e^{R(T-t)}\ell(T) + 2c\int_t^T e^{R(s-t)}\varphi\, ds$, where $R = [R_{ij}]$ is the $N \times N$ matrix with entries

$$R_{ij} = \begin{cases} 2c, & i = j, \\ \lambda_{ij}, & i \neq j, \end{cases}$$

and φ is an N-dimensional vector formed by 1 in all elements. Since $\ell_i(T) \geq 0$ and all elements in R are positive, we have that $\ell_i(t) \geq 0$. Clearly, we can find $a \geq 0$ such that $\ell_i(t) \leq a$ for all $i = 1, \dots, N$ and all $t \in [0, T]$, so that (A.9) follows for $k = 0$ since $Y_i^k(0) = 0$. Suppose that (A.9) holds for k. From (A.8) we have that

$$\ell_i(t) = e^{2c(T-t)}\ell_i(T) + \int_t^T e^{2c(s-t)}\Big[\sum_{j\neq i}\lambda_{ij}\ell_j(s) + 2c\Big]\,ds, \tag{A.10}$$

and from (A.7) and (A.10), due to $\|Y_i^k(t)\| \leq \ell_i(t)$, it follows that

$$\begin{aligned}\big\|Y_j^{k+1}(t)\big\| &\leq \big\|\Phi_i(T,t)\big\|^2\|X_{fi}\| + \int_t^T \big\|\Phi_i(s,t)\big\|^2\Big[\sum_{j\neq i}\lambda_{ij}\big\|Y_j^k(s)\big\| \\ &\quad + \big\|C_i^*(s)C_i(s)\big\| + \big\|\Psi_i^*(s)D_i^*(s)D_i(s)\Psi_i(s)\big\|\Big]ds \\ &\leq e^{2c(T-t)}\ell_i(T) + \int_t^T e^{2c(s-t)}\Big[\sum_{j\neq i}\lambda_{ij}\ell_j(s) + 2c\Big]\,ds = \ell_i(t),\end{aligned}$$

showing (A.9) for $k+1$. Since $0 \leq Y_i^k(t) \leq Y_i^{k+1}(t)$ and $\|Y_i^k(t)\| \leq a$, we get from the monotone convergence result (Lemma 2.17) that there exists $\mathbf{Y}(t) = (Y_1(t), \dots, Y_N(t)) \in \mathbb{H}^{n+}$ such that $\lim_{k\to\infty}\mathbf{Y}^k(t) = \mathbf{Y}(t)$ and $\|Y_i(t)\| \leq a$. From the bounded convergence theorem, taking the limit as $k \to \infty$ in (A.7), it follows that

$$\begin{aligned}Y_i(t) &= \Phi_i^*(T,t)X_{fi}\Phi_i(T,t) + \int_t^T \Phi_i^*(s,t)\Big[\sum_{j\neq i}\lambda_{ij}Y_j(s) + C_i^*(s)C_i(s) \\ &\quad + \Psi_i^*(s)D_i^*(s)D_i(s)\Psi_i(s)\Big]\Phi_i(s,t)\,ds.\end{aligned} \tag{A.11}$$

From (A.11) and $\|Y_i(t)\| \leq a$ it follows that $\mathbf{Y}(t)$ is continuous at all points $t \in [0, T]$ and at the points $t \in [0, T] \setminus \mathbb{D}$ the derivative of (A.11) exists and satisfies $\dot{\mathbf{Y}}(t) + \mathcal{T}(\mathbf{\Psi}(t), \mathbf{Y}(t), t) = 0$. Thus, from the uniqueness of the solution of (A.6) we get that $\widetilde{\mathbf{X}}(t) = \mathbf{Y}(t)$, showing that indeed $\widetilde{\mathbf{X}}(t) \in \mathbb{H}^{n+}$. □

Before proceeding to the proof of Theorem A.1, we notice the following identity valid for any $\widetilde{\mathbf{X}}(t) = (\widetilde{X}_1(t), \dots, \widetilde{X}_N(t)) \in \mathbb{H}^n$:

$$\begin{aligned}&\big(A_i(t) - B_i(t)\mathcal{K}_i(\widetilde{\mathbf{X}}, t)\big)^*\widetilde{X}_i + \widetilde{X}_i^*\big(A_i(t) - B_i(t)\mathcal{K}_i(\widetilde{\mathbf{X}}, t)\big) \\ &\quad + \mathcal{K}_i^*(\widetilde{\mathbf{X}}, t)D_i^*(t)D_i(t)\mathcal{K}_i(\widetilde{\mathbf{X}}, t)\end{aligned}$$

$$
\begin{aligned}
&= \big(A_i(t) - B_i(t)\Psi_i(t)\big)^* \widetilde{X}_i + \widetilde{X}_i^* \big(A_i(t) - B_i(t)\Psi_i(t)\big) \\
&\quad + \Psi_i^*(t) D_i^*(t) D_i(t) \Psi_i(t) \\
&\quad - \big(\Psi_i(t) - \mathcal{K}_i(\widetilde{\mathbf{X}}, t)\big)^* \big(D_i^*(t) D_i(t)\big) \big(\Psi_i(t) - \mathcal{K}_i(\widetilde{\mathbf{X}}, t)\big). \qquad \text{(A.12)}
\end{aligned}
$$

We have the following lemma, showing the minimality of the solution of (A.4).

Lemma A.3 *Consider $\mathbf{\Psi}(t) = (\Psi_1(t), \dots, \Psi_N(t))$, $\Psi_i(t)$ in the class* **PC** *with discontinuity points $t \in \mathbb{D}$ and let $\widetilde{\mathbf{X}}(t) = (\widetilde{X}_1(t), \dots, \widetilde{X}_N(t)) \in \mathbb{H}^n$, $0 \le t \le T$, be the unique continuous solution satisfying* (A.6) *with boundary conditions $\widetilde{\mathbf{X}}(T) = \mathbf{L}$ for all $t \in [0, T] \setminus \mathbb{D}$. If $\mathbf{X}(t) = (X_1(t), \dots, X_N(t)) \in \mathbb{H}^n$ is continuous for $0 \le t \le T$ and satisfies* (A.4) *for all $t \in [0, T] \setminus \mathbb{D}$ with boundary condition $\mathbf{X}(T) = \mathbf{L}$, then $\mathbf{X}(t) \le \widetilde{\mathbf{X}}(t)$ for all $0 \le t \le T$.*

Proof Setting $\widehat{\mathbf{\Psi}}(t) = \mathcal{K}(\mathbf{X}(t), t)$ and $\mathbf{R}(t) = (R_1(t), \dots, R_N(t)) \in \mathbb{H}^{n+}$, $R_i(t) = (\Psi_i(t) - \widehat{\Psi}_i(t))^* (D_i^*(t) D_i(t))(\Psi_i(t) - \widehat{\Psi}_i(t))$, we have from (A.12) that we can rewrite (A.4), for all $t \in [0, T] \setminus \mathbb{D}$, as

$$
0 = \dot{\mathbf{X}}(t) + \mathcal{T}\big(\widehat{\mathbf{\Psi}}(t), \mathbf{X}(t), t\big) = \dot{\mathbf{X}}(t) + \mathcal{T}\big(\mathbf{\Psi}(t), \mathbf{X}(t), t\big) - \mathbf{R}(t). \qquad \text{(A.13)}
$$

Setting $\mathbf{Q}(t) = \widetilde{\mathbf{X}}(t) - \mathbf{X}(t)$, $\mathbf{Q}(T) = 0$, it follows from (A.6) and (A.13) that for all $t \in [0, T] \setminus \mathbb{D}$,

$$
\dot{\mathbf{Q}}(t) + \mathcal{T}\big(\mathbf{\Psi}(t), \mathbf{Q}(t), t\big) + \mathbf{R}(t) = 0, \quad \mathbf{Q}(T) = 0. \qquad \text{(A.14)}
$$

By Lemma A.2, $\mathbf{Q}(t) \ge 0$, showing the result. □

Proof of Theorem A.1 Starting with $\mathbf{\Psi}^1(t) = (\Psi_1^1(t), \dots, \Psi_N^1(t))$, $\Psi_i^1(t)$ in the class **PC** with discontinuity points $t \in \mathbb{D}$, we define recursively, as in Sect. 3 of [302], for $k = 1, 2, \dots$, $\mathbf{X}^k(t) = (X_1^k(t), \dots, X_N^k(t)) \in \mathbb{H}^{n+}$, $0 \le t \le T$, as the unique continuous solution (see Lemma A.2) satisfying for all $t \in [0, T] \setminus \mathbb{D}$,

$$
\dot{\mathbf{X}}^k(t) + \mathcal{T}\big(\mathbf{\Psi}^k(t), \mathbf{X}^k(t), t\big) = 0, \quad \mathbf{X}^k(T) = \mathbf{L}, \qquad \text{(A.15)}
$$

where

$$
\mathbf{\Psi}^{k+1}(t) = \mathcal{K}\big(\mathbf{X}^k(t), t\big). \qquad \text{(A.16)}
$$

Notice that since $\mathbf{X}^k(t)$ is continuous for $t \in [0, T]$, the discontinuity points of the differential equation (A.15) remain unchanged along the iterations on k. From (A.12), (A.15), and (A.16) it follows that for all $t \in [0, T] \setminus \mathbb{D}$,

$$
\begin{aligned}
\dot{\mathbf{X}}^k(t) + \mathcal{T}\big(\mathbf{\Psi}^{k+1}(t), \mathbf{X}^k(t), t\big) &\le \dot{\mathbf{X}}^k(t) + \mathcal{T}\big(\mathbf{\Psi}^k(t), \mathbf{X}^k(t), t\big) \\
&= 0 \\
&= \dot{\mathbf{X}}^{k+1}(t) + \mathcal{T}\big(\mathbf{\Psi}^{k+1}(t), \mathbf{X}^{k+1}(t), t\big), \qquad \text{(A.17)}
\end{aligned}
$$

and thus, writing $\mathbf{Q}(t) = \mathbf{X}^k(t) - \mathbf{X}^{k+1}(t)$, we have from (A.17) that for all $t \in [0, T] \setminus \mathbb{D}$, $\dot{\mathbf{Q}}(t) + \mathcal{T}(\boldsymbol{\Psi}^{k+1}(t), \mathbf{Q}(t), t) + \mathbf{R}(t) = 0$ for some $\mathbf{R}(t) \in \mathbb{H}^{n+}$ with discontinuity points in $\mathbb{D}$. From Lemma A.2 it follows that $\mathbf{Q}(t) \in \mathbb{H}^{n+}$, that is, $\mathbf{X}^k(t) \geq \mathbf{X}^{k+1}(t) \geq 0$, and by the monotone convergence result again (Lemma 3.1 in [302]) there exists $\mathbf{S}(t) = (S_1(t), \ldots, S_N(t)) \in \mathbb{H}^{n+}$ such that $\lim_{k\to\infty} \mathbf{X}^k(t) = \mathbf{S}(t)$ and, by (A.16), $\lim_{k\to\infty} \boldsymbol{\Psi}^k(t) = \boldsymbol{\Psi}(t)$, where $\boldsymbol{\Psi}(t) = \mathcal{K}(\mathbf{S}(t), t)$. Moreover, as seen in Lemma A.2, there exists $a \geq 0$ such that $\|S_i(t)\| \leq \|X_i^1(t)\| \leq a$ for all $i = 1, \ldots, N$ and $0 \leq t \leq T$. Setting $\widehat{\Phi}_i(s,t)$ as the fundamental matrix associated to the matrix $A_i(t)$, it follows from (A.15) that

$$\begin{aligned} X_i^k(t) &= \widehat{\Phi}_i^*(T,t) L_i \widehat{\Phi}_i(T,t) \\ &\quad + \int_t^T \widehat{\Phi}_i^*(s,t) \Bigg[\sum_{j=1}^N \lambda_{ij} X_j^k(s) - \big(\Psi_i^k(s)\big)^* B_i^*(s) X_i^k(s) - X_i^k(s) B_i(s) \Psi_i^k(s) \\ &\quad + C_i^*(s) C_i(s) + \big(\Psi_i^k(s)\big)^* D_i^*(s) D_i(s) \Psi_i^k(s) \Bigg] \widehat{\Phi}_i(s,t)\, ds. \end{aligned} \tag{A.18}$$

From the bounded convergence theorem, taking the limit in (A.18) as $k \to \infty$, we get that

$$\begin{aligned} S_i(t) &= \widehat{\Phi}_i^*(T,t) L_i \widehat{\Phi}_i(T,t) \\ &\quad + \int_t^T \widehat{\Phi}_i^*(s,t) \Bigg[\sum_{j=1}^N \lambda_{ij} S_j(s) - \big(\Psi_i(s)\big)^* B_i^*(s) S_i(s) - S_i(s) B_i(s) \Psi_i(s) \\ &\quad + C_i^*(s) C_i(s) + \Psi_i^*(s) D_i^*(s) D_i(s) \Psi_i(s) \Bigg] \widehat{\Phi}_i(s,t)\, ds. \end{aligned} \tag{A.19}$$

From (A.19) we have that $\mathbf{S}(t)$ is continuous for $t \in [0, T]$ and for all $t \in [0, T] \setminus \mathbb{D}$, and the derivative of (A.19) exists and satisfies (A.4). Notice that by (A.12) we can rewrite (A.4) as $\dot{\mathbf{S}}(t) + \mathcal{T}(\boldsymbol{\Psi}(t), \mathbf{S}(t), t) = 0$, $\mathbf{S}(T) = \mathbf{L}$. Suppose that $\mathbf{Y}(t) = (Y_1(t), \ldots, Y_N(t)) \in \mathbb{H}^n$ is continuous for $t \in [0, T]$ and satisfies

$$\begin{aligned} &\dot{Y}_i(t) + A_i^*(t) Y_i(t) + Y_i(t) A_i(t) - Y_i(t) B_i(t) \big[D_i^*(t) D_i(t)\big]^{-1} B_i^*(t) Y_i(t) \\ &\quad + \sum_{j=1}^N \lambda_{ij} Y_j(t) + C_i^*(t) C_i(t) = 0, \quad i \in \mathcal{S}, \end{aligned}$$

with boundary condition $\mathbf{Y}(T) = \mathbf{L}$. Then we have that

$$\dot{\mathbf{Y}}(t) + \mathcal{T}\big(\widehat{\boldsymbol{\Psi}}(t), \mathbf{Y}(t), t\big) = 0, \quad \mathbf{Y}(T) = \mathbf{L},$$

where $\widehat{\boldsymbol{\Psi}}(t) = \mathcal{K}(\mathbf{Y}, t)$. From the minimality property, Lemma A.3, we have that $\mathbf{Y}(t) \leq \mathbf{S}(t)$ and similarly $\mathbf{S}(t) \leq \mathbf{Y}(t)$, which shows the uniqueness. □

A.3 Maximal Solution

In the next sections we consider all matrices in (A.1) time invariant and want to study conditions for the existence of solutions of (4.28) and asymptotic convergence of $\mathbf{X}^T(0)$ to a solution $\mathbf{X}$ of (4.28) as $T \to \infty$. Parallel to the classical LQ problem, when dealing with the infinite-time optimal control problem (infinite horizon), two structural concepts turn out to be essential: *mean-square stabilizability* and *mean-square detectability*, defined in Sect. 3.6.

The purpose of this section is to prove the existence of the maximal solution for the CARE (see Definition 4.10), under the hypothesis of mean-square stabilizability. This result is summarized in Theorem A.5 below. But first we need the following auxiliary result. For $\mathbf{G} = (G_1, \dots, G_N) \in \mathbb{H}^{n,m}$, $\mathbf{K} = (K_1, \dots, K_N) \in \mathbb{H}^{n,m}$, and $\widetilde{\mathbf{X}} = (\widetilde{X}_1, \dots, \widetilde{X}_N) \in \mathbb{H}^n$, define the operators $\mathcal{T}^{\mathbf{G}}, \mathcal{T}^{\mathbf{K}} \in \mathbb{B}(\mathbb{H}^n)$ as

$$\mathcal{T}_i^{\mathbf{G}}(\widetilde{\mathbf{X}}) = (A_i - B_i G_i)^* \widetilde{X}_i + \widetilde{X}_i (A_i - B_i G_i) + \sum_{j=1}^{N} \lambda_{ij} \widetilde{X}_j,$$

$$\mathcal{T}_i^{\mathbf{K}}(\widetilde{\mathbf{X}}) = (A_i - B_i K_i)^* \widetilde{X}_i + \widetilde{X}_i (A_i - B_i K_i) + \sum_{j=1}^{N} \lambda_{ij} \widetilde{X}_j,$$

and $\mathcal{L}^{\mathbf{G}} = (\mathcal{T}^{\mathbf{G}})^*$, $\mathcal{L}^{\mathbf{K}} = (\mathcal{T}^{\mathbf{K}})^*$.

Proposition A.4 *Suppose that for some* $\mathbf{G} = (G_1, \dots, G_N) \in \mathbb{H}^{n,m}$,

$$\operatorname{Re}\{\lambda(\mathcal{L}^{\mathbf{G}})\} < 0$$

and for some $\widetilde{\mathbf{X}} = (\widetilde{X}_1, \dots, \widetilde{X}_N) \in \mathbb{H}^{n+}$, $\mathbf{K} = (K_1, \dots, K_N) \in \mathbb{H}^{n,m}$, *and* $\delta > 0$,

$$\mathcal{T}_i^{\mathbf{K}}(\widetilde{\mathbf{X}}) + \delta (K_i - G_i)^* (K_i - G_i) \le 0, \quad i \in \mathcal{S}. \tag{A.20}$$

Then $\operatorname{Re}\{\lambda(\mathcal{L}^{\mathbf{K}})\} < 0$.

Proof Consider $x(t)$ given by (4.1) with $t \in \mathbb{R}^+$ and $u(t) = -K_{\theta(t)} x(t)$, $\mathbf{K} = (K_1, \dots, K_N) \in \mathbb{H}^{n,m}$ and arbitrary initial data (x_0, θ_0). Let us consider $\mathbf{Q}(t)$, $t \in \mathbb{R}^+$, given by (3.9) and (3.12), so that, by Proposition 3.7(b),

$$\dot{\mathbf{Q}}(t) = \mathcal{L}^{\mathbf{K}}\big(\mathbf{Q}(t)\big), \tag{A.21}$$

and thus,

$$\dot{Q}_i(t) = \mathcal{L}_i^{\mathbf{G}}\big(\mathbf{Q}(t)\big) + \Delta_i Q_i(t) + Q_i(t) \Delta_i^*, \tag{A.22}$$

where

$$\Delta_i = B_i (G_i - K_i). \tag{A.23}$$

Now, for arbitrary $\varepsilon > 0$,

$$0 \leq \left(\varepsilon I - \frac{1}{\varepsilon}\Delta_i\right) Q_i(t) \left(\varepsilon I - \frac{1}{\varepsilon}\Delta_i\right)^*,$$

so that

$$\Delta_i Q_i(t) + Q_i(t)\Delta_i^* \leq \varepsilon^2 Q_i(t) + \frac{1}{\varepsilon^2}\Delta_i Q_i(t)\Delta_i^*.$$

Thus, by (A.22),

$$\dot{Q}_i(t) \leq \mathcal{L}_i^{\mathbf{G}}(\mathbf{Q}(t)) + \varepsilon^2 Q_i(t) + \frac{1}{\varepsilon^2}\Delta_i Q_i(t)\Delta_i^*. \tag{A.24}$$

Now, for $\mathbf{H} = (H_1, \dots, H_N) \in \mathbb{H}^n$, let us define the operators

$$\begin{aligned}
\widetilde{\mathcal{L}}(\mathbf{H}) &= \big(\widetilde{\mathcal{L}}_1(\mathbf{H}), \dots, \widetilde{\mathcal{L}}_N(\mathbf{H})\big), \\
\Gamma(\mathbf{H}) &= \big(\Gamma_1(\mathbf{H}), \dots, \Gamma_N(\mathbf{H})\big), \\
\mathcal{V}(\mathbf{H}) &= \big(\mathcal{V}_1(\mathbf{H}), \dots, \mathcal{V}_N(\mathbf{H})\big)
\end{aligned}$$

in $\mathbb{B}(\mathbb{H}^n)$ such that

$$\widetilde{\mathcal{L}}_i(\mathbf{H}) = \mathcal{L}_i^{\mathbf{G}}(\mathbf{H}) + \varepsilon^2 H_i, \quad \Gamma_i(\mathbf{H}) = \Delta_i H_i \Delta_i^*, \tag{A.25}$$

and

$$\mathcal{V}_i(\mathbf{H}) = \left(\varepsilon I - \frac{1}{\varepsilon}\Delta_i\right) H_i \left(\varepsilon I - \frac{1}{\varepsilon}\Delta_i^*\right), \quad i \in \mathcal{S}. \tag{A.26}$$

From (A.25) we have that

$$\widetilde{\mathcal{L}} = \mathcal{L}^{\mathbf{G}} + \varepsilon^2 I \tag{A.27}$$

with I being the identity operator associated to $\mathbb{H}^n$. We can rewrite (A.24) as

$$\dot{Q}_i(t) \leq \widetilde{\mathcal{L}}_i\big(\mathbf{Q}(t)\big) + \frac{1}{\varepsilon^2}\Gamma_i\big(\mathbf{Q}(t)\big). \tag{A.28}$$

In order to use a comparison theorem, we consider now the following nonhomogeneous differential equation:

$$\begin{cases}
\dot{R}_i(t) = \widetilde{\mathcal{L}}_i\big(\mathbf{R}(t)\big) + \dfrac{1}{\varepsilon^2}\Gamma_i\big(\mathbf{Q}(t)\big), \\
R_i(0) = Q_i(0), \quad i \in \mathcal{S},
\end{cases} \tag{A.29}$$

or equivalently, the differential equations

$$\begin{cases}
\dot{\mathbf{R}}(t) = \widetilde{\mathcal{L}}\big(\mathbf{R}(t)\big) + \dfrac{1}{\varepsilon^2}\Gamma\big(\mathbf{Q}(t)\big), \\
\mathbf{R}(0) = \mathbf{Q}(0) \in \mathbb{H}^{n+},
\end{cases} \tag{A.30}$$

with $\mathbf{R}(t) = (R_1(t), \dots, R_N(t))$. Now, for each finite T and time interval $[0, T]$, $\mathbf{Q}$ and consequently $\frac{1}{\varepsilon^2}\Gamma(\mathbf{Q})$ belong to $L_1([0,T], \mathbb{H}_1^n)$ and are continuously differentiable. Thus, the unique solution $\mathbf{R}(t) \in \mathbb{H}^n$ to (A.30) is given by

$$\mathbf{R}(t) = e^{\widetilde{\mathcal{L}}t}\big(\mathbf{Q}(0)\big) + \frac{1}{\varepsilon^2}\int_0^t e^{\widetilde{\mathcal{L}}(t-s)}\big(\Gamma\big(\mathbf{Q}(s)\big)\big)\,ds, \quad t \in [0, T], \tag{A.31}$$

for any $\mathbf{Q}(0) \in \mathbb{H}^{n+}$. Let us now define

$$U_i(t) = R_i(t) - Q_i(t), \quad i \in \mathcal{S}. \tag{A.32}$$

Then $\mathbf{U}(t) = (U_1(t), \dots, U_N(t))$ belongs to $\mathbb{H}^n$ and satisfies the differential equation

$$\begin{cases} \dot{\mathbf{U}}(t) = \widetilde{\mathcal{L}}\big(\mathbf{U}(t)\big) + \mathcal{V}\big(\mathbf{Q}(t)\big), \\ \mathbf{U}(0) = 0, \end{cases} \tag{A.33}$$

as shown next:

$$\begin{aligned} \dot{U}_i(t) = \dot{R}_i(t) - \dot{Q}_i(t) &= \widetilde{\mathcal{L}}_i\big(\mathbf{R}(t)\big) + \frac{1}{\varepsilon^2}\Delta_i Q_i(t)\Delta_i^* - \mathcal{L}_i^{\mathbf{G}}\big(\mathbf{Q}(t)\big) \\ &= \mathcal{L}_i^{\mathbf{G}}\big(\mathbf{R}(t)\big) + \varepsilon^2 R_i(t) + \frac{1}{\varepsilon^2}\Delta_i Q_i(t)\Delta_i^* \\ &\quad - \big(\mathcal{L}_i^{\mathbf{G}}\big(\mathbf{Q}(t)\big) + \Delta_i Q_i(t) + Q_i(t)\Delta_i^*\big) \\ &= \widetilde{\mathcal{L}}_i\big(\mathbf{U}(t)\big) + \left(\varepsilon I - \frac{1}{\varepsilon}\Delta_i\right) Q_i(t) \left(\varepsilon I - \frac{1}{\varepsilon}\Delta_i^*\right) \\ &= \widetilde{\mathcal{L}}_i\big(\mathbf{U}(t)\big) + \mathcal{V}_i\big(\mathbf{Q}(t)\big). \end{aligned}$$

Now, $\mathbf{U}(t) \in \mathbb{H}^n$ defined in (A.32) is a unique solution to (A.33) and is given by

$$\mathbf{U}(t) = \int_0^t e^{\widetilde{\mathcal{L}}(t-s)}\big(\mathcal{V}\big(\mathbf{Q}(s)\big)\big)\,ds, \quad t \in [0, T].$$

Since $\mathbf{Q}(s)$ and consequently $\mathcal{V}(\mathbf{Q}(s))$ belongs to $\mathbb{H}^{n+}$, it follows from Lemma 3.10(b) that $e^{\widetilde{\mathcal{L}}(t-s)}(\mathcal{V}(\mathbf{Q}(s)))$ belongs to $\mathbb{H}^{n+}$. Hence, $\mathbf{U}(t) \in \mathbb{H}^{n+}$, which, together with (A.32), sets our comparison result, i.e., $0 \le \mathbf{Q}(t) \le \mathbf{R}(t) \in \mathbb{H}^{n+}$, $t \in [0, T]$, for arbitrary $\mathbf{Q}(0) \in \mathbb{H}^{n+}$ and each finite T. Now, using (A.31) and (2.31), we have that

$$\|\mathbf{Q}(t)\|_1 \le \|e^{\widetilde{\mathcal{L}}(t)}\big(\mathbf{Q}(0)\big)\|_1 + \frac{1}{\varepsilon^2}\int_0^t \|e^{\widetilde{\mathcal{L}}(t-s)}\big(\Gamma\big(\mathbf{Q}(s)\big)\big)\|_1\,ds.$$

Hence, integration on $[0, T]$ yields

$$\int_0^T \|\mathbf{Q}(t)\|_1\,dt \le \int_0^T \|e^{\widetilde{\mathcal{L}}(t)}\big(\mathbf{Q}(0)\big)\|_1 dt$$

$$+\frac{1}{\varepsilon^2}\int_0^T\int_0^t \left\| e^{\widetilde{\mathcal{L}}(t-s)}\big(\Gamma(\mathbf{Q}(s))\big)\right\|_1 ds\,dt. \tag{A.34}$$

Referring to the last term of (A.34), let us define $l = t - s$ and

$$T^E_{\widetilde{\mathcal{L}}}(r) = \begin{cases} e^{\widetilde{\mathcal{L}}(r)} & \text{if } r \geq 0,\\ 0 & \text{if } r < 0,\end{cases}$$

so that

$$\begin{aligned}
&\int_0^T\int_0^t \left\| e^{\widetilde{\mathcal{L}}(t-s)}\big(\Gamma(\mathbf{Q}(s))\big)\right\|_1 ds\,dt\\
&\quad = \int_0^T\int_0^T \left\| T^E_{\widetilde{\mathcal{L}}}(t-s)\big(\Gamma(\mathbf{Q}(s))\big)\right\|_1 dt\,ds\\
&\quad \leq \int_0^T \left\|\Gamma(\mathbf{Q}(s))\right\|_1 \int_0^{T-s} \left\| e^{\widetilde{\mathcal{L}}(l)}\right\| dl\,ds\\
&\quad \leq \int_0^T \left\|\Gamma(\mathbf{Q}(s))\right\|_1 ds \int_0^T \left\| e^{\widetilde{\mathcal{L}}(l)}\right\| dl.
\end{aligned} \tag{A.35}$$

Hence,

$$\int_0^T \left\|\mathbf{Q}(t)\right\|_1 dt \leq \left\{ \left\|\mathbf{Q}(0)\right\|_1 + \frac{1}{\varepsilon^2}\int_0^T \left\|\Gamma(\mathbf{Q}(s))\right\|_1 ds\right\} \int_0^T \left\| e^{\widetilde{\mathcal{L}}(s)}\right\| ds. \tag{A.36}$$

By (A.36) we end up with

$$\int_0^T \left\|\mathbf{Q}(t)\right\|_1 dt \leq \left\{ \left\|\mathbf{Q}(0)\right\|_1 + \frac{\|\mathbf{B}\|^2_{\max}}{\varepsilon^2}\int_0^T \left\|\widetilde{\Gamma}(\mathbf{Q}(s))\right\|_1 ds\right\} \int_0^T \left\| e^{\widetilde{\mathcal{L}}(s)}\right\| ds, \tag{A.37}$$

where for $\widetilde{\mathbf{Y}} = (\widetilde{Y}_1, \ldots, \widetilde{Y}_N) \in \mathbb{H}^n$, $\widetilde{\Gamma}(\widetilde{\mathbf{Y}}) = (\widetilde{\Gamma}_1(\widetilde{\mathbf{Y}}), \ldots, \widetilde{\Gamma}_N(\widetilde{\mathbf{Y}}))$ with $\widetilde{\Gamma}_i(\widetilde{\mathbf{Y}}) = (K_i - G_i)\widetilde{Y}_i(K_i - G_i)^*$, recalling that $\widetilde{\mathcal{L}} = \mathcal{L}^{\mathbf{G}} + \varepsilon^2 I$. Let us show that

$$\int_0^\infty \left\|\widetilde{\Gamma}(\mathbf{Q}(s))\right\|_1 ds < \infty. \tag{A.38}$$

To show (A.38), we notice from (A.20) and (A.21) that

$$\begin{aligned}
\left\|\widetilde{\Gamma}(\mathbf{Q}(s))\right\|_1 &\leq \langle \widetilde{\Gamma}(\mathbf{Q}(s)); I\rangle \leq -\frac{1}{\delta}\langle \mathbf{Q}(s); \mathcal{T}^{\mathbf{K}}(\widetilde{\mathbf{X}})\rangle\\
&= -\frac{1}{\delta}\langle \mathcal{L}^{\mathbf{K}}(\mathbf{Q}(s)); \widetilde{\mathbf{X}}\rangle = -\frac{1}{\delta}\langle \dot{\mathbf{Q}}(s); \widetilde{\mathbf{X}}\rangle.
\end{aligned} \tag{A.39}$$

Integrating (A.39), we get that

$$\int_0^T \left\|\widetilde{\Gamma}(\mathbf{Q}(s))\right\|_1 ds \leq \frac{1}{\delta}\big(\langle \mathbf{Q}(0); \widetilde{\mathbf{X}}\rangle - \langle \mathbf{Q}(T); \widetilde{\mathbf{X}}\rangle\big) \leq \frac{1}{\delta}\langle \mathbf{Q}(0); \widetilde{\mathbf{X}}\rangle,$$

showing (A.38). Thus, for some constant $c > 0$, passing in (A.37) to the limit, we have that

$$\lim_{T\to\infty}\int_0^T \|\mathbf{Q}(t)\|_1\,dt \le \left\{\|(\mathbf{Q}(0))\|_1 + \frac{c}{\varepsilon^2}\right\}\lim_{T\to\infty}\int_0^T \|e^{\widetilde{\mathcal{L}}(t)}\|\,dt. \tag{A.40}$$

By construction, $\mathrm{Re}\{\lambda(\mathcal{L}^{\mathbf{G}})\} < 0$, and since by (A.27), $\sigma(\widetilde{\mathcal{L}}) = \sigma(\mathcal{L}^{\mathbf{G}}) + \varepsilon^2$, we have that $\mathrm{Re}\{\lambda(\widetilde{\mathcal{L}})\} < 0$ for some $\varepsilon > 0$ sufficiently small. Thus, by Proposition 2.23, $\lim_{T\to\infty}\int_0^T \|e^{\widetilde{\mathcal{L}}(t)}\|\,dt < \infty$, or else (see (A.40)),

$$\lim_{T\to\infty}\int_0^T \|\mathbf{Q}(t)\|_1\,dt < \infty$$

whenever $\mathbf{Q}(0) \in \mathbb{H}^{n+}$. Moreover, $E[\|x(t)\|^2] \le n\|\mathbf{Q}(t)\|_1$, so that

$$\lim_{T\to\infty}\int_0^T E\big[\|x(t)\|^2\big]\,dt < \infty$$

for any initial condition (x_0, θ_0). Hence, $\mathrm{Re}\{\lambda(\mathcal{L}^{\mathbf{K}})\} < 0$, completing the proof. □

We can now prove the main result of this section.

Theorem A.5 (Maximal solution) *Suppose that* $(\mathbf{A}, \mathbf{B}, \Pi)$ *is SS. Then for* $\ell = 0, 1, 2, \ldots$, *there exists* $\mathbf{X}^\ell = (X_1^\ell, \ldots, X_N^\ell) \in \mathbb{H}^{n+}$ *that satisfies the following properties*:

(a) $\mathbf{X}^0 \ge \mathbf{X}^1 \ge \cdots \ge \mathbf{X}^\ell \ge \widetilde{\mathbf{X}}$ *for arbitrary* $\widetilde{\mathbf{X}} \in \mathbb{H}^{n*}$ *such that* $\mathcal{R}(\widetilde{\mathbf{X}}) \ge 0$;
(b) $\mathrm{Re}\{\lambda(\mathcal{L}^\ell)\} < 0$, *where* $\mathcal{L}^\ell = (\mathcal{L}_1^\ell, \ldots, \mathcal{L}_N^\ell)$, *and for* $i = 1, \ldots, N$, $\mathbf{P} = (P_1, \ldots, P_N) \in \mathbb{H}^n$,

$$\mathcal{L}_i^\ell(\mathbf{P}) := A_i^\ell P_i + P_i\big(A_i^\ell\big)^* + \sum_{j\in\mathcal{S}} \lambda_{ji} P_j,$$

$$A_i^\ell := A_i - B_i K_i^\ell,$$

$$K_i^\ell := \mathcal{K}_i\big(\mathbf{X}^{\ell-1}\big) \quad \textit{for } \ell = 1, 2, \ldots,$$

$$\mathbf{K}^0 \in \mathbb{K} \textit{ arbitrary}.$$

(c) $\mathbf{X}^\ell$ *is the unique solution of the set of coupled linear equations*

$$\big(A_i - B_i K_i^\ell\big)^* X_i^\ell + X_i^\ell\big(A_i - B_i K_i^\ell\big) + \sum_{j\in\mathcal{S}} \lambda_{ij} X_j^\ell + C_i^* C_i + K_i^{\ell*} D_i^* D_i K_i^\ell = 0, \quad i = 1, \ldots, N. \tag{A.41}$$

Moreover, $\mathbf{X}^\ell \to \mathbf{X}^+$ *as* $\ell \to \infty$, *where* $\mathbf{X}^+ = (X_1^+, \ldots, X_N^+) \in \mathbb{H}^{n+}$ *satisfies* $\mathcal{R}(\mathbf{X}^+) = 0$ *and* $\mathbf{X}^+ \ge \widetilde{\mathbf{X}}$ *for any* $\widetilde{\mathbf{X}} \in \mathbb{H}^{n*}$ *such that* $\mathcal{R}(\widetilde{\mathbf{X}}) \ge 0$. *Furthermore,*

$\mathrm{Re}\{\lambda(\mathcal{L}^+)\} \leq 0$, *where* $\mathcal{L}^+ = (\mathcal{L}_1^+, \ldots, \mathcal{L}_N^+)$ *is defined for* $\mathbf{P} = (P_1, \ldots, P_N)$ *as* $\mathcal{L}_i^+(\mathbf{P}) = A^+ P_i + P_i (A_i^+)^* + \sum_{j\in\mathcal{S}} \lambda_{ji} P_j$ *for* $i = 1, \ldots, N$, *and*

$$A_i^+ = A_i - B_i K_i^+, \qquad K_i^+ = \mathcal{K}_i(\mathbf{X}^+).$$

Proof Let us apply induction on ℓ. By the hypothesis that $(\mathbf{A}, \mathbf{B}, \Pi)$ is SS we can find $\mathbf{K}^0 \in \mathbb{K}$ such that $\mathrm{Re}\{\lambda(\mathcal{L}^0)\} < 0$ and thus $\mathrm{Re}\{\lambda(\mathcal{T}^0)\} < 0$ where $\mathcal{T}^0 = \mathcal{L}^{0*}$. Hence, by Proposition 3.20, there exists a unique solution $\mathbf{X}^0$ for (A.41), and moreover $\mathbf{X}^0 \in \mathbb{H}^{n+}$. Suppose that at iteration ℓ we have $\mathbf{X}^{\ell-1} \in \mathbb{H}^{n+}$ satisfying (A.41) and $\mathrm{Re}\{\lambda(\mathcal{L}^{\ell-1})\} < 0$. From the identity (A.12), (A.41), and recalling that $\mathbf{K}^\ell = \mathcal{K}(\mathbf{X}^{\ell-1})$, we have that for some $\delta > 0$ (in fact, $\delta > 0$ such that $D_i^* D_i \geq \delta I$ for all $i \in \mathcal{S}$),

$$\begin{aligned}
0 &= \big(A_i - B_i K_i^\ell\big)^* X_i^{\ell-1} + X_i^{\ell-1}\big(A_i - B_i K_i^\ell\big) + \sum_{j\in\mathcal{S}} \lambda_{ij} X_j^{\ell-1} \\
&\quad + C_i^* C_i + K_i^{\ell*} D_i^* D_i K_i^\ell + \big(K_i^\ell - K_i^{\ell-1}\big)^* D_i^* D_i \big(K_i^\ell - K_i^{\ell-1}\big) \\
&\geq \big(A_i - B_i K_i^\ell\big)^* X_i^{\ell-1} + X_i^{\ell-1}\big(A_i - B_i K_i^\ell\big) + \sum_{j\in\mathcal{S}} \lambda_{ij} X_j^{\ell-1} \\
&\quad + \delta\big(K_i^\ell - K_i^{\ell-1}\big)^*\big(K_i^\ell - K_i^{\ell-1}\big),
\end{aligned} \tag{A.42}$$

so that by (A.42) and Proposition A.4, $\mathrm{Re}\{\lambda(\mathcal{L}^\ell)\} < 0$. From this and from Proposition 3.20 it follows that there exists a unique solution $\mathbf{X}^\ell$ satisfying (A.41) and moreover $\mathbf{X}^\ell \in \mathbb{H}^{n+}$. Writing $\mathbf{Q} = \mathbf{X}^{\ell-1} - \mathbf{X}^\ell$, we get, taking (A.42) minus (A.41), that

$$\begin{aligned}
0 &= \big(A_i - B_i K_i^\ell\big)^* Q_i + Q_i\big(A_i - B_i K_i^\ell\big) + \sum_{j\in\mathcal{S}} \lambda_{ij} Q_j \\
&\quad + \big(K_i^\ell - K_i^{\ell-1}\big)^* D_i^* D_i \big(K_i^\ell - K_i^{\ell-1}\big),
\end{aligned}$$

so that by Proposition 3.20, $\mathbf{Q} \in \mathbb{H}^{n+}$, that is, $\mathbf{X}^{\ell-1} \geq \mathbf{X}^\ell$. Similarly, for arbitrary $\widetilde{\mathbf{X}} \in \mathbb{H}^{n*}$ such that $\mathcal{R}(\widetilde{\mathbf{X}}) \geq 0$, we have from the identity (A.12) and some $\mathbf{R} \in \mathbb{H}^{n+}$ that

$$\begin{aligned}
0 &= \big(A_i - B_i K_i^\ell\big)^* \widetilde{X}_i + \widetilde{X}_i\big(A_i - B_i K_i^\ell\big) + \sum_{j\in\mathcal{S}} \lambda_{ij} \widetilde{X}_j + C_i^* C_i \\
&\quad + K_i^{\ell*} D_i^* D_i K_i^\ell - \big(K_i^\ell - \bar{K}_i\big)^* D_i^* D_i \big(K_i^\ell - \bar{K}_i\big) - R_i,
\end{aligned} \tag{A.43}$$

where $\bar{\mathbf{K}} = \mathcal{K}(\widetilde{\mathbf{X}})$. Taking (A.41) minus (A.43) and $\mathbf{Q} = \mathbf{X}^\ell - \widetilde{\mathbf{X}}$, we obtain that

$$\begin{aligned}
0 &= \big(A_i - B_i K_i^\ell\big)^* Q_i + Q_i\big(A_i - B_i K_i^\ell\big) + \sum_{j\in\mathcal{S}} \lambda_{ij} Q_j \\
&\quad + \big(K_i^\ell - \bar{K}_i\big)^* D_i^* D_i \big(K_i^\ell - \bar{K}_i\big) + R_i,
\end{aligned} \tag{A.44}$$

and from Proposition 3.20 and (A.44) it follows that $\mathbf{Q} \geq 0$, that is, $\mathbf{X}^\ell \geq \widetilde{\mathbf{X}}$. Since $\{\mathbf{X}^\ell\}_{\ell=0}^\infty$ is a decreasing sequence with $\mathbf{X}^\ell \geq 0$ for all $\ell = 0, 1, \dots,$ we get from Lemma 2.17 that there exists $\mathbf{X}^+ \in \mathbb{H}^{n+}$ such that $\mathbf{X}^\ell \downarrow \mathbf{X}^+$ as $\ell \to \infty$. Clearly, $\mathbf{X}^+ \geq \widetilde{\mathbf{X}}$ for any $\widetilde{\mathbf{X}} \in \mathbb{H}^{n*}$ such that $\mathcal{R}(\widetilde{\mathbf{X}}) \geq 0$. Moreover, substituting $\mathbf{K}^\ell = \mathcal{K}(\mathbf{X}^{\ell-1})$ into (A.41) and taking the limit as $\ell \to \infty$, we get that

$$\left(A_i - B_i\mathcal{K}_i(\mathbf{X}^+)\right)^* X_i^+ + X_i^+\left(A_i - B_i\mathcal{K}_i(\mathbf{X}^+)\right) + \sum_{j\in\mathcal{S}} \lambda_{ij} X_j^+$$
$$+ C_i^* C_i + \mathcal{K}_i(\mathbf{X}^+)^* D_i^* D_i \mathcal{K}_i(\mathbf{X}^+) = 0, \quad i = 1, \dots, N.$$

Rearranging the terms, we obtain, for $i = 1, \dots, N$, that $\mathcal{R}(\mathbf{X}^+) = 0$. Finally, notice that since $\operatorname{Re}\{\lambda(\mathcal{L}^\ell)\} < 0$, we get that (see [264], p. 328, for the continuity of the eigenvalues on finite-dimensional linear operator entries) $\operatorname{Re}\{\lambda(\mathcal{L}^+)\} \leq 0$. □

The next result establishes a link between an LMIs optimization problem and the maximal solution $\mathbf{X}^+$. Consider the following convex optimization programming problem:

$$\begin{aligned}
&\max \operatorname{tr}\left(\sum_{i=1}^N X_i\right) \\
&\text{subject, for } i = 1, \dots, N, \text{ to} \\
&\begin{bmatrix} A_i^* X_i + X_i A_i + \sum_{j=1}^N \lambda_{ij} X_j + C_i^* C_i & X_i B_i \\ B_i^* X_i & D_i^* D_i \end{bmatrix} \geq 0, \\
&X_i = X_i^*.
\end{aligned} \tag{A.45}$$

Lemma A.6 *Suppose that* $(\mathbf{A}, \mathbf{B}, \Pi)$ *is SS. Then the maximal solution* $\mathbf{X}^+ \in \mathbb{H}^{n+}$ *for the CARE* (*which exists from Theorem* A.5) *is the unique solution of the convex programming problem* (A.45).

Proof First of all, notice that, from Schur's complement (see Lemma 2.26), $\widetilde{\mathbf{X}} = (\widetilde{X}_1, \dots, \widetilde{X}_N)$ satisfies the constraints (A.45) if and only if $\widetilde{\mathbf{X}} \in \mathbb{H}^{n*}$ and

$$A_i^* \widetilde{X}_i + \widetilde{X}_i A_i + \sum_{j=1}^N \lambda_{ij} \widetilde{X}_j + C_i^* C_i - \widetilde{X}_i B_i \left(D_i^* D_i\right)^{-1} B_i^* \widetilde{X}_i \geq 0,$$

that is, if and only if $\widetilde{\mathbf{X}} \in \mathbb{H}^{n*}$ and $\mathcal{R}(\widetilde{\mathbf{X}}) \geq 0$. By Theorem A.5, $\mathbf{X}^+ \in \mathbb{H}^{n+}$, $\mathcal{R}(\mathbf{X}^+) = 0$ (thus, the constraints (A.45) are satisfied for $\mathbf{X}^+$), and $\mathbf{X}^+ \geq \widetilde{\mathbf{X}}$ for any $\widetilde{\mathbf{X}}$ satisfying the constraints (A.45), which implies that

$$\sum_{i=1}^N \operatorname{tr}\left(X_i^+\right) \geq \sum_{i=1}^N \operatorname{tr}(\widetilde{X}_i),$$

and so the optimal solution is given by $\mathbf{X}^+$. □

A.4 Stabilizing Solution

This section presents results concerning the existence of a stabilizing solution for the CARE and its relation with the maximal solution. As we are going to see in this section, the maximal solution coincides with the stabilizing solution whenever the latter exists. Also in this section we present a necessary and sufficient condition for the existence of a stabilizing solution.

Proposition A.7 *There exists at most one mean-square stabilizing solution* $\mathbf{X} = (X_1, \ldots, X_N) \in \mathbb{H}^{n+}$ *for the CARE* (4.28), *which coincides with the maximal solution.*

Proof Suppose that $\mathbf{X} = (X_1, \ldots, X_N)$ is a mean-square stabilizing solution for the CARE (4.28). Clearly $(\mathbf{A}, \mathbf{B}, \Pi)$ is SS, and thus the maximal solution $\mathbf{X}^+$ exists. From the identity (A.12) we have

$$\begin{aligned}&\big(A_i - B_i\mathcal{K}_i(\mathbf{X})\big)^*\big(X_i - X_i^+\big) + \big(X_i - X_i^+\big)\big(A_i - B_i\mathcal{K}_i(\mathbf{X})\big)\\&\quad + \sum_{j=1}^{N}\lambda_{ij}\big(X_j - X_j^+\big) + \big(\mathcal{K}_i(\mathbf{X}^+) - \mathcal{K}_i(\mathbf{X})\big)^*\big(D_i^*D_i\big)\big(\mathcal{K}_i(\mathbf{X}^+) - \mathcal{K}_i(\mathbf{X})\big) = 0,\end{aligned} \tag{A.46}$$

and from Proposition 3.20 and (A.46) it follows that $\mathbf{X} - \mathbf{X}^+ \geq 0$, that is, $\mathbf{X} \geq \mathbf{X}^+$. But from Theorem A.5 and $\mathcal{R}(\mathbf{X}) = 0$ we have that $\mathbf{X}^+ \geq \mathbf{X}$, showing that $\mathbf{X} = \mathbf{X}^+$. □

The next theorem presents a necessary and sufficient condition for the existence of the mean-square stabilizing solution for the CARE (4.28). In what follows we use the following notation: for any $\mathbf{V} = (V_1, \ldots, V_N) \in \mathbb{H}^{n*}$ such that $\mathcal{R}(\mathbf{V}) \geq 0$, we write $\mathcal{R}(\mathbf{V})^{1/2} = (\mathcal{R}_1(\mathbf{V})^{1/2}, \ldots, \mathcal{R}_N(\mathbf{V})^{1/2})$ and $\mathbf{A} - \mathbf{B}\mathcal{K}(\mathbf{V}) = (A_1 - B_1\mathcal{K}_1(\mathbf{V}), \ldots, A_N - B_N\mathcal{K}_N(\mathbf{V}))$.

Theorem A.8 *The following statements are equivalent:*

(i) $(\mathbf{A}, \mathbf{B}, \Pi)$ *is SS, and* $(\mathcal{R}(\widetilde{\mathbf{X}})^{1/2}, \mathbf{A} - \mathbf{B}\mathcal{K}(\widetilde{\mathbf{X}}), \Pi)$ *is SD for some* $\widetilde{\mathbf{X}} = (\widetilde{X}_1, \ldots, \widetilde{X}_N) \in \mathbb{H}^{n*}$ *such that* $\mathcal{R}(\widetilde{\mathbf{X}}) \geq 0$.
(ii) *There exists the mean-square stabilizing solution for the CARE* (4.28).

Moreover, if (i) *is satisfied, then the class* $\mathcal{C}(\widetilde{\mathbf{X}}) := \{\mathbf{X} \in \mathbb{H}^{n+}; \mathcal{R}(\mathbf{X}) = 0, \mathbf{X} \geq \widetilde{\mathbf{X}}\}$ *contains only the mean-square stabilizing solution for the CARE* (4.28).

Proof Let us show that (i) implies that $\mathcal{C}(\widetilde{\mathbf{X}})$ contains only the mean-square stabilizing solution for the CARE (4.28). Since $(\mathbf{A}, \mathbf{B}, \Pi)$ is SS, we have from Theorem A.5 that the maximal solution $\mathbf{X}^+$ exists, and since $\mathcal{R}(\widetilde{\mathbf{X}}) \geq 0$, we have from

Theorem A.5 that $\mathbf{X}^+ \geq \widetilde{\mathbf{X}}$, and thus $\mathbf{X}^+ \in \mathcal{C}(\widetilde{\mathbf{X}})$. Consider any $\mathbf{X} \in \mathcal{C}(\widetilde{\mathbf{X}})$ and set $\mathbf{R} = \mathbf{X} - \widetilde{\mathbf{X}} \geq 0$. From identity (A.12) we have that

$$\begin{aligned}&\big(A_i - B_i\mathcal{K}_i(\mathbf{X})\big)^* R_i + R_i\big(A_i - B_i\mathcal{K}_i(\mathbf{X})\big) + \sum_{j=1}^N \lambda_{ij} R_j\\&\quad + \mathcal{R}_i(\widetilde{\mathbf{X}}) + \big(\mathcal{K}_i(\mathbf{X}) - \mathcal{K}_i(\widetilde{\mathbf{X}})\big)^* D_i^* D_i\big(\mathcal{K}_i(\mathbf{X}) - \mathcal{K}_i(\widetilde{\mathbf{X}})\big) = 0. \qquad (A.47)\end{aligned}$$

We can find $0 < \delta \leq 1$ such that $D_i^* D_i > \delta I$ for each $i = 1, \ldots, N$, and thus, by (A.47),

$$\begin{aligned}&\big(A_i - B_i\mathcal{K}_i(\mathbf{X})\big)^* R_i + R_i\big(A_i - B_i\mathcal{K}_i(\mathbf{X})\big) + \sum_{j=1}^N \lambda_{ij} R_j\\&\quad + \delta\big(\mathcal{R}_i(\widetilde{\mathbf{X}}) + \big(\mathcal{K}_i(\mathbf{X}) - \mathcal{K}_i(\widetilde{\mathbf{X}})\big)^* \big(\mathcal{K}_i(\mathbf{X}) - \mathcal{K}_i(\widetilde{\mathbf{X}})\big)\big) \leq 0. \qquad (A.48)\end{aligned}$$

From the hypothesis that $(\mathcal{R}(\widetilde{\mathbf{X}})^{1/2}, \mathbf{A} - \mathbf{B}\mathcal{K}(\widetilde{\mathbf{X}}), \Pi)$ is SD we can find $\mathbf{V} = (V_1, \ldots, V_N)$ such that the operator $\mathcal{T}^{\mathbf{V}}$, defined as

$$\begin{aligned}\mathcal{T}_i^{\mathbf{V}}(\mathbf{Z}) &= \big(A_i - B_i\mathcal{K}_i(\widetilde{\mathbf{X}}) - V_i\mathcal{R}_i(\widetilde{\mathbf{X}})^{1/2}\big)^* Z_i\\&\quad + Z_i\big(A_i - B_i\mathcal{K}_i(\widetilde{\mathbf{X}}) - V_i\mathcal{R}_i(\widetilde{\mathbf{X}})^{1/2}\big) + \sum_{j=1}^N \lambda_{ij} Z_j, \quad i \in \mathcal{S},\end{aligned}$$

is such that $\mathrm{Re}\{\lambda(\mathcal{T}^{\mathbf{V}})\} < 0$. We set now $\widehat{\mathbf{B}} = (\widehat{B}_1, \ldots, \widehat{B}_N)$, $\widehat{\mathbf{G}} = (\widehat{G}_1, \ldots, \widehat{G}_N)$, and $\widehat{\mathbf{K}} = (\widehat{K}_1, \ldots, \widehat{K}_N)$ as follows:

$$\widehat{B}_i = \begin{bmatrix} V_i & B_i \end{bmatrix}, \qquad \widehat{G}_i = \begin{bmatrix} \mathcal{R}_i(\widetilde{\mathbf{X}})^{1/2} \\ \mathcal{K}_i(\widetilde{\mathbf{X}}) \end{bmatrix}, \qquad \widehat{K}_i = \begin{bmatrix} 0 \\ \mathcal{K}_i(\mathbf{X}) \end{bmatrix}.$$

Then it is easy to see that

$$\begin{aligned}&A_i - \widehat{B}_i\widehat{G}_i = A_i - B_i\mathcal{K}_i(\widetilde{\mathbf{X}}) - V_i\mathcal{R}_i(\widetilde{\mathbf{X}})^{1/2},\\&A_i - \widehat{B}_i\widehat{K}_i = A_i - B_i\mathcal{K}_i(\mathbf{X}),\\&(\widehat{G}_i - \widehat{K}_i)^*(\widehat{G}_i - \widehat{K}_i) = \mathcal{R}_i(\widetilde{\mathbf{X}}) + \big(\mathcal{K}_i(\widetilde{\mathbf{X}}) - \mathcal{K}_i(\mathbf{X})\big)^*\big(\mathcal{K}_i(\widetilde{\mathbf{X}}) - \mathcal{K}_i(\mathbf{X})\big).\end{aligned}$$

In other words, (A.48) can be rewritten as

$$\mathcal{T}_i^{\widehat{\mathbf{K}}}(\mathbf{R}) + \delta(\widehat{K}_i - \widehat{G}_i)^*(\widehat{K}_i - \widehat{G}_i) \leq 0, \quad i \in \mathcal{S},$$

where $\mathcal{T}^{\widehat{\mathbf{K}}}$ is as in Proposition A.4, replacing $\mathbf{B}$ by $\widehat{\mathbf{B}}$. Since $\mathcal{T}^{\widehat{\mathbf{G}}} = \mathcal{T}^{\mathbf{V}}$ and $\mathrm{Re}\{\lambda(\mathcal{T}^{\mathbf{V}})\} < 0$, we get from Proposition A.4 that $\mathrm{Re}\{\lambda(\mathcal{T}^{\widehat{\mathbf{K}}})\} < 0$, showing (from Proposition A.7) that $\mathbf{X}$ is the mean-square stabilizing solution for the CARE (4.28).

The proof of (ii) implies (i) is immediate, since in this case clearly $(\mathbf{A}, \mathbf{B}, \Pi)$ is SS, and considering the mean-square stabilizing solution $\mathbf{X}$ for the CARE (4.28), we have that $(\mathcal{R}(\mathbf{X})^{1/2}, \mathbf{A} - \mathbf{B}\mathcal{K}(\mathbf{X}), \Pi) = (0, \mathbf{A} - \mathbf{B}\mathcal{K}(\mathbf{X}), \Pi)$ is SD. □

From Theorem A.8 we have the following corollary.

Corollary A.9 *If* $(\mathbf{A}, \mathbf{B}, \Pi)$ *is SS and* $(\mathbf{C}, \mathbf{A}, \Pi)$ *is SD, then there exists a unique solution for the CARE* (4.28) *in* $\mathbb{H}^{n+}$, *which coincides with the mean-square stabilizing solution.*

Proof This follows from Theorem A.8 taking in (i) $\widetilde{\mathbf{X}} = 0$, so that $\mathcal{R}_i(0) = C_i^* C_i \geq 0$ and $\mathcal{K}_i(0) = 0$, and recalling that $C_i = U_i(C_i^* C_i)^{1/2}$ for an orthogonal matrix U_i (see Theorem 7.5 in [298]). □

Remark A.10 The application of the iterative technique presented in Theorem A.5, also known as Newton's method (see, for instance, [95, 96, 161, 177, 178]), requires an initial mean-square stabilizing gain $\mathbf{K}^0$, which is not in general easily obtained. Thus, some kind of numerical procedure would be required to derive this initial mean-square stabilizing gain $\mathbf{K}^0$, and after that we could apply the iterative technique (which corresponds to solving a sequence of linear equations) to get an approximation for the solution of the CARE. On the other hand, the reformulation of the problem in terms of LMIs optimization is more advantageous since it provides directly a numerical technique for obtaining the maximal solution. Moreover, there are nowadays very efficient numerical algorithms for solving LMIs optimization problems with the global optimum found in polynomial time [49]. Once a solution $\mathbf{X}$ for (A.45) is found, one can check whether it is the mean-square stabilizing solution by verifying if $\mathrm{Re}\{\lambda(\mathcal{L})\} < 0$, where for $\mathbf{P} \in \mathbb{H}^n$, $\mathcal{L}(\mathbf{P}) = (\mathcal{L}_1(\mathbf{P}), \dots, \mathcal{L}_N(\mathbf{P}))$ is $\mathcal{L}_j(\mathbf{P}) = (A_j - B_j\mathcal{K}_j(\mathbf{X}))P_j + P_j(A_j - B_j\mathcal{K}_j(\mathbf{X}))^* + \sum_{i\in\mathcal{S}} \lambda_{ij} P_i$. If so, then clearly $(\mathbf{A}, \mathbf{B}, \Pi)$ is SS, and by Lemma A.6, $\mathbf{X}$ is the maximal solution and, in fact, by Proposition A.7, is the mean-square stabilizing solution. Notice also that by using the Kronecker product we can rewrite the operator $\mathcal{L}$ as an $Nn^2 \times Nn^2$ matrix (see Chap. 3), so that the test $\mathrm{Re}\{\lambda(\mathcal{L})\} < 0$ can be easily performed.

A.5 Filtering Coupled Algebraic Riccati Equations

The results presented in Sects. A.2, A.3, and A.4 are for the CARE related to the optimal control problem. For the CARE related to the filtering problem, the results are similar, and for this reason, we only state the main results in this section. For the CDRE, we will consider a time-varying model defined by the following jump controlled system $\mathcal{G}$:

$$\mathcal{G} = \begin{cases} dx(t) = A_{\theta(t)}(t)x(t)\,dt + J_{\theta(t)}(t)\,dw(t), \\ dy(t) = H_{\theta(t)}(t)x(t)\,dt + G_{\theta(t)}(t)\,dw(t), \end{cases} \tag{A.49}$$

where $A_i(t)$, $J_i(t)$, $H_i(t)$, $G_i(t)$ are real matrices of class **PC** (see Definition 2.3). We also assume that $G_i(t)G_i^*(t) > 0$ and $J_i(t)G_i^*(t) = 0$ for all $i \in \mathcal{S}$ and $t \in \mathbb{R}^+$ and that the matrices $(G_i(t)G_i^*(t))^{-1}$ are of class **PC**. As before, we write $\mathbb{D} \subset \mathbb{R}^+$ as the union of the discontinuity points of A_i, J_i, H_i, G_i, $(G_iG_i^*)^{-1}$, $i \in \mathcal{S}$. In what follows we recall that, under our assumptions, $p_i(t) > 0$ for all $i \in \mathcal{S}$ and $t \in \mathbb{R}^+$, and $p_i(t) \to \pi_i > 0$ as $t \to \infty$. We define the linear operator $\mathcal{K}^f(\cdot, t) \in \mathbb{B}(\mathbb{H}^n)$ and the nonlinear operator $\mathcal{R}^f(\cdot, t) : \mathbb{H}^n \to \mathbb{H}^n$ in the following way: for any $\mathbf{Z} = (Z_1, \dots, Z_N) \in \mathbb{H}^n$, $\mathcal{K}^f(\mathbf{Z}, t) = (\mathcal{K}_1(\mathbf{Z}, t), \dots, \mathcal{K}_N(\mathbf{Z}, t))$ and $\mathcal{R}^f(\mathbf{Z}, t) = (\mathcal{R}_1(\mathbf{Z}, t), \dots, \mathcal{R}_N(\mathbf{Z}, t))$ are defined as

$$\mathcal{K}_i^f(\mathbf{Z}, t) = Z_i H_i^*(t)\big[p_i(t)G_i(t)G_i^*(t)\big]^{-1}, \quad i \in \mathcal{S}, \tag{A.50}$$

and

$$\begin{aligned}\mathcal{R}_i^f(\mathbf{Z}, t) = {} & A_i(t)Z_i + Z_iA_i^*(t) - Z_iH_i^*(t)\big(p_i(t)G_i(t)G_i^*(t)\big)^{-1}H_i(t)Z_i \\ & + \sum_{j=1}^{N} \lambda_{ji}Z_j + J_i^*(t)J_i(t)p_i(t), \quad i \in \mathcal{S}.\end{aligned} \tag{A.51}$$

The set of filtering CDRE is defined as

$$-\dot{\mathbf{Y}}(t) + \mathcal{R}^f\big(\mathbf{Y}(t), t\big) = 0, \quad \mathbf{Y}(0) \in \mathbb{H}^{n+}, \tag{A.52}$$

where $\mathbf{Y}(t) = (Y_1(t), \dots, Y_N(t))$. As in Sect. A.2, we have the following result:

Theorem A.11 *There exists a unique set of N positive semi-definite and continuous $n \times n$ matrices $\mathbf{Y}(t) = (Y_1(t), \dots, Y_N(t)) \in \mathbb{H}^{n+}$, $t \in \mathbb{R}^+$, satisfying* (A.52) *for each $t \in \mathbb{R}^+ \setminus \mathbb{D}$.*

Consider now all matrices in (A.49) time invariant with $G_iG_i^* > 0$ and $J_iG_i^* = 0$ for each $i \in \mathcal{S}$. The linear operator $\mathcal{K}^f \in \mathbb{B}(\mathbb{H}^n)$ and the nonlinear operator $\mathcal{R}^f : \mathbb{H}^n \to \mathbb{H}^n$ are defined in the following way: for any $\mathbf{Z} = (Z_1, \dots, Z_N) \in \mathbb{H}^n$, $\mathcal{K}^f(\mathbf{Z}) = (\mathcal{K}_1^f(\mathbf{Z}), \dots, \mathcal{K}_N^f(\mathbf{Z}))$ and $\mathcal{R}^f(\mathbf{Z}) = (\mathcal{R}_1^f(\mathbf{Z}), \dots, \mathcal{R}_N^f(\mathbf{Z}))$ are defined as

$$\mathcal{K}_i^f(\mathbf{Z}) = Z_iH_i^*\big[\pi_iG_iG_i^*\big]^{-1}, \quad i \in \mathcal{S}, \tag{A.53}$$

and

$$\begin{aligned}\mathcal{R}_i(\mathbf{Z}) = {} & A_iZ_i + Z_iA_i^* - Z_iH_i^*\big(\pi_iG_iG_i^*\big)^{-1}H_iZ_i \\ & + \sum_{j=1}^{N} \lambda_{ji}Z_j + J_i^*(t)J_i(t)\pi_i, \quad i \in \mathcal{S}.\end{aligned} \tag{A.54}$$

We have the filtering CARE

$$\mathcal{R}^f(\mathbf{S}) = 0 \tag{A.55}$$

and the following definitions.

Definition A.12 We say that $\mathbf{Y} = (Y_1, \dots, Y_N) \in \mathbb{H}^{n+}$ is a maximal solution of (A.55) if $\mathcal{R}^f(\mathbf{Y}) = 0$ and, for any $\mathbf{V} = (V_1, \dots, V_N) \in \mathbb{H}^{n*}$ such that $\mathcal{R}^f(\mathbf{V}) \geq 0$, we have that $\mathbf{Y} \geq \mathbf{V}$.

Definition A.13 We say that $\mathbf{Y} = (Y_1, \dots, Y_N) \in \mathbb{H}^{n}$ is a positive semi-definite solution to the CARE if $\mathbf{Y} \in \mathbb{H}^{n+}$ and satisfies (A.55). Furthermore, $\mathbf{Y}$ is a (mean-square) stabilizing solution of (4.28) if $\mathcal{R}^f(\mathbf{Y}) = 0$ and for $\mathbf{K} = (K_1, \dots, K_N) \in \mathbb{H}^{n,m}$ defined as $\mathbf{K} = \mathcal{K}^f(\mathbf{Y})$, we have that $\mathbf{K}$ stabilizes $(\mathbf{H}, \mathbf{A}, \Pi)$.

The following results come from the same arguments as those presented in Sects. A.3 and A.4.

Theorem A.14 (Maximal solution) *Suppose that* $(\mathbf{H}, \mathbf{A}, \Pi)$ *is SD. Then there exists* $\mathbf{Y}^+ = (Y_1^+, \dots, Y_N^+) \in \mathbb{H}^{n+}$ *such that* $\mathcal{R}^f(\mathbf{Y}^+) = 0$ *and* $\mathbf{Y}^+ \geq \widetilde{\mathbf{Y}}$ *for any* $\widetilde{\mathbf{Y}} \in \mathbb{H}^{n*}$ *such that* $\mathcal{R}^f(\widetilde{\mathbf{Y}}) \geq 0$. *Furthermore,* $\operatorname{Re}\{\lambda(\mathcal{T}^+)\} \leq 0$, *where* $\mathcal{T}^+ = (\mathcal{T}_1^+, \dots, \mathcal{T}_N^+)$ *is defined for* $\mathbf{P} = (P_1, \dots, P_N)$ *as* $\mathcal{T}_i^+(\mathbf{P}) = (A_i^+)^* P_i + P_i A_i^+ + \sum_{j \in \mathcal{S}} \lambda_{ij} P_j$, *for* $i = 1, \dots, N$, *and*

$$A_i^+ = A_i - K_i^+ H_i, \qquad K_i^+ = \mathcal{K}_i^f(\mathbf{Y}^+).$$

In what follows we use the following notation: for any $\mathbf{V} = (V_1, \dots, V_N) \in \mathbb{H}^{n*}$ such that $\mathcal{R}^f(\mathbf{V}) \geq 0$, we write $\mathcal{R}^f(\mathbf{V})^{1/2} = (\mathcal{R}_1^f(\mathbf{V})^{1/2}, \dots, \mathcal{R}_N^f(\mathbf{V})^{1/2})$ and $\mathbf{A} - \mathcal{K}^f(\mathbf{V})\mathbf{H} = (A_1 - \mathcal{K}_1^f(V)H_1, \dots, A_N - \mathcal{K}_N^f(\mathbf{V})H_N)$.

Theorem A.15 *The following statements are equivalent:*

(i) $(\mathbf{H}, \mathbf{A}, \Pi)$ *is SD, and* $(\mathbf{A} - \mathcal{K}^f(\widetilde{\mathbf{Y}})\mathbf{H}, \mathcal{R}^f(\widetilde{\mathbf{Y}})^{1/2}, \Pi)$ *is SS for some* $\widetilde{\mathbf{Y}} = (\widetilde{Y}_1, \dots, \widetilde{Y}_N) \in \mathbb{H}^{n*}$ *such that* $\mathcal{R}^f(\widetilde{\mathbf{Y}}) \geq 0$.
(ii) *There exists the mean-square stabilizing solution for the CARE* (A.55).

Moreover, if (i) *is satisfied, then the class* $\mathcal{C}(\widetilde{\mathbf{Y}}) := \{\mathbf{Y} \in \mathbb{H}^{n+}; \mathcal{R}^f(\mathbf{Y}) = 0, \mathbf{Y} \geq \widetilde{\mathbf{Y}}\}$ *contains only the mean-square stabilizing solution for the CARE* (A.55).

From Theorem A.15 we have the following corollary.

Corollary A.16 *If* $(\mathbf{H}, \mathbf{A}, \Pi)$ *is SD and* $(\mathbf{A}, \mathbf{J}, \Pi)$ *is SS, then there exists a unique solution for the CARE* (A.55) *in* $\mathbb{H}^{n+}$, *which coincides with the mean-square stabilizing solution.*

A.6 Asymptotic Convergence

In this section we deal with the convergence of the solution of the CDRE to the mean-square stabilizing solution of the CARE. It will be more interesting to consider the filtering case, due to the presence of the time-varying terms $p_i(t)$. For this case, we will need to assume that the Markov process $\{\theta(t)\}$ is irreducible. As seen in Sect. 2.5, there will exist limit probabilities $\{\pi_i; i \in \mathcal{S}\}$, which do not depend on the initial distribution, with $\sum_{i\mathcal{S}} \pi_i = 1$, satisfying (2.21). Therefore, $p_{ij}(t) \to \pi_j$ and $p_j(t) \to \pi_j$, exponentially fast, as $t \to \infty$. We have the following result.

Theorem A.17 *Suppose that* $(\mathbf{H}, \mathbf{A}, \Pi)$ *is SD and* $(\mathbf{A}, \mathbf{J}, \Pi)$ *is SS. Suppose also that* (2.21) *holds. Let* $\mathbf{Y}(t) = (Y_1(t), \dots, Y_N(t)) \in \mathbb{H}^{n+}$, $t \in \mathbb{R}^+$, *be the solution of* (A.52), *and* $\mathbf{Y} = (Y_1, \dots, Y_N) \in \mathbb{H}^{n+}$ *the mean-square stabilizing solution for the CARE* (A.55). *Then* $\mathbf{Y}(t) \to \mathbf{Y}$ *as* $t \to \infty$.

Proof The idea of the proof is to obtain lower and upper bound functions $\mathbf{P}_\star(t)$ and $\mathbf{P}^\star(t)$ for $\mathbf{Y}(t)$ that converge to $\mathbf{Y}$, i.e., $\mathbf{P}_\star(t) \le \mathbf{Y}(t) \le \mathbf{P}^\star(t)$, and these functions squeeze asymptotically $\mathbf{Y}(t)$ to $\mathbf{Y}$. First, we define the function $\mathbf{P}^\star(t) = (P_1^\star(t), \dots, P_N^\star(t))$ as follows:

$$\dot{P}_i^\star(t) = A_i^\star P_i^\star(t) + P_i^\star(t)\big(A_i^\star\big)^* + K_i^\star\big(p_i(t)G_iG_i^*\big)\big(K_i^\star\big)^* \\ + \sum_{j=1}^N \lambda_{ji} P_j^\star(t) + J_iJ_i^* p_i(t), \quad P_i^\star(0) = S_i(0), \tag{A.56}$$

where $\mathbf{K}^\star = (K_1^\star, \dots, K_N^\star)$ and $\mathbf{A}^\star = (A_1^\star, \dots, A_N^\star)$ are defined as

$$K_i^\star = \mathcal{K}_i^f(\mathbf{Y}) = Y_iH_i^*\big[\pi_iG_iG_i^*\big]^{-1}, \tag{A.57}$$

$$A_i^\star = A_i - K_i^\star H_i. \tag{A.58}$$

We notice now that, by identity (A.12),

$$\dot{Y}_i(t) = A_i^\star Y_i(t) + Y_i(t)\big(A_i^\star\big)^* + K_i^\star\big(p_i(t)G_iG_i^*\big)\big(K_i^\star\big)^* + \sum_{j=1}^N \lambda_{ji} Y_j(t) \\ + J_iJ_i^* p_i(t) - \big(K_i^\star - K_i(t)\big)\big(p_i(t)G_iG_i^*\big)\big(K_i^\star - K_i(t)\big)^*, \tag{A.59}$$

where $K_i(t) = Y_i(t)H_i^*[p_i(t)G_iG_i^*]^{-1}$. Defining $R_i(t) = P_i^\star(t) - Y_i(t)$, we get from (A.56) and (A.59) that

$$\dot{R}_i(t) = A_i^\star R_i(t) + R_i(t)\big(A_i^\star\big)^* + \big(K_i^\star - K_i(t)\big)\big(p_i(t)G_iG_i^*\big)\big(K_i^\star - K_i(t)\big)^* \\ + \sum_{j=1}^N \lambda_{ji} R_j(t), \quad R_i(0) = 0. \tag{A.60}$$

Then by (A.60) and Lemma A.2, $\mathbf{R}(t) = (R_1(t), \dots, R_N(t)) \ge 0$, and thus $\mathbf{P}^\star(t) \ge \mathbf{Y}(t)$. Moreover, from (A.56), Corollary 3.31, and that $\mathbf{K}^\star = (K_1^\star, \dots, K_N^\star)$ stabilizes $(\mathbf{H}, \mathbf{A}, \Pi)$, we have that $\mathbf{P}^\star(t) \to \bar{\mathbf{P}}$, which is the unique solution of the coupled equations

$$(A_i - K_i^\star H_i)\bar{P}_i + \bar{P}_i(A_i - K_i^\star H_i)^* + K_i^\star(\pi_i G_i G_i^*)(K_i^\star)^* \\ + \sum_{j=1}^{N} \lambda_{ji}\bar{P}_j + J_i J_i^* \pi_i = 0. \tag{A.61}$$

But we notice that $\mathbf{Y}$ is also a solution for (A.61), and thus $\bar{\mathbf{P}} = \mathbf{Y}$. Define now for $i \in \mathcal{S}$, $\alpha_i(t) = \inf_{s\in\mathbb{R}^+}\{p_i(t+s)\}$ and

$$\dot{P}_{\star,i}(t) = A_{\star,i}(t)P_{\star,i}(t) + P_{\star,i}(t)\big(A_{\star,i}(t)\big)^* + K_{\star,i}(t)\big(\alpha_i(t)G_iG_i^*\big)\big(K_{\star,i}(t)\big)^* \\ + \sum_{j=1}^{N}\lambda_{ji}P_{\star,j}(t) + J_iJ_i^*\alpha_i(t), \quad P_{\star,i}(0) = 0, \tag{A.62}$$

where $\mathbf{K}_\star(t) = (K_{\star,1}(t), \ldots, K_{\star,N}(t))$ and $\mathbf{A}_\star(t) = (A_{\star,1}(t), \ldots, A_{\star,N}(t))$ are defined as

$$K_{\star,i}(t) = P_{\star,i}(t)H_i^*\big[\alpha_i(t)G_iG_i^*\big]^{-1}, \tag{A.63}$$

$$A_{\star,i}(t) = A_i - K_{\star,i}(t)H_i. \tag{A.64}$$

By considering $\Phi_i(t,s)$ the fundamental matrix associated to the matrix $A_{\star,i}(t) + \frac{\lambda_{ii}}{2}I$, we have from (A.62) that

$$P_{\star,i}(t) = \int_0^t \Phi_i(t,s)\bigg[\sum_{j\neq i}\lambda_{ij}P_{\star,j}(s) + K_{\star,i}(s)\big(\alpha_i(s)G_iG_i^*\big)K_{\star,i}(s) \\ + J_iJ_i^*\alpha_i(s)\bigg]\Phi_i^*(t,s)\,ds, \tag{A.65}$$

and from the fact that all terms inside the integral of (A.65) are positive semi-definite we have that $\mathbf{P}_\star(t) \geq \mathbf{P}_\star(s)$ whenever $t \geq s$. Using identity (A.12), we can write (A.62) as

$$\dot{P}_{\star,i}(t) = \bar{A}_i(t)P_{\star,i}(t) + P_{\star,i}(t)\big(\bar{A}_i(t)\big)^* + K_i(t)\big(\alpha_i(t)G_iG_i^*\big)\big(K_i(t)\big)^* \\ + J_iJ_i^*\alpha_i(t) + \sum_{j=1}^{N}\lambda_{ji}P_{\star,j}(t) \\ - \big(K_{\star,i}(t) - K_i(t)\big)\big(\alpha_i(t)G_iG_i^*\big)\big(K_{\star,i}(t) - K_i(t)\big)^*, \tag{A.66}$$

where

$$K_i(t) = \mathcal{K}_i^f\big(\mathbf{Y}(t)\big) = Y_i(t)H_i^*\big[p_i(t)G_iG_i^*\big]^{-1}, \tag{A.67}$$

$$\bar{A}_i(t) = A_i - K_i(t)H_i. \tag{A.68}$$

Again by identity (A.12),

$$\dot{Y}_i(t) = \bar{A}_i(t)Y_i(t) + Y_i(t)\big(\bar{A}_i(t)\big)^* + K_i(t)\big(p_i(t)G_iG_i^*\big)\big(K_i(t)\big)^* + J_iJ_i^*p_i(t) + \sum_{j=1}^{N}\lambda_{ji}Y_j(t), \tag{A.69}$$

and defining $R_i(t) = Y_i(t) - P_{\star,i}(t)$, we get from (A.66) and (A.69) that

$$\begin{aligned}\dot{R}_i(t) &= \bar{A}_i(t)R_i(t) + R_i(t)\big(A_i(t)\big)^* \\ &\quad + \big(p_i(t) - \alpha_i(t)\big)\big(K_i(t)\big(G_iG_i^*\big)\big(K_i(t)\big)^* + J_iJ_i^*\big) \\ &\quad + \sum_{j=1}^{N}\lambda_{ji}R_j(t) + \big(K_{\star,i}(t) - K_i(t)\big)\big(\alpha_i(t)G_iG_i^*\big)\big(K_{\star,i}(t) - K_i(t)\big)^*, \end{aligned} \tag{A.70}$$

$$R_i(0) = Y_i(0).$$

Then from (A.70), $(p_i(t) - \alpha_i(t)) \geq 0$, and Lemma A.2 we have that $\mathbf{R}(t) = (R_1(t), \dots, R_N(t)) \geq 0$, and thus $\mathbf{Y}(t) \geq \mathbf{P}_\star(t)$. Now, since $\mathbf{P}_\star(t)$ is a nondecreasing function of t and is bounded above by $\mathbf{Y}$, because $\mathbf{P}_\star(t) \leq \mathbf{Y}(t) \leq \mathbf{P}^\star(t)$ and $\lim_{t\to\infty}\mathbf{P}^\star(t) = \mathbf{Y}$, we have that $\lim_{t\to\infty}\mathbf{P}_\star(t)$ exists. So, there exists a matrix $\bar{\mathbf{P}}_\star$ such that $\mathbf{P}_\star(t) \to \bar{\mathbf{P}}_\star$ as $t \to \infty$. Since $\mathbf{P}_\star(t)$ is bounded, so is $\dot{\mathbf{P}}_\star(t)$ and $\ddot{\mathbf{P}}_\star(t)$ by (A.66), and from the convergence of the integral $\int_0^\infty \dot{\mathbf{P}}_\star(t)\,dt = \bar{\mathbf{P}}_\star$ it follows by Lemma 2.12 in [233] that $\dot{\mathbf{P}}_\star(t) \to 0$ as $t \to \infty$. Therefore, from (A.62), (A.63), (A.64), and that $\alpha_i(t) \to \pi_i$ as $t \to \infty$ it follows that $\bar{\mathbf{P}}_\star$ is the solution of the coupled algebraic Riccati equation given by

$$A_{\star,i}P_{\star,i} + P_{\star,i}(A_{\star,i})^* + K_{\star,i}\big(\pi_iG_iG_i^*\big)(K_{\star,i})^* + \sum_{j=1}^{N}\lambda_{ji}P_{\star,j} + J_iJ_i^*\pi_i = 0, \tag{A.71}$$

where $K_{\star,i} = P_{\star,i}H_i^*[\pi_iG_iG_i^*]^{-1}$ and $A_{\star,i} = A_i - K_{\star,i}H_i$. But (A.71) is equivalent to $\mathcal{R}^f(\mathbf{P}_\star) = 0$, and, since $\mathbf{P}_\star \in \mathbb{H}^{n+}$, we have from the uniqueness of a solution of the CARE in $\mathbb{H}^{n+}$ (see Corollary A.16) that $\mathbf{P}_\star = \mathbf{Y}$. Therefore, $\lim_{t\to\infty}\mathbf{P}_\star(t) = \mathbf{Y}$, completing the proof. □

For the control CARE, we have an equivalent result. Notice however that in this case we do not need the ergodicity assumption since in this case the time-varying terms $p_i(t)$ are not present in the CDRE (A.4).

Theorem A.18 *Suppose that* $(\mathbf{A}, \mathbf{B}, \Pi)$ *is SS and* $(\mathbf{C}, \mathbf{A}, \Pi)$ *is SD. Let* $\mathbf{X}^T(t) = (X_1^T(t), \dots, X_N^T(t)) \in \mathbb{H}^{n+}$, $t \in [0, T]$, *be the solution of* (A.4), *and* $\mathbf{X} = (X_1, \dots, X_N) \in \mathbb{H}^{n+}$ *the mean-square stabilizing solution for the CARE* (4.28). *Then* $\mathbf{X}^T(0) \to \mathbf{X}$ *as* $T \to \infty$.

A.7 Filtering Differential and Algebraic Riccati Equation for Unknown $\theta(t)$

The main purpose of this section is to present a proof of Theorem 7.15. This proof follows the same ideas as the proof of Theorem A.17. In order to prove Theorem 7.15, we first need several intermediate results. We assume from now on the hypothesis of Theorem 7.15. In addition, we assume the results regarding the existence and uniqueness of a positive semi-definite solution, $\tilde{Z}$, for the algebraic Riccati equation (7.38), including the fact that $F - \tilde{Z}H^*(G^pG^{p*})^{-1}H$ is stable. In order to prove the following results, we rewrite the matrix Riccati equation (7.36) in a more convenient way. Defining $T_t := \tilde{Z}(t)H^*K_t^{-1}$ and $K_t := G_t^pG_t^{p*}$, (7.36) is given by

$$\dot{\tilde{Z}}(t) = (F - T_tH)\tilde{Z}(t) + \tilde{Z}(t)(F - T_tH)^* + T_tK_tT_t^* + J_t^pJ_t^{p*} + \mathcal{V}(\mathbf{Q}(t)),$$
$$\tilde{Z}(0) = E[\tilde{z}_0\tilde{z}_0^*] \geq 0.$$

Since $p_j(t) \to \pi_j$ exponentially fast as $t \to \infty$, we have that

$$G_t^p = [\sqrt{p_1(t)}G_1 \quad \dots \quad \sqrt{p_N(t)}G_N] \to [\sqrt{\pi_1}G_1 \quad \dots \quad \sqrt{\pi_N}G_N] = G^p$$

and

$$J_t^p = \operatorname{diag}(\sqrt{p_i(t)}J_i) \to \operatorname{diag}(\sqrt{\pi_i}J_i) = J^p.$$

So, as $t \to \infty$, we have $G_t^pG_t^{p*} \to G^pG^{p*}$ and $J_t^pJ_t^{p*} + \mathcal{V}(\mathbf{Q}(t)) \to J^pJ^{p*} + \mathcal{V}(\mathbf{Q})$ exponentially fast.

Lemma A.19 *Let $P^\star(t)$ be the solution of the matrix differential equation given by*

$$\begin{aligned} \dot{P}^\star(t) &= \mathfrak{A}P^\star(t) + P^\star(t)\mathfrak{A}^* + T_\infty K_tT_\infty^* + J_t^pJ_t^{p*} + \mathcal{V}(\mathbf{Q}(t)), \\ P^\star(0) &= \tilde{Z}(0) = E[\tilde{z}_0\tilde{z}_0^*] \geq 0, \end{aligned} \tag{A.72}$$

where $\mathfrak{A} := F - T_\infty H$, $T_\infty := \tilde{Z}H^*K^{-1}$ with $\tilde{Z}$ the positive semi-definite solution of (7.38), and $K := G^pG^{p*}$ with $G^p := [\sqrt{\pi_1}G_1 \dots \sqrt{\pi_N}G_N]$. Then $P^\star(t) \geq \tilde{Z}(t)$ for any $t \in \mathbb{R}^+$, and $\lim_{t\to\infty} P^\star(t) = \tilde{Z}$.

Proof Define $\tilde{P}^\star(t) := P^\star(t) - \tilde{Z}(t)$. Then

$$\begin{aligned} \dot{\tilde{P}}^\star(t) &= \mathfrak{A}\tilde{P}^\star(t) + \tilde{P}^\star(t)\mathfrak{A}^* + (T_t - T_\infty)K_t(T_t - T_\infty)^*, \\ \tilde{P}^\star(0) &= 0. \end{aligned} \tag{A.73}$$

Let $\Phi^\star(t,s)$ be the transition matrix associated with $\mathfrak{A}$, i.e., $\Phi^\star(t,s) = e^{\mathfrak{A}(t-s)}$. Then, the solution of (A.73) is given by

$$\tilde{P}^\star(t) = \int_0^t \Phi^\star(t,s)(T_s - T_\infty)K_s(T_s - T_\infty)^*\Phi^\star(t,s)^*\,ds.$$

Since $K_t > 0$ for all $t \in \mathbb{R}^+$, we have $\tilde{P}^\star(t) \geq 0$ and, consequently, $P^\star(t) \geq \tilde{Z}(t)$. From Propositions 3.28 and 3.8(a) we have

$$\hat{\varphi}(\dot{\mathbf{Q}}(t)) = \mathcal{A}\hat{\varphi}(\mathbf{Q}(t)) + \hat{\varphi}(\mathbf{R}(t)), \tag{A.74}$$

where $\mathbf{R}(t) := (R_1(t), \dots, R_N(t))$ with $R_i(t) := J_i J_i^* p_i(t)$. Now, let us consider the limit solution of (A.72) and (A.74). Since $\mathfrak{A}$ and $\mathcal{A}$ are stable, we have from Proposition 2.22 that $\lim_{t\to\infty} P^\star(t) := \bar{P}^\star$ exists and satisfies

$$(F - T_\infty H)\bar{P}^\star + \bar{P}^\star(F - T_\infty H)^* + T_\infty K T_\infty^* + J^p J^{p*} + \mathcal{V}(\mathbf{Q}) = 0. \tag{A.75}$$

Notice that $\tilde{Z}$ also is a solution of (A.75) because, replacing $\bar{P}^\star$ by $\tilde{Z}$ in (A.75) and taking into account that $T_\infty = \tilde{Z}H^*K^{-1}$, we have

$$F\tilde{Z} + \tilde{Z}F^* - \tilde{Z}H^*\left(G^p G^{p*}\right)^{-1}H\tilde{Z} + J^p J^{p*} + \mathcal{V}(\mathbf{Q}) = 0,$$

which is (7.38). Since $F - T_\infty H$ is stable, the algebraic Riccati equation (A.75) admits a unique solution. Therefore, we must have $\bar{P}^\star = \tilde{Z}$. In short, we have $\lim_{t\to\infty} P^\star(t) = \tilde{Z}$ and $P^\star(t) \geq \tilde{Z}(t)$. □

Lemma A.20 *Let $Q_i(t)$ and $\bar{Q}_i(t)$ be solutions of matrix differential equations given, respectively, by*

$$\begin{aligned} \dot{Q}_i(t) &= A_i Q_i(t) + Q_i(t)A_i^* + \sum_{j=1}^N \lambda_{ji} Q_j(t) + J_i J_i^* p_i(t), \\ Q_i(0) &= V_i \geq 0, \end{aligned} \tag{A.76}$$

and

$$\begin{aligned} \dot{\bar{Q}}_i(t) &= A_i \bar{Q}_i(t) + \bar{Q}_i(t)A_i^* + \sum_{j=1}^N \lambda_{ji} \bar{Q}_j(t) + J_i J_i^* \alpha_i(t), \\ \bar{Q}_i(0) &= V_i \geq 0, \end{aligned} \tag{A.77}$$

where $\alpha_i(t) = \inf_{s\in\mathbb{R}^+}\{p_i(t+s)\}$. Then, for all $t > 0$, we have $Q_i(t) \geq \bar{Q}_i(t)$.

Proof First observe that $p_i(t) \geq \alpha_i(t)$ and that for $0 \leq s \leq t$, $\alpha_i(t) \geq \alpha_i(s)$, i.e., $\alpha_i(t)$ is a nondecreasing function of t. Moreover, $\alpha_i(t) \to \pi_i$ exponentially fast as $t \to \infty$ because $p_i(t) \to \pi_i$ exponentially fast. Next, by Lemma A.2 under positive semi-definite initial conditions, (A.76) and (A.77) admit positive semi-definite solutions $Q_i(t) \geq 0$ and $\bar{Q}_i(t) \geq 0$ for all $t \in \mathbb{R}^+$. We now show that $Q_i(t) \geq \bar{Q}_i(t)$. (A.76) can be rewritten as

$$\dot{Q}_i(t) = \left(A_i + \frac{1}{2}\lambda_{ii}I\right)Q_i(t) + Q_i(t)\left(A_i + \frac{1}{2}\lambda_{ii}I\right)^*$$

$$+\sum_{j=1,\ j\neq i}^{N}\lambda_{ji}Q_j(t)+J_iJ_i^*p_i(t),$$
$$Q_i(0)=V_i\geq 0.$$

Define $R_i(t):=Q_i(t)-\bar{Q}_i(t)$. Then, we get

$$\dot{R}_i(t)=\left(A_i+\frac{1}{2}\lambda_{ii}I\right)R_i(t)+R_i(t)\left(A_i+\frac{1}{2}\lambda_{ii}I\right)^*$$
$$+\sum_{j=1,\ j\neq i}^{N}\lambda_{ji}R_j(t)+J_iJ_i^*\big[p_i(t)-\alpha_i(t)\big],$$
$$R_i(0)=0.$$

Since $J_iJ_i^*[p_i(t)-\alpha_i(t)]\geq 0$, because $p_i(t)\geq\alpha_i(t)$ for all t, the solution $R_i(t)$ is obtained in the same way as we did for the solution of $\bar{Q}_i(t)$ and possesses the same properties as those of $\bar{Q}_i(t)$. So, $R_i(t)\geq 0$, which proves that $Q_i(t)\geq\bar{Q}_i(t)$, completing the proof of the lemma. □

Lemma A.21 *Let* $\mathbf{Q}(t)=(Q_1(t),\dots,Q_N(t))\in\mathbb{H}^{n+}$ *and* $\bar{\mathbf{Q}}(t)=(\bar{Q}_1(t),\dots,\bar{Q}_N(t))\in\mathbb{H}^{n+}$ *be as in Lemma* A.20. *Then* $\mathcal{V}(\mathbf{Q}(t))\geq\mathcal{V}(\bar{\mathbf{Q}}(t))$ *for all* $t\geq 0$, *where* $\mathcal{V}(\mathbf{Q}(t))$ *is defined by* (7.37).

Proof We have shown that $\mathcal{V}(\mathbf{Q}(t))$ is a linear operator and that $\mathcal{V}(\mathbf{Q}(t))\geq 0$ for all $\mathbf{Q}(t)=(Q_1(t),\dots,Q_N(t))\in\mathbb{H}^{n+}$. By Lemma A.20, $\mathbf{Q}(t)-\bar{\mathbf{Q}}(t)\geq 0$. Then, $0\leq\mathcal{V}(\mathbf{Q}(t)-\bar{\mathbf{Q}}(t))=\mathcal{V}(\mathbf{Q}(t))-\mathcal{V}(\bar{\mathbf{Q}}(t))$, which is equivalent to saying that $\mathcal{V}(\mathbf{Q}(t))\geq\mathcal{V}(\bar{\mathbf{Q}}(t))$, completing the proof. □

Lemma A.22 *Let* $P_\star(t)$ *be the solution of the Riccati differential equation given by*

$$\dot{P}_\star(t)=\big(F-T_\star(t)H\big)P_\star(t)+P_\star(t)\big(F-T_\star(t)H\big)^*$$
$$+T_\star(t)\bar{K}_tT_\star^*(t)+\bar{J}_t\bar{J}_t^*+\mathcal{V}\big(\bar{\mathbf{Q}}(t)\big),\qquad\text{(A.78)}$$
$$P_\star(0)=0,$$

where $T_\star(t):=P_\star(t)H^*\bar{K}_t^{-1}$, $\bar{K}_t:=\bar{G}_t\bar{G}_t^*$, $\bar{G}_t:=[\sqrt{\alpha_1(t)}G_1\dots\sqrt{\alpha_N(t)}G_N]$, $\bar{J}_t:=\operatorname{diag}(\sqrt{\alpha_i(t)}J_i)$, $\mathcal{V}(\bar{\mathbf{Q}}(t))$ *is the linear operator defined by* (7.37) *applied to* $\bar{\mathbf{Q}}(t)=(\bar{Q}_1(t),\dots,\bar{Q}_N(t))$, *the solution of the matrix differential equation given by* (A.77), *and* $\alpha_i(t)=\inf_{s\in\mathbb{R}^+}\{p_i(t+s)\}$. *Then, for* $0\leq s\leq t$, $P_\star(s)\leq P_\star(t)$. *In addition,* $P_\star(t)\leq\tilde{Z}(t)$ *for all* $t\in\mathbb{R}^+$ *and* $\lim_{t\to\infty}P_\star(t)=\tilde{Z}$, *where* $\tilde{Z}$ *is the solution of* (7.38).

Proof First, observe that, for all t, $J_t^pJ_t^{p*}\geq\bar{J}_t\bar{J}_t^*\geq 0$ and $K_t\geq\bar{K}_t\geq 0$. Also by the previous lemma, $\mathcal{V}(\mathbf{Q}(t))\geq\mathcal{V}(\bar{\mathbf{Q}}(t))\geq 0$. In addition, from the exponential speed of convergence of $\alpha_i(t)$ to π_i we have that $\bar{J}_t\bar{J}_t^* \to J^p J^{p*}$, $\mathcal{V}(\bar{\mathbf{Q}}(t)) \to \mathcal{V}(\mathbf{Q})$, and $\bar{K}_t \to K = G^p G^{p*}$ exponentially fast as $t \to \infty$. Also, for $0 \le s \le t$, $P_\star(s) \le P_\star(t)$, by Lemma 5.2 in [302]. Let us now prove that $P_\star(t) \le \tilde{Z}(t)$ for all $t \in \mathbb{R}^+$. In order to do that, we rewrite (A.78) in the following way:

$$\begin{aligned}\dot{P}_\star(t) &= (F - T_t H)P_\star(t) + P_\star(t)(F - T_t H)^* \\ &\quad + \bar{J}_t\bar{J}_t^* + \mathcal{V}(\bar{\mathbf{Q}}(t)) + T_t\bar{K}_t T_t^* - \big(T_t - T_\star(t)\big)\bar{K}_t\big(T_t - T_\star(t)\big)^*.\end{aligned}$$

Defining $\tilde{P}_\star(t) := \tilde{Z}(t) - P_\star(t)$, we have

$$\begin{aligned}\dot{\tilde{P}}_\star(t) &= (F - T_t H)\tilde{P}_\star(t) + \tilde{P}_\star(t)(F - T_t H)^* \\ &\quad + T_t[K_t - \bar{K}_t]T_t^* + \big[J_t^p J_t^{p*} - \bar{J}_t\bar{J}_t^*\big] + \big[\mathcal{V}(\mathbf{Q}(t)) - \mathcal{V}(\bar{\mathbf{Q}}(t))\big] \\ &\quad + \big(T_t - T_\star(t)\big)\bar{K}_t\big(T_t - T_\star(t)\big)^*, \\ \tilde{P}_\star(0) &= \tilde{Z}(0) \ge 0.\end{aligned}$$

Let $\Phi(t,s)$ be the transition matrix associated with $F - T_t H$. Then

$$\begin{aligned}\tilde{P}_\star(t) &= \Phi(t,0)\tilde{Z}(0)\Phi^*(t,0) \\ &\quad + \int_0^t \Phi(t,s)\big\{T_s[K_s - \bar{K}_s]T_s^* + \big[J_s^p J_s^{p*} - \bar{J}_s\bar{J}_s^*\big] \\ &\quad + \big[\mathcal{V}(\mathbf{Q}(s)) - \mathcal{V}(\bar{Q}(s))\big] + \big(T_s - T_\star(s)\big)\bar{K}_s\big(T_s - T_\star(s)\big)^*\big\}\Phi^*(t,s)\,ds.\end{aligned}$$

Since $\tilde{Z}(0) \ge 0$, $K_t - \bar{K}_t \ge 0$, $J_t^p J_t^{p*} - \bar{J}_t\bar{J}_t^* \ge 0$, $\mathcal{V}(\mathbf{Q}(t)) - \mathcal{V}(\bar{\mathbf{Q}}(t)) \ge 0$, and $(T_t - T_\star(t))\bar{K}_t(T_t - T_\star(t))^* \ge 0$ because $\bar{K}_t \ge 0$ for all $t \ge 0$, we have $\tilde{P}_\star(t) \ge 0$, which proves that $\tilde{Z}(t) \ge P_\star(t)$.

Now, since $P_\star(t)$ is a nondecreasing function of t and is bounded above by $\tilde{Z}$ because $P_\star(t) \le \tilde{Z}(t) \le P^\star(t)$ and $\lim_{t\to\infty} P^\star(t) = \tilde{Z}$, we have that $\lim_{t\to\infty} P_\star(t)$ exists. So, there exists a matrix $\bar{P}_\star$ such that $P_\star(t) \to \bar{P}_\star$ as $t \to \infty$. Now, due to the monotonicity and convergence of $P_\star(t)$, it follows that $\dot{P}_\star(t) \to 0$ as $t \to \infty$. Therefore, $\bar{P}_\star$ is the solution of the algebraic Riccati equation given by

$$(F - \bar{T}_\star H)\bar{P}_\star + \bar{P}_\star(F - \bar{T}_\star H)^* + \bar{T}_\star K\bar{T}_\star^* + J^p J^{p*} + \mathcal{V}(\mathbf{Q}) = 0,$$

where $\bar{T}_\star = \bar{P}_\star H^*(\mathcal{G}\mathcal{G}^*)^{-1}$. But the above equation can be rewritten as

$$F\bar{P}_\star + \bar{P}_\star F^* - \bar{P}_\star H^*\big(G^p G^{p*}\big)^{-1}H\bar{P}_\star + J^p J^{p*} + \mathcal{V}(\mathbf{Q}) = 0. \qquad \text{(A.79)}$$

Since (A.79) is equivalent to (7.38), we have, from the uniqueness of solution of (7.38), that $\bar{P}_\star = \tilde{Z}$. Therefore, $\lim_{t\to\infty} P_\star(t) = \tilde{Z}$. □

We can now proceed to the proof of Theorem 7.15. The idea of the proof runs as in the proof of Theorem A.17. First, notice that from standard results on the theory of ARE (see, for instance, [53], or Lemma 5.1 in [156]), (7.36) has a unique positive semi-definite solution. We have to prove then the existence and uniqueness of a positive semi-definite solution, $\tilde{Z}$, for the algebraic Riccati equation (7.38). Now, proving that $\tilde{Z}(t) \to \tilde{Z}$ is tantamount to proving that there exist lower and upper bound functions $P_\star(t)$ and $P^\star(t)$ for $\tilde{Z}(t)$, i.e., $P_\star(t) \le \tilde{Z}(t) \le P^\star(t)$, and these functions squeeze asymptotically $\tilde{Z}(t)$ to $\tilde{Z}$.

Proof of Theorem 7.15 By Theorem 3.33, if system (7.2) is MSS, then $\mathrm{Re}\{\lambda(\mathcal{A})\} < 0$, where $\mathcal{A} = \Pi' \otimes I_{n^2} + \mathrm{diag}(A_i \oplus A_i)$. But by Proposition 3.13, if $\mathrm{Re}\{\lambda(\mathcal{A})\} < 0$, then $\mathrm{Re}\{\lambda(F)\} < 0$. Therefore, we conclude that the matrix F in (7.36) is stable. Now, from standard results on the theory of ARE (see, for instance, [53], or Lemma 5.1 in [156]) it follows that there exists a unique positive semi-definite solution $\tilde{Z}$ to (7.38) and, furthermore, $F - \tilde{Z}H^*(G^p G^{p*})^{-1}H$ is stable. Now, by Proposition 3.29, under the assumption of mean-square stability of system (7.2), $\|Q_i(t) - Q_i\| \to 0$ as $t \to \infty$, where $Q_i = \hat{\varphi}_i^{-1}(-\mathcal{A}^{-1}\hat{\varphi}(\mathbf{R}))$ with $R_i = J_i J_i^* \pi_i$, $\mathbf{R} = (R_1, \ldots, R_N)$. In addition, notice that $\mathcal{V}(\mathbf{Q}(t)) \to \mathcal{V}(\mathbf{Q})$ as $t \to \infty$, with $\mathcal{V}(\mathbf{Q}(t))$ defined by (7.37), since $\mathcal{V}(\mathbf{Q}(t))$ is a linear bounded operator. Finally, by Lemmas A.19 and A.22 there exist matrices $P_\star(t)$ and $P^\star(t)$ such that

$$P_\star(t) \le \tilde{Z}(t) \le P^\star(t)$$

and

$$\lim_{t\to\infty} P_\star(t) = \lim_{t\to\infty} P^\star(t) = \tilde{Z},$$

and, consequently, we have $\lim_{t\to\infty} \tilde{Z}(t) = \tilde{Z}$, which completes the proof. □

Appendix B
The Adjoint Operator and Some Auxiliary Results

B.1 Outline of the Appendix

The separation principle presented in Chap. 6 is based on some results related to an adjoint operator for MJLS. We derive in this appendix these results.

B.2 Preliminaries

Consider the MJLS given by

$$\dot{x}(t) = A_{\theta(t)}x(t) + B_{\theta(t)}u(t), \quad t \in \mathbb{R}^+, \qquad x(0) = x_0, \tag{B.1}$$

with $\{u(t); t \in \mathbb{R}^+\}$ a deterministic input signal, and write $\vartheta_t = (x(t), \theta(t))$. In what follows we use the same notation and definitions as in Sect. 4.3. For $\mathbf{S} = (S_1, \ldots, S_N) \in \mathbb{H}^{n+}$ and $h(t)$ a bounded continuously differentiable random function for $t \in [0, T)$, taking values in $\mathbb{R}^n$, define $g \in \mathcal{C}^1(\mathcal{X})$ as

$$g\big(t, x(t), \theta(t)\big) = x(t)^* S_{\theta(t)} h(t). \tag{B.2}$$

Then, as in (4.8), it follows that

$$\begin{aligned} \mathfrak{L}^u g\big(t, x(t), \theta(t)\big) &= x(t)^* A^*_{\theta(t)} S_{\theta(t)} h(t) + x(t)^* S_{\theta(t)} \dot{h}(t) \\ &\quad + x(t)^* \sum_{j \in \mathcal{S}} \lambda_{\theta(t)j} S_j h(t) + u(t)^* B^*_{\theta(t)} S_{\theta(t)} h(t). \end{aligned} \tag{B.3}$$

Now, for $x_0 = 0$ and $u(t) = \delta(t)e$, where $e \in \mathbb{R}^m$, and $\delta(t)$ is the usual delta of Dirac, we have from (B.1) that $x(t) = \Phi(t, 0)B_{\theta_0}e$ for $t > 0$. It follows then from Dynkin's formula (4.16) and (B.3) that, letting

$$\Sigma(\ell) := A^*_{\theta(\ell)} S_{\theta(\ell)} h(\ell) + S_{\theta(\ell)} \dot{h}(\ell) + \sum_{j \in \mathcal{S}} \lambda_{\theta(\ell)j} S_j h(\ell), \tag{B.4}$$

O.L.V. Costa et al., *Continuous-Time Markov Jump Linear Systems*,
Probability and Its Applications, DOI 10.1007/978-3-642-34100-7,

we have

$$\begin{aligned} &E\big(e^* B_{\theta_0}^* \Phi(t,0)^* S_{\theta(t)} h(t) \big| x(s), \theta(s)\big) \\ &\quad = x(s)^* S_{\theta(s)} h(s) \\ &\qquad + E\left(\int_s^t e^* B_{\theta_0}^* \Phi(\ell,0)^* \Sigma(\ell)\, d\ell + e^* B_{\theta_0}^* S_{\theta_0} h(0) \,\middle|\, x(s), \theta(s) \right), \end{aligned} \tag{B.5}$$

or yet, for $t > 0$, we have

$$\begin{aligned} &E\big(e^* B_{\theta_0}^* \Phi(t,0)^* S_{\theta(t)} h(t) \big| x_0 = 0, \theta_0\big) \\ &\quad = e^* B_{\theta_0}^* S_{\theta_0} h(0) + E\left(\int_0^t e^* B_{\theta_0}^* \Phi(\ell,0)^* \Sigma(\ell)\, d\ell \,\middle|\, x_0 = 0, \theta_0 \right). \end{aligned} \tag{B.6}$$

If we assume now that θ_0 is a random variable such that $\theta_0 = i$ with probability π_i, then we have

$$\begin{aligned} &\sum_{i\in\mathcal{S}} \pi_i \big\{ E\big(e^* B_i^* \Phi(t,0)^* S_{\theta(t)} h(t) \big| x_0 = 0, \theta_0 = i\big) - e^* B_i^* S_i h(0) \big\} \\ &\quad = \sum_{i\in\mathcal{S}} \pi_i E\left(\int_0^t e^* B_i^* \Phi(\ell,0)^* \Sigma(\ell)\, d\ell \,\middle|\, x_0 = 0, \theta_0 = i \right). \end{aligned} \tag{B.7}$$

B.3 Main Results

We consider now the MJLS

$$\mathcal{G} = \begin{cases} \dot{x}(t) = A_{\theta(t)} x(t) + B_{\theta(t)} u(t), \\ z(t) = C_{\theta(t)} x(t) + D_{\theta(t)} u(t), \end{cases} \tag{B.8}$$

with the time running in $\mathbb{R}$, that is, $t \in \mathbb{R}$. We can write $x(t)$ as

$$x(t) = \int_{-\infty}^t \Phi(t,s) B_{\theta(s)} u(s)\, ds,$$

and thus, $z(t)$ as

$$z(t) = C_{\theta(t)} \left\{ \int_{\infty}^t \Phi(t,s) B_{\theta(s)} u(s)\, ds \right\} + D_{\theta(t)} u(t). \tag{B.9}$$

Let the operator $\mathcal{G}$ from $L_2^m(\Omega, \mathcal{F}, P)$ to $L_2^p(\Omega, \mathcal{F}, P)$ be defined as $\mathcal{G}(u) = \{z(t); t \in \mathbb{R}\}$, so that $\mathcal{G}(u)(t) = z(t)$. From Theorem 3.27 we have that if system (B.8) is MSS, then $\mathcal{G} \in \mathbb{B}(L_2^m(\Omega, \mathcal{F}, P), L_2^p(\Omega, \mathcal{F}, P))$.

We define next the adjoint operator $\mathcal{G}^*$ of $\mathcal{G}$, which is such that, for any $u \in L_2^m(\Omega, \mathcal{F}, P)$ and any $v \in L_2^p(\Omega, \mathcal{F}, P)$,

$$\langle \mathcal{G}(u); v \rangle := \int_{-\infty}^{\infty} E\big(\mathcal{G}(u)(t)^* v(t)\big)\, dt = \int_{-\infty}^{\infty} E\big(u(t)^* \mathcal{G}^*(v)(t)\big)\, dt =: \langle u; \mathcal{G}^*(v) \rangle$$

is satisfied. Thus,

$$\begin{aligned}
\langle \mathcal{G}(u); v \rangle &= \int_{-\infty}^{\infty} E\big(z(t)^* v(t)\big)\, dt \\
&= E\bigg(\int_{-\infty}^{\infty} \bigg\{ C_{\theta(t)} \bigg\{ \int_{-\infty}^{\infty} \Phi(t,s) B_{\theta(s)} u(s) 1_{\{s \le t\}}\, ds \bigg\} + D_{\theta(t)} u(t) \bigg\}^* v(t)\, dt \bigg) \\
&= E\bigg(\int_{-\infty}^{\infty} \int_s^{\infty} E\big(\{C_{\theta(t)} \Phi(t,s) B_{\theta(s)} u(s)\}^* v(t) \big| \mathcal{F}_s\big)\, dt\, ds \bigg) \\
&\quad + E\bigg(\int_{-\infty}^{\infty} \{D_{\theta(t)} u(t)\}^* v(t)\, dt \bigg) \\
&= E\bigg(\int_{-\infty}^{\infty} u(s)^* \bigg\{ B_{\theta(s)}^* \int_s^{\infty} E\big(\Phi(t,s)^* C_{\theta(t)}^* v(t) \big| \mathcal{F}_s\big)\, dt \\
&\quad + D_{\theta(s)}^* v(s) \bigg\}\, ds \bigg) \\
&= \langle u; \mathcal{G}^*(v) \rangle,
\end{aligned}$$

and therefore,

$$\mathcal{G}^*(v)(s) = B_{\theta(s)}^* \int_s^{\infty} E\big(\Phi(t,s)^* C_{\theta(t)}^* v(t) \big| \mathcal{F}_s\big)\, dt + D_{\theta(s)}^* v(s). \tag{B.10}$$

Proposition B.1 *Suppose that system* (B.8) *is MSS and that* $\mathbf{X} = (X_1, \ldots, X_N) \in \mathbb{H}^n$, $X_i = X_i^*$, *satisfies, for each* $i \in \mathcal{S}$,

$$A_i^* X_i + X_i A_i + C_i^* C_i + \sum_{j \in \mathcal{S}} \lambda_{ij} X_j = 0, \tag{B.11}$$

$$D_i^* C_i + B_i^* X_i = 0. \tag{B.12}$$

Then

$$\mathcal{G}^* \mathcal{G}(u)(t) = D_{\theta(t)}^* D_{\theta(t)} u(t). \tag{B.13}$$

Proof From (B.9) and (B.10) we have

$$\mathcal{G}^* \mathcal{G}(u)(s) = \int_s^{\infty} E\big(B_{\theta(s)}^* \Phi(t,s)^* C_{\theta(t)}^* \mathcal{G}(u)(t) \big| \mathcal{F}_s\big)\, dt + D_{\theta(s)}^* \mathcal{G}(u)(s)$$

$$= \int_s^\infty E\bigg(B^*_{\theta(s)} \Phi(t,s)^* C^*_{\theta(t)}$$
$$\times \bigg(C_{\theta(t)} \int_{-\infty}^t \Phi(t,\ell) B_{\theta(\ell)} u(\ell)\, d\ell + D_{\theta(t)} u(t) \bigg) \bigg| \mathcal{F}_s \bigg)\, dt$$
$$+ D^*_{\theta(s)} \bigg(C_{\theta(s)} \int_{-\infty}^s \Phi(s,\ell) B_{\theta(\ell)} u(\ell)\, d\ell \bigg) + D^*_{\theta(s)} D_{\theta(s)} u(s).$$

Thus, defining

$$h(t) = \int_{-\infty}^t \Phi(t,\ell) B_{\theta(\ell)} u(\ell)\, d\ell,$$

we obtain from (B.11) and (B.12) that

$$\mathcal{G}^*\mathcal{G}(u)(s)$$
$$= -B^*_{\theta(s)} \bigg\{ \int_s^\infty E\bigg(\Phi(t,s)^* \bigg[\bigg\{ A^*_{\theta(t)} X_{\theta(t)} + X_{\theta(t)} A_{\theta(t)} + \sum_{j\in\mathcal{S}} \lambda_{\theta(t)j} X_j \bigg\} h(t)$$
$$+ X_{\theta(t)} B_{\theta(t)} u(t) \bigg] \bigg| \mathcal{F}_s \bigg) dt + X_{\theta(s)} h(s) \bigg\} + D^*_{\theta(s)} D_{\theta(s)} u(s).$$

Noticing also that

$$\dot{h}(t) = A_{\theta(t)} h(t) + B_{\theta(t)} u(t),$$

we further obtain

$$\mathcal{G}^*\mathcal{G}(u)(s) = -B^*_{\theta(s)} \bigg\{ \int_s^\infty E\bigg(\Phi(t,s)^* \bigg[A^*_{\theta(t)} X_{\theta(t)} h(t) + X_{\theta(t)} \dot{h}(t)$$
$$+ \sum_{j\in\mathcal{S}} \lambda_{\theta(t)j} X_j h(t) \bigg] \bigg| \mathcal{F}_s \bigg)\, dt + X_{\theta(s)} h(s) \bigg\} + D^*_{\theta(s)} D_{\theta(s)} u(s).$$

If we show that

$$E\big(\Phi(\tau,s)^* X_{\theta(\tau)} h(\tau) \big| \mathcal{F}_s \big) \to 0$$

as $\tau \to \infty$, then, by (B.6),

$$\int_s^\infty E\bigg(\Phi(t,s)^* \bigg[A^*_{\theta(t)} X_{\theta(t)} h(t) + X_{\theta(t)} \dot{h}(t) + \sum_{j\in\mathcal{S}} \lambda_{\theta(t)j} X_j h(t) \bigg] \bigg| \mathcal{F}_s \bigg)\, dt$$
$$= -X_{\theta(s)} h(s),$$

showing the desired result. As shown in Theorem 3.27, $\|h(t)\|_2 \to 0$ as $t \to \infty$. Writing $\xi(t) = \Phi(t,s)\xi$ for any $\xi \in \mathbb{C}^n$, we have from Lemma 3.12 that $\|\xi(t)\|_2 \to 0$

as $t \to \infty$. Therefore,

$$\begin{aligned}
&\left\| \xi^* E\big(\Phi(\tau,s)^* X_{\theta(\tau)} h(\tau) \big| \mathcal{F}_s\big) \right\| \\
&\quad = \left\| E\big(\xi(\tau)^* X_{\theta(\tau)} h(\tau) \big| \mathcal{F}_s\big) \right\| \\
&\quad \le \|\mathbf{X}\|_{\max} E\big(\|\xi(\tau)\| \, \|h(\tau)\| \, |\mathcal{F}_s\big) \\
&\quad \le \|\mathbf{X}\|_{\max} E\big(\|\xi(\tau)\|^2 |\mathcal{F}_s\big)^{1/2} E\big(\|h(\tau)\|^2 |\mathcal{F}_s\big)^{1/2} \to 0
\end{aligned}$$

as $\tau \to \infty$, completing the proof of the proposition. □

We conclude this appendix by presenting a proof of Proposition 6.25.

Proof of Proposition 6.25 (a) From the control CARE (6.52) we have that

$$\widetilde{A}_i^* X_i + X_i \widetilde{A}_i + \widetilde{C}_i^* \widetilde{C}_i + \sum_{j \in \mathcal{S}} \lambda_{ij} X_j = 0, \tag{B.14}$$

and since $R_i F_i = D_i^* D_i F_i = B_i^* X_i$, we have that

$$\begin{aligned}
R_i^{-1/2}\big(D_i^* \widetilde{C}_i + B_i^* X_i\big) &= R_i^{-1/2}\big(D_i^*(C_i - D_i F_i) + B_i^* X_i\big) \\
&= R_i^{-1/2}\big(-D_i^* D_i F_i + B_i^* X_i\big) = 0.
\end{aligned} \tag{B.15}$$

Thus, from (B.14), (B.15), and Proposition B.1 we have that

$$\mathcal{G}_U^* \mathcal{G}_U(v)(t) = \big(R_{\theta(t)}^{-1/2} D_{\theta(t)}^* D_{\theta(t)} R_{\theta(t)}^{-1/2}\big) v(t) = \big(R_{\theta(t)}^{-1/2} R_{\theta(t)} R_{\theta(t)}^{-1/2}\big) v(t) = v(t),$$

completing the proof of (a) of the proposition.

(b) Let us calculate now $\mathcal{G}_U^* \mathcal{G}_c(w)(t)$. We set $\widetilde{B}_i = B_i R_i^{-1/2}$, $\widetilde{D}_i = D_i R_i^{-1/2}$. We have that

$$\begin{aligned}
\mathcal{G}_c(w)(t) &= \widetilde{C}_{\theta(t)} \left\{ \int_{-\infty}^{t} \widetilde{\Phi}(t,s) J_{\theta(s)} w(s)\, ds \right\}, \\
\mathcal{G}_U^*(v)(s) &= \widetilde{B}_{\theta(s)}^* \int_s^{\infty} E\big(\widetilde{\Phi}(t,s)^* \widetilde{C}_{\theta(t)}^* v(t) \big| \mathcal{F}_s\big)\, dt + \widetilde{D}_{\theta(s)}^* v(s),
\end{aligned}$$

and thus,

$$\begin{aligned}
&\mathcal{G}_U^* \mathcal{G}_c(w)(s) \\
&\quad = \widetilde{B}_{\theta(s)}^* \int_s^{\infty} E\left(\widetilde{\Phi}(t,s)^* \widetilde{C}_{\theta(t)}^* \left(\widetilde{C}_{\theta(t)} \left\{ \int_{-\infty}^{t} \widetilde{\Phi}(t,s) J_{\theta(s)} w(s)\, ds \right\} \right) \middle| \mathcal{F}_s \right) dt \\
&\qquad + \widetilde{D}_{\theta(s)}^* \left(\widetilde{C}_{\theta(s)} \left\{ \int_{-\infty}^{s} \widetilde{\Phi}(t,\ell) J_{\theta(\ell)} w(\ell)\, d\ell \right\} \right).
\end{aligned}$$

From (B.14) and (B.15), writing

$$\widetilde{h}(t) = \int_{-\infty}^{t} \widetilde{\Phi}(t,s) J_{\theta(s)} w(s)\, ds,$$

we have

$$\begin{aligned}
\mathcal{G}_U^* \mathcal{G}_c(w)(s) &= -\widetilde{B}_{\theta(s)}^* \biggl\{ \int_s^{\infty} E\biggl(\widetilde{\Phi}(t,s)^* \biggl(\widetilde{A}_{\theta(t)}^* X_{\theta(t)} + X_{\theta(t)} \widetilde{A}_{\theta(t)} \\
&\quad + \sum_{j\in\mathcal{S}} \lambda_{\theta(t)j} X_j \biggr) \widetilde{h}(t) \Big| \mathcal{F}_s \biggr)\, dt \\
&\quad + X_{\theta(s)} \widetilde{h}(s) + \int_s^{\infty} E\bigl(\widetilde{\Phi}(t,s)^* X_{\theta(t)} J_{\theta(t)} w(t) \big| \mathcal{F}_s \bigr)\, dt \\
&\quad - \int_s^{\infty} E\bigl(\widetilde{\Phi}(t,s)^* X_{\theta(t)} J_{\theta(t)} w(t) \big| \mathcal{F}_s \bigr)\, dt \biggr\} \\
&= -\widetilde{B}_{\theta(s)}^* \biggl\{ \int_s^{\infty} E\biggl(\widetilde{\Phi}(t,s)^* \biggl(\widetilde{A}_{\theta(t)}^* X_{\theta(t)} \widetilde{h}(t) + X_{\theta(t)} \dot{\widetilde{h}}(t) \\
&\quad + \sum_{j\in\mathcal{S}} \lambda_{\theta(t)j} X_j \widetilde{h}(t) \biggr) \Big| \mathcal{F}_s \biggr)\, dt \\
&\quad + X_{\theta(s)} \widetilde{h}(s) - \int_s^{\infty} E\bigl(\widetilde{\Phi}(t,s)^* X_{\theta(t)} J_{\theta(t)} w(t) \big| \mathcal{F}_s \bigr)\, dt \biggr\}
\end{aligned}$$

since

$$\dot{\widetilde{h}}(t) = \widetilde{A}_{\theta(t)} \widetilde{h}(t) + J_{\theta(t)} w(t).$$

By (B.6),

$$\begin{aligned}
&\int_s^{\infty} E\biggl(\widetilde{\Phi}(t,s)^* \biggl[\widetilde{A}_{\theta(t)}^* X_{\theta(t)} \widetilde{h}(t) + X_{\theta(t)} \dot{\widetilde{h}}(t) + \sum_{j\in\mathcal{S}} \lambda_{\theta(t)j} X_j \widetilde{h}(t) \biggr] \Big| \mathcal{F}_s \biggr)\, dt \\
&\quad = -X_{\theta(s)} \widetilde{h}(s)
\end{aligned} \tag{B.16}$$

since from MSS (see the proof of Proposition B.1) we have

$$E\bigl(\widetilde{\Phi}(\tau,s)^* X_{\theta(\tau)} \widetilde{h}(\tau) \big| \mathcal{F}_s \bigr) \to 0$$

as $\tau \to \infty$. Therefore,

$$\mathcal{G}_U^* \mathcal{G}_c(w)(s) = \widetilde{B}_{\theta(s)}^* \int_s^{\infty} E\bigl(\widetilde{\Phi}(t,s)^* X_{\theta(t)} J_{\theta(t)} w(t) \big| \mathcal{F}_s \bigr)\, dt,$$

completing the proof of (b) of the proposition. □

Notation and Conventions

As a general rule, lowercase greek and roman letters are used for vector, scalar variables and functions, while uppercase greek and roman letters are used for matrix variables and functions as well as sequences of matrices. Sets and spaces are denoted by blackboard uppercase characters (such as $\mathbb{R}$, $\mathbb{C}$), operators by calligraphic characters (such as $\mathcal{L}$, $\mathcal{T}$), and sequences of N matrices are indicated in boldface roman (such as $\mathbf{A} = (A_1, \ldots, A_N)$). Sometimes it is not possible or convenient to adhere completely to this rule, but the exceptions should be clearly perceived based on their specific context.

The following lists present the main symbols and general notation used throughout the book, followed by a brief explanation and the number of the page of their definition or first appearance.

Symbol	Description	Page
$\square$	End of proof.	
$\emptyset$	Empty set.	89
$1_{\{\cdot\}}$	Indicator function.	35
$\\|\cdot\\|$	Any norm.	16
$\\|\cdot\\|_1$	1-norm of a sequence of N matrices.	23
$\\|\cdot\\|_2$	$L_2(\Omega, \mathcal{F}, P)$-norm of random object.	16
	2-norm of a sequence of N matrices.	23
$\\|\cdot\\|_{\max}$	max-norm of matrices.	21
	max-norm of a sequence of N matrices.	23
$\\|\cdot\\|_{2,T}$	Finite-horizon energy norm of a signal.	153
$\\|\mathcal{G}\\|_2^2$	H_2-norm (sometimes H_2-cost) of system $\mathcal{G}$.	85, 88
$\\|\mathcal{G}\\|_\infty$	H_∞-norm of system $\mathcal{G}$.	154
$\\|\mathbb{L}\\|$	L_2-induced norm of operator $\mathbb{L}$.	154
$\langle\cdot;\cdot\rangle$	N-sequence inner product.	24
$\langle\cdot;\cdot\rangle_L$	Vector inner product, $\langle x; y\rangle_L = x^* L y$.	15

O.L.V. Costa et al., *Continuous-Time Markov Jump Linear Systems*, Probability and Its Applications, DOI 10.1007/978-3-642-34100-7,

Symbol	Description	Page
$\|\cdot\|_L$	Norm induced by the inner product $\langle\cdot;\cdot\rangle_L$.	16
$\otimes$	Kronecker product.	25
$\oplus$	Kronecker sum.	25
Z'	Transpose of Z.	15
$\bar{Z}$	Complex conjugate of complex matrix Z.	15
Z^*	Conjugate transpose of Z.	15
∇_x	Gradient with respect to x.	74
α	Disturbance gain.	183
β	Transition rate of 2-state Markov chain.	46
	Guaranteed H_2 cost.	88, 193
γ	H_∞ performance level.	154
Γ_ℓ	An $N \times N$ real matrix.	141
$\mathbf{\Gamma}$	Sequence of dynamic controller $n \times n$ matrices.	65, 168
$\delta(t)$	Unit impulse.	85
$\mathbf{\Delta}$	Sequence of matrices with uncertain entries.	183
ζ_i	ith operation point of robotic manipulator.	224
η	A random vector in $\mathbb{R}^n$.	116
$\theta(t)$	Markov state (operation mode) at time t.	2
θ_0	Initial Markov state, $\theta_0 = \theta(0)$.	34
$\hat{\theta}(t)$	Estimate for $\theta(t)$.	61
Θ	A Markov chain with perturbed transition rates.	190
$\Theta_j(\cdot)$	A linear operator.	89
ϑ_t	Augmented Markov state, $\vartheta_t = (x(t), \theta(t))$.	23
ϑ_0	Initial augmented state, $\vartheta_0 = (x(0), \theta(0))$.	23
κ	Vertex index.	21
$\lambda_i(\cdot)$	ith eigenvalue.	16
$\lambda_{\max}(\cdot)$	Maximal eigenvalue.	51
$\mathbf{\Lambda}$	Sequence of dynamic controller $p \times n$ matrices.	168
μ	Convex upper bound to H_2 cost.	89
	An expected value, $\mu = E[x(0)]$.	98
μ_i	An expected value, $\mu_i = E[x_0 1_{\{\theta_0=i\}}]$.	129
$\mu(t)$	An expected value, $\mu(t) = E[x(t)]$.	136
ν	Initial distribution of the Markov chain.	19
ν_i	Component of distribution, $\nu_i = P(\theta(0) = i)$.	19
$\nu(t)$	Innovations process.	132
$\varpi(t)$	Additive feedback disturbance at time t.	193
π, π_i	Stationary probabilities.	20
Π	Transition rate matrix of the Markov chain.	20
Π^κ	Vertex of transition rates polytope.	84

Symbol	Description	Page
ρ, $\rho^\star$	Robustness margins.	185
ρ^κ	Element of unit simplex.	21
$\sigma(\cdot)$	Spectrum of a matrix/operator.	16
$\sigma\{\cdot\}$	Sigma-field generated by a random object.	99
Φ	State transition matrix.	18
ϕ	Lyapunov function.	49
Υ	An operator.	24
	Set of all well-behaved realizations of the Markov chain.	36
Υ_ℓ	A matrix.	142
$\varphi(\cdot)$	Linear operator mapping a matrix to a vector.	24
$\varphi^{-1}(\cdot)$	Linear operator mapping a vector to a matrix, inverse of φ.	25
$\hat{\varphi}(\cdot)$	Linear operator mapping a sequence of N matrices to a vector.	25
$\hat{\varphi}^{-1}(\cdot)$	Linear operator mapping a vector to a sequence of N matrices, inverse of $\hat{\varphi}$.	25
$\hat{\varphi}_j(\cdot)$	Linear operator mapping the jth matrix of a sequence of N matrices to a vector.	24
$\hat{\varphi}_j^{-1}(\cdot)$	Linear operator mapping a vector to the jth matrix of a sequence of N matrices, inverse of $\hat{\varphi}_j$.	25
$\mathbf{\Phi}$	Sequence of dynamic controller $p \times r$ matrices.	168
χ	Fixed initial state in $\mathbb{R}^n$.	156
$\Psi = [\alpha_{ij}]$	$N \times N$ transition rate matrix associated with the estimation of the Markov chain.	62
Ψ_ℓ	A matrix.	142
$\mathbf{\Psi}$	Sequence of dynamic controller $n \times r$ matrices.	168
Ω	Sample space.	16
ω	A random outcome, $\omega \in \Omega$.	16
$(\Omega, \mathcal{F}, P)$	Probability space.	16
ζ	Robustness output.	193
$\mathbf{A}$	Sequence of $n \times n$ system matrices.	34
$\widehat{\mathbf{A}}$	Sequence of $n \times n$ controller matrices.	64
$\tilde{\mathbf{A}}$	Sequence of $n \times n$ system matrices.	40
$\widetilde{\mathbf{A}}$	Sequence of $n \times n$ closed-loop matrices.	60, 84
$\mathcal{A}$	$Nn^2 \times Nn^2$ matrix associated to the second moment of x.	38
A_f	Filter matrix.	140
$A_{f,\mathrm{op}}$	Optimal filter matrix.	147
$\widehat{\mathbf{B}}$	Sequence of $n \times p$ controller matrices.	64
$\mathbf{B}$	Sequence of $n \times m$ system matrices.	59
$\mathbb{B}(\mathbb{X}, \mathbb{Y})$	Banach space of linear maps from $\mathbb{X}$ into $\mathbb{Y}$.	15
$\mathbb{B}(\mathbb{X})$	Particular case, $\mathbb{B}(\mathbb{X}) = \mathbb{B}(\mathbb{X}, \mathbb{X})$.	
$\mathbb{B}(\mathbb{C}^n, \mathbb{C}^m)$	Linear space of all $m \times n$ complex matrices.	15

Symbol	Description	Page
$\mathbb{B}(\mathbb{R}^n, \mathbb{R}^m)$	Linear space of all $m \times n$ real matrices.	15
B_f	Filter matrix.	140
$B_{f,\text{op}}$	Optimal filter matrix.	147
$\mathbb{C}^n$	n-dimensional complex Euclidean space.	16
$\mathbf{C}$	Sequence of $p \times n$ system matrices.	60
$\widehat{\mathbf{C}}$	Sequence of $p \times n$ controller matrices.	64
$\widetilde{\mathbf{C}}$	Sequence of $p \times n$ closed-loop matrices.	120
$\mathcal{C}$	A feasible set of LMIs.	89
$\mathcal{C}^1(\mathcal{X})$	A set.	74
$\text{cov}(\cdot)$	Covariance.	128
det	Determinant function.	
$D, \mathbb{D}$	Set of discontinuity points.	18, 76
$\mathbf{D}$	Sequence of $p \times m$ system matrices.	60
$\mathfrak{D}(\mathcal{L})$	Domain of operator $\mathcal{L}$.	17
$\mathcal{D}_i(\cdot)$	Linear mapping.	65
$\text{diag}(\cdot)$	Block diagonal matrix with distinct blocks.	24
$\text{dg}(\cdot)$	Block diagonal matrix with repeated blocks.	141
$e(t)$	Error output.	140
e_s	Vector in $\mathbb{R}^r$, with 1 in the sth position and zero elsewhere.	85
$\mathbf{e}$	Vector in $\mathbb{R}^N$, with all components equal to 1.	20
$E(\cdot), E[\cdot]$	Expectation with respect to P.	16
$E_{\vartheta_t}(\cdot)$	Expectation conditioned to $(x(t), \theta(t))$.	72
F	$Nn \times Nn$ matrix associated to the first moment of the state process.	38
$\mathcal{F}$	σ-algebra.	16
$\mathcal{F}(\cdot)$	Linear operator.	37
$\mathcal{F}_t$	Filtration at time t.	16
$\mathbf{F}$	Sequence of $m \times n$ decision variables.	63
	Sequence of $m \times n$ controller gains.	115, 166
$\mathbf{G}$	Sequence of $q \times r$ system matrices.	98
	Sequence of $n \times r$ observer gains.	60
$\mathcal{G}$	A system.	3
$\mathcal{G}^*$	Adjoint of $\mathcal{G}$.	259
$\mathcal{G}_{\text{cl}}$	A closed-loop system.	112
$\mathcal{G}_u$	A control system.	64
$\mathcal{G}_c$	A decomposed system, associated to control.	120
$\mathcal{G}_U$	A decomposed system, associated to the separation of the cost of estimation with the cost of control.	120
$\mathcal{G}_v$	System associated with the optimal filtering problem.	115
$\mathcal{G}_{\mathbf{K}}, \mathcal{G}_{\widehat{\mathbf{K}}}$	Closed-loop H_2 control systems.	88, 90

Symbol	Description	Page
$\mathcal{G}_{\mathbf{K}}^{\Pi}$	Closed-loop H_2 control system with uncertain transition rates.	88
$\mathcal{G}_{w}^{\mathcal{K}}$	Closed-loop H_∞ control system.	167
$\mathbf{H}$	Sequence of $q \times n$ system matrices.	98
$\mathbb{H}_{\mathbb{C}}^{n,m}$	Linear space of sequences of N matrices with $m \times n$ entries.	23
$\mathbb{H}_{\mathbb{C}}^{n}$	Particular case, $\mathbb{H}_{\mathbb{C}}^{n} = \mathbb{H}_{\mathbb{C}}^{n,n}$.	
$\mathbb{H}^{n,m}$	Real linear space of sequences of N matrices with $m \times n$ entries.	23
$\mathbb{H}^{n}$	Particular case, $\mathbb{H}^{n} = \mathbb{H}^{n,n}$.	
$\mathbb{H}^{n*}$	Set of sequences of symmetric N matrices.	23
$\mathbb{H}^{n+}$	Cone of positive semidefinite sequences of N matrices.	23
$\mathrm{Her}(\cdot)$	Hermitian sum of matrices.	16
iff	If and only if.	
i, j	Indices (usually operation modes).	19
I, I_ℓ	$\ell \times \ell$ identity matrix.	16
Im	Imaginary part.	16
$\mathcal{I}$	$n \times 2n$ matrix.	174
$\mathfrak{I}(t)$	A martingale.	23
$\mathfrak{I}$	$2n \times n$ matrix.	174
$\mathbf{J}$	Sequence of $n \times r$ matrices.	34
J_t^p	$Nn \times Np$ matrix.	130
$\mathcal{J}$	$Nn \times Nn$ matrix.	130
	Quadratic cost functional.	73
$\mathcal{J}^{\mathrm{op}}, \mathcal{J}_{\mathrm{op}}, \mathcal{J}_{\mathrm{op}}^{\mathrm{op}}$	Optimal cost.	99, 108
$\mathcal{J}_T^{\gamma}$	Functional related to H_∞ control.	156
$\mathbb{K}$	Set of stabilizing controllers.	60
$\mathbb{K}_{\mathrm{r}}$	Set of robustly stabilizing controllers.	84
$\mathbb{K}_q$	Set of quadratically stabilizing controllers.	84
$\mathbf{K}$	Sequence of $m \times n$ controller gains.	59
$\mathbf{K}^f$	Innovations gain.	99
$\mathcal{K}, \boldsymbol{\mathcal{K}}$	Dynamic output feedback controller.	64, 168
$\mathcal{K}(\cdot), \mathcal{K}(\cdot, t)$	Linear operators.	232, 115
$L_2^n(\Omega, \mathcal{F}, P)$	Space of square integrable stochastic processes.	16
$\mathbb{L}$	Input–output operator.	154
$\mathbb{L}^{c\ell}$	Input–output closed-loop operator.	165
$\mathcal{L}(\cdot)$	A linear operator.	37
$\mathcal{L}^{\ell}(\cdot)$	A linear operator.	241
$\mathcal{L}^{+}(\cdot)$	A linear operator.	242
$\mathfrak{L}^{u}(\cdot)$	Infinitesimal operator.	74
$\mathfrak{L}_K(t)$	A filter with kernel K.	130
L_f	Filter matrix.	140

Symbol	Description	Page
$L_{f,\text{op}}$	Optimal filter matrix.	147
ℓ	Number of vertices (sometimes just an index).	21
$\mathbf{L}$	Sequence of $p \times r$ matrices.	153
$\mathbf{M}$	Sequence of $n \times n$ matrices.	86
N	Number of operation modes.	19
$\mathcal{N}$	Kernel of a matrix.	16
$\nu\mathbf{N}$	Sequence of $n \times n$ matrices.	86
$o(\cdot)$	Small functional dependence, $o(h)/h \to 0$.	19
P	Probability measure.	16
$P(t)$	Probability vector in $\mathbb{R}^N$.	19
$p_i(t)$	Probability distribution of the Markov chain at time t.	19
$p_{ij}(t)$	Probability of transition from mode i to mode j at time t.	19
$\mathbf{P}$	Observability Gramian.	86
$\mathbf{Q}(t)$	Sequence of N second moment matrices.	35
$q(t)$	Vector of first moments at time t.	35
$q_i(t)$	First moment of $x(t)$ at mode i.	35
$\hat{q}(t)$	Column vector obtained by stacking $q_i(t)$.	35
$Q(t)$	Second moment of $x(t)$.	35
$Q_i(t)$	Second moment of $x(t)1_{\{\theta(t)=i\}}$.	35
$Q(\tau, t)$	Autocorrelation matrix of x at times τ, t.	35
$\mathcal{Q}(\tau)$	Stationary autocorrelation matrix of x at lag τ.	36
$\mathbb{R}^n$	n-dimensional real Euclidean space.	15
$\mathbb{R}^+$	Set of positive reals, $\mathbb{R}^+ = [0, \infty)$.	15
$\mathbf{R}$	Sequence of $m \times m$ weights.	119
Re	Real part.	16
$\text{Re}\{\lambda(\cdot)\}$	Largest real part of the spectrum of $(\cdot)$.	16
$\mathcal{R}(\cdot)$	Range of a matrix.	16
	Nonlinear Riccati operator in $\mathbb{H}^n$.	79
$\mathcal{R}_i(\cdot)$	Linear mapping.	65
$\mathbf{r}_{\mathbb{C}}$	Complex stability radius.	185
$\mathbf{r}_{\mathbb{R}}$	Real stability radius.	185
$\widehat{\mathbf{r}}_{\mathbb{C}}, \widehat{\mathbf{r}}_{\mathbb{R}}$	Unstructured stability radii.	186
$\mathcal{S}$	Set of operation modes, $\mathcal{S} = \{1, \dots, N\}$.	19
$\mathbf{S}$	Controllability Gramian.	86
$T(t)$	Probability transition matrix in $\mathbb{R}^{N \times N}$.	19
$\text{tr}(\cdot)$	Trace operator.	16
t	Time.	2
dt	Infinitesimal time interval.	16
T_k	kth jump time of the Markov chain.	36
$\mathcal{T}(\cdot)$	A linear operator.	37

Symbol	Description	Page
$u(t)$	Control variable at time t.	59
$\hat{u}(t)$	An optimal control law.	73
$\mathcal{U}^T, \mathcal{U}$	Sets of admissible controllers.	72, 73
$\mathcal{U}_i(\cdot)$	An affine mapping.	154
$\mathbf{v}(t)$	Closed-loop state at time t.	65
$\mathbb{V}$	Polytope of transition rate matrices.	84
$\mathcal{V}(\cdot)$	A linear operator.	137
	A mapping.	137
$\mathcal{V}_i^\gamma$	A matrix.	154
$\widehat{v}(t)$	Dynamic estimate for the output $v(t)$.	140
W, W_0	Wiener processes.	34, 129
$w(t)$	Additive disturbance (sometimes Wiener process) at time t.	34
$x(t)$	System state at time t.	34
x_0	Initial state, $x_0 = x(0)$.	34
$\dot{x}(t)$	System state's derivative at time t.	34
$dx(t)$	System state's differential at time t.	34
$x_{\mathrm{zi}}(t)$	Zero-input response at time t.	153
$x_{\mathrm{zs}}(t)$	Zero-state response at time t.	153
$\widehat{x}(t)$	Filter's state at time t.	60, 64, 99
$\widetilde{x}(t)$	Estimation error at time t.	60
$\widehat{x}_{\mathrm{op}}(t)$	Filter's state at time t.	99
$\widetilde{x}_{\mathrm{op}}(t)$	Estimation error at time t.	107
$\widehat{x}_e(t)$	Filter's state at time t.	116
$\widetilde{x}_e(t)$	Estimation error at time t.	116
$\mathbb{X}$	Banach space.	15
$\mathcal{X}$	A set.	74
$\mathbf{X}$	Solution of the control CARE.	251
	Stabilizing solution of the control CARE.	244
$\mathbf{X}^+$	Maximal solution of the control CARE.	241
$\mathbb{Y}$	Banach space.	15
$\mathbf{Y}$	Solution to the filtering CARE.	115
$y(t)$	Measured variable at time t.	60
$dy(t)$	Incremental observation at time t.	98
$z(t)$	System output at time t.	71
	Vector which stacks up $z_1(t), \ldots, z_N(t)$.	129
$z_i(t)$	Product of the form $z_i(t) = x(t)1_{\{\theta(t)=i\}}$.	129
$\tilde{z}(t)$	Estimation error.	60
$\tilde{Z}(t)$	Error covariance matrix.	132
$\widehat{z}(t)$	Dynamic filter state.	140
$\mathfrak{Z}(\cdot)$	A mapping.	195
$\mathcal{Z}$	A subset of operation modes, $\mathcal{Z} \subseteq \mathcal{S}$.	190

Abbreviation	Description	Page
APV	Analytical Point of View.	6
a.a.	Almost all.	18
a.s.	Almost surely.	
ARE	Algebraic Riccati Equation.	136
AWSS	Asymptotic Wide Sense Stationarity (also Asymptotically Wide Sense Stationary).	36
CARE	Coupled Algebraic Riccati Equations.	79
CDRE	Coupled Differential Riccati Equations.	76
COF	Complete Observations Filter.	227
HMM	Hidden Markov Model.	6
JLQ	Jump Linear Quadratic.	13
LMI, LMIs	Linear Matrix Inequality/Inequalities.	29
LQG	Linear Quadratic Gaussian.	11
MJLS	Markov Jump Linear System(s).	2
MSS	Mean Square Stability (also Mean Square Stable).	33
MM	Multiple Models.	6
o.i.	Orthogonal increments.	128
PC	Piecewise Constant.	17
q.m.	Quadratic mean.	127
r.v.	Random variable.	
StS	Stochastic Stability (also Stochastically Stable).	36
SS	Stochastic/Mean Square Stabilizability (also Stochastically/Mean Square Stabilizable).	59
SD	Stochastic Detectability (also Stochastically Detectable).	60

References

1. F. Abdollahi, K. Khorasani, A decentralized $\mathcal{H}_\infty$ routing control strategy for mobile networked multi-agents, in *Proceedings of the 2009 American Control Conference*, St. Louis, USA, June 2009, pp. 1555–1560
2. F. Abdollahi, K. Khorasani, A decentralized Markovian jump $\mathcal{H}_\infty$ control routing strategy for mobile multi-agent networked systems. IEEE Transactions on Control Systems Technology **19**, 269–283 (2011)
3. S. Aberkane, J.C. Ponsart, M. Rodrigues, D. Sauter, Output feedback control of a class of stochastic hybrid systems. Automatica **44**, 1325–1332 (2008)
4. H. Abou-Kandil, G. Freiling, V. Ionescu, G. Jank, *Matrix Riccati Equations in Control and Systems Theory (Systems and Control: Foundations and Applications)* (Birkhäuser, Basel, 2003)
5. H. Abou-Kandil, G. Freiling, G. Jank, Solution and asymptotic behaviour of coupled Riccati equations in jump linear systems. IEEE Transactions on Automatic Control **39**, 1631–1636 (1994)
6. H. Abou-Kandil, G. Freiling, G. Jank, On the solution of discrete-time Markovian jump linear quadratic control problems. Automatica **31**, 765–768 (1995)
7. G.A. Ackerson, K.S. Fu, On state estimation in switching environments. IEEE Transactions on Automatic Control **15**(1), 10–17 (1970)
8. R. Akella, P.R. Kumar, Optimal control of production rate in a failure prone manufacturing system. IEEE Transactions on Automatic Control **31**, 116–126 (1986)
9. S. Allam, F. Dufour, P. Bertrand, Discrete-time estimation of a Markov chain with marked point process observations. Application to Markovian jump filtering. IEEE Transactions on Automatic Control **46**, 903–908 (2001)
10. W.J. Anderson, *Continuous-Time Markov Chains; An Application Oriented Approach* (Springer, New York, 1991)
11. P. Apkarian, R.J. Adams, Advanced gain-scheduling techniques for uncertain systems. IEEE Transactions on Control Systems Technology **6**, 21–32 (1998)
12. H. Arai, S. Tachi, Position control of a manipulator with passive joints using dynamic coupling. IEEE Transactions on Robotics and Automation **7**, 528–534 (1991)
13. K.J. Åström, *Introduction to Stochastic Control Theory* (Academic Press, New York, 1970)
14. M. Athans, Command and control (C2) theory: A challenge to control science. IEEE Transactions on Automatic Control **32**, 286–293 (1987)
15. R.E. Azouzi, M. Abbad, E. Altman, Perturbation of multivariable linear quadratic systems with jump parameters. IEEE Transactions on Automatic Control **46**, 1666–1671 (2001)
16. J. Baczynski, M.D. Fragoso, Maximal solution to algebraic Riccati equations linked to infinite Markov jump linear systems. Mathematics of Control. Signals and Systems **20**, 157–172 (2008)

O.L.V. Costa et al., *Continuous-Time Markov Jump Linear Systems*, Probability and Its Applications, DOI 10.1007/978-3-642-34100-7,

17. J. Baczynski, M.D. Fragoso, Maximal versus strong solution of algebraic Riccati equations arising in infinite Markov jump linear systems. Systems & Control Letters **57**, 246–254 (2008)
18. J. Baczynski, M.D. Fragoso, E.P. Lopes, On a discrete time linear jump stochastic dynamic game. International Journal of Systems Science **32**, 979–988 (2001)
19. A.V. Balakrishnan, A note on the structure of optimal stochastic control. Applied Mathematics & Optimization **1**, 87–94 (1974)
20. Y. Bar-Shalom, Tracking methods in a multitarget environment. IEEE Transactions on Automatic Control **23**, 618–626 (1978)
21. Y. Bar-Shalom, K. Birmiwal, Variable dimension for maneuvering target tracking. IEEE Transactions on Automatic Control **18**, 621–628 (1982)
22. Y. Bar-Shalom, X.R. Li, *Estimation and Tracking. Principles, Techniques, and Software.* (Artech House, Norwood, 1993)
23. R. Bellman, *Dynamic Programming* (Princeton University Press, Princeton, 1957)
24. K. Benjelloun, E.K. Boukas, Mean square stochastic stability of linear time-delay systems with Markovian jump parameters. IEEE Transactions on Automatic Control **43**, 1456–1460 (1998)
25. A. Bensoussan, *Stochastic Control of Partially Observable Systems* (Cambridge University Press, Cambridge, 1992)
26. F. Bernard, F. Dufour, P. Bertrand, On the JLQ problem with uncertainty. IEEE Transactions on Automatic Control **42**, 869–872 (1997)
27. S.P. Bhattacharyya, H. Chapellat, L.H. Keel, *Robust Control: The Parametric Approach*. Information and System Sciences (Prentice Hall, Upper Saddle River, 1995)
28. T. Bielecki, P.R. Kumar, Necessary and sufficient conditions for a zero inventory policy to be optimal in an unreliable manufacturing system, in *Proceedings of the 25th IEEE Conference on Decision and Control*, Athens, 1986, pp. 248–250
29. P. Billingsley, *Probability and Measure* (Wiley, New York, 1986)
30. T. Björk, Finite dimensional optimal filters for a class of Itô processes with jumping parameters. Stochastics **4**, 167–183 (1980)
31. T. Björk, Finite optimal filters for a class of nonlinear diffusions with jumping parameters. Stochastics **6**, 121–138 (1982)
32. W.P. Blair, D.D. Sworder, Continuous-time regulation of a class of econometric models. IEEE Transactions on Systems, Man and Cybernetics **SMC-5**, 341–346 (1975)
33. W.P. Blair, D.D. Sworder, Feedback control of a class of linear discrete systems with jump parameters and quadratic cost criteria. International Journal of Control **21**, 833–844 (1975)
34. A.P. Blake, F. Zampolli, Optimal policy in Markov-switching rational expectations models. Journal of Economic Dynamics & Control **35**, 1626–1651 (2011)
35. H.A.P. Blom, An efficient filter for abruptly changing systems, in *Proceedings of the 23rd IEEE Conference on Decision and Control*, Las Vegas, NV, 1984, pp. 656–658
36. H.A.P. Blom, Continuous-discrete filtering for the systems with Markovian switching coefficients and simultaneous jump, in *Proceedings of the 21st Asilomar Conference on Signals, Systems and Computers*, Pacific Grove, 1987, pp. 244–248
37. H.A.P. Blom, Y. Bar-Shalom, The interacting multiple model algorithm for systems with Markovian jumping parameters. IEEE Transactions on Automatic Control **33**, 780–783 (1988)
38. S. Bohacek, E. Jonckheere, Nonlinear tracking over compact sets with linear dynamically varying H^{∞} control. SIAM Journal on Control and Optimization **40**, 1042–1071 (2001)
39. S. Bohacek, E.A. Jonckheere, Relationships between linear dynamically varying systems and jump linear systems. Mathematics of Control. Signals and Systems **16**, 207–224 (2003)
40. E.K. Boukas, *Stochastic Switching Systems: Analysis and Design* (Birkhäuser, Basel, 2006)
41. E.K. Boukas, *Control of Singular Systems with Random Abrupt Changes* (Springer, Berlin, 2008)
42. E.K. Boukas, K. Benjelloun, On stabilization of uncertain linear systems with jump parameters. International Journal of Control **72**, 842–850 (2001)

43. E.K. Boukas, A. Benzaouia, Stability of discrete-time linear systems with Markovian jumping parameters and constrained control. IEEE Transactions on Automatic Control **47**, 516–521 (2002)
44. E.K. Boukas, Z.K. Liu, Jump linear quadratic regulator with controlled jump rates. IEEE Transactions on Automatic Control **46**, 301–305 (2001)
45. E.K. Boukas, Z.K. Liu, Robust H_∞-control of discrete-time Markovian jump linear systems with mode-dependent time-delays. IEEE Transactions on Automatic Control **46**, 1918–1924 (2001)
46. E.K. Boukas, Z.K. Liu, G.X. Liu, Delay-dependent robust stability and H_∞-control of jump linear systems with time-delay. International Journal of Control **74**, 329–340 (2001)
47. E.K. Boukas, P. Shi, H_∞-control for discrete-time linear systems with Markovian jump parameters, in *Proceedings of the 36th IEEE Conference on Decision and Control*, 1997, pp. 4143–4148
48. E.K. Boukas, H. Yang, Stability of discrete-time linear systems with Markovian jumping parameters. Mathematics of Control. Signals and Systems **8**, 390–402 (1995)
49. S. Boyd, L.E. Ghaoui, E. Feron, V. Balakrishnan, *Linear Matrix Inequalities in System and Control Theory* (SIAM, Philadelphia, 1994)
50. J.W. Brewer, Kronecker products and matrix calculus in system theory. IEEE Transactions on Circuits and Systems **CAS-25**, 772–781 (1978)
51. P.E. Caines, J. Zhang, On the adaptive control of jump parameter systems via nonlinear filtering. SIAM Journal on Control and Optimization **33**, 1758–1777 (1995)
52. D.O. Cajueiro, Stochastic optimal control of jumping Markov parameter processes with applications to finance, PhD Thesis, Instituto Tecnológico de Aeronáutica—ITA, Brazil (2002)
53. F.M. Callier, C.A. Desoer, *Linear System Theory* (Springer, Berlin, 1999)
54. S.L. Campbell, C.D. Meyer Jr., *Generalized Inverses of Linear Transformations* (Dover, New York, 1991)
55. Y.Y. Cao, J. Lam, Robust H_∞ control of uncertain Markovian jump systems with time-delay. IEEE Transactions on Automatic Control **45**, 77–83 (2000)
56. C.G. Cassandras, J. Lygeros, *Stochastic Hybrid Systems* (Taylor & Francis, Boca Raton, 2007)
57. H. Chan, Ü. Özgüner, Optimal control of systems over a communication network with queues via a jump system approach, in *Proceedings of the 4th IEEE Conference on Control Applications*, Albany, NY, 1995, pp. 1148–1153
58. C.G. Chang, M. Athans, State estimation for discrete systems with switching parameters. IEEE Transactions on Aerospace and Electronic Systems **14**, 418–424 (1978)
59. H.J. Chizeck, Y. Ji, Optimal quadratic control of jump linear systems with Gaussian noise in discrete-time, in *Proceedings of the 27th Conference on Decision and Control*, 1988, pp. 1989–1992
60. H.J. Chizeck, A.S. Willsky, D. Castañon, Discrete-time Markovian jump linear quadratic optimal control. International Journal of Control **43**, 197–216 (1986)
61. E.F. Costa, J.B.R. do Val, On the detectability and observability of discrete-time Markov jump linear systems. Systems & Control Letters **44**, 135–145 (2001)
62. E.F. Costa, J.B.R. do Val, On the detectability and observability of continuous-time Markov jump linear systems. SIAM Journal on Control and Optimization **41**, 1295–1314 (2002)
63. E.F. Costa, J.B.R. do Val, Weak detectability and the linear-quadratic control problem of discrete-time Markov jump linear systems. International Journal of Control **75**, 1282–1292 (2002)
64. E.F. Costa, J.B.R. do Val, M.D. Fragoso, On a detectability concept of discrete-time infinite Markov jump linear systems. Stochastic Analysis and Applications **23**, 1–14 (2005)
65. O.L.V. Costa, Linear minimum mean square error estimation for discrete-time Markovian jump linear systems. IEEE Transactions on Automatic Control **39**, 1685–1689 (1994)
66. O.L.V. Costa, Discrete-time coupled Riccati equations for systems with Markov switching parameters. Journal of Mathematical Analysis and Applications **194**, 197–216 (1995)

67. O.L.V. Costa, Mean square stabilizing solutions for discrete-time coupled algebraic Riccati equations. IEEE Transactions on Automatic Control **41**, 593–598 (1996)
68. O.L.V. Costa, M.V. Araujo, A generalized multi-period mean-variance portfolio optimization with Markov switching parameters. Automatica **44**, 2487–2497 (2008)
69. O.L.V. Costa, E.O. Assumpção, E.K. Boukas, R.P. Marques, Constrained quadratic state feedback control of discrete-time Markovian jump linear systems. Automatica **35**, 617–626 (1999)
70. O.L.V. Costa, J.C.C. Aya, Monte-Carlo TD(λ)-methods for the optimal control of discrete-time Markovian jump linear systems. Automatica **38**, 217–225 (2002)
71. O.L.V. Costa, E.K. Boukas, Necessary and sufficient conditions for robust stability and stabilizability of continuous-time linear systems with Markovian jumps. Journal of Optimization Theory and Applications **99**, 359–379 (1998)
72. O.L.V. Costa, W.L. de Paulo, Indefinite quadratic with linear costs optimal control of Markov jump with multiplicative noise systems. Automatica **43**, 587–597 (2007)
73. O.L.V. Costa, J.B.R. do Val, Full information H_∞-control for discrete-time infinite Markov jump parameter systems. Journal of Mathematical Analysis and Applications **202**, 578–603 (1996)
74. O.L.V. Costa, J.B.R. do Val, Jump LQ-optimal control for discrete-time Markovian systems with l_2 stochastic inputs. Stochastic Analysis and Applications **16**, 843–858 (1998)
75. O.L.V. Costa, J.B.R. do Val, J.C. Geromel, A convex programming approach to H_2-control of discrete-time Markovian jump linear systems. International Journal of Control **66**, 557–579 (1997)
76. O.L.V. Costa, J.B.R. do Val, J.C. Geromel, Continuous-time state-feedback H_2-control of Markovian jump linear systems via convex analysis. Automatica **35**, 259–268 (1999)
77. O.L.V. Costa, M.D. Fragoso, Stability results for discrete-time linear systems with Markovian jumping parameters. Journal of Mathematical Analysis and Applications **179**, 154–178 (1993)
78. O.L.V. Costa, M.D. Fragoso, Discrete-time LQ-optimal control problems for infinite Markov jump parameter systems. IEEE Transactions on Automatic Control **40**, 2076–2088 (1995)
79. O.L.V. Costa, M.D. Fragoso, Comments on stochastic stability of jump linear systems. IEEE Transactions on Automatic Control **49**, 1414–1416 (2004)
80. O.L.V. Costa, M.D. Fragoso, A separation principle for the H_2-control of continuous-time infinite Markov jump linear systems with partial observations. Journal of Mathematical Analysis and Applications **331**, 97–120 (2007)
81. O.L.V. Costa, M.D. Fragoso, R.P. Marques, *Discrete-Time Markov Jump Linear Systems* (Springer, Berlin, 2004)
82. O.L.V. Costa, M.D. Fragoso, M.G. Todorov, On the filtering problem for continuous-time Markov jump linear systems with no observation of the Markov chain. European Journal of Control **17**, 339–354 (2011)
83. O.L.V. Costa, S. Guerra, Robust linear filtering for discrete-time hybrid Markov linear systems. International Journal of Control **75**, 712–727 (2002)
84. O.L.V. Costa, S. Guerra, Stationary filter for linear minimum mean square error estimator of discrete-time Markovian jump systems. IEEE Transactions on Automatic Control **47**, 1351–1356 (2002)
85. O.L.V. Costa, R.P. Marques, Mixed H_2/H_∞-control of discrete-time Markovian jump linear systems. IEEE Transactions on Automatic Control **43**, 95–100 (1998)
86. O.L.V. Costa, R.P. Marques, Maximal and stabilizing Hermitian solutions for discrete-time coupled algebraic Riccati equations. Automatica **12**, 167–195 (1999)
87. O.L.V. Costa, R.P. Marques, Robust H_2-control for discrete-time Markovian jump linear systems. International Journal of Control **73**, 11–21 (2000)
88. O.L.V. Costa, R.T. Okimura, Discrete-time mean variance optimal control of linear systems with Markovian jumps and multiplicative noise. International Journal of Control **2**, 256–267 (2009)

89. O.L.V. Costa, E.F. Tuesta, Finite horizon quadratic optimal control and a separation principle for Markovian jump linear systems. IEEE Transactions on Automatic Control **48**, 1836–1842 (2003)
90. O.L.V. Costa, E.F. Tuesta, H_2-control and the separation principle for discrete-time Markovian jump linear systems. Mathematics of Control. Signals and Systems **16**, 320–350 (2004)
91. D.R. Cox, H.D. Miller, *The Theory of Stochastic Processes* (Chapman & Hall, London, 1965)
92. J.J. Craig, *Introduction to Robotics: Mechanics and Control* (Addison-Wesley, Reading, 1986)
93. A. Czornik, *On Control Problems for Jump Linear Systems* (Wydawnictwo Politechniki Śląskiej, Gliwice, 2003)
94. A. Czornik, A. Swierniak, On the discrete JLQ and JLQG problems. Nonlinear Analysis **47**, 423–434 (2001)
95. T. Damm, *Rational Matrix Equations in Stochastic Control*. Lecture Notes in Control and Information Sciences, vol. 297 (Springer, Berlin, 2004). ISBN 3-540-20516-0
96. T. Damm, D. Hinrichsen, Newton's method for concave operators with resolvent positive derivatives in ordered Banach spaces. Linear Algebra and Its Applications **363**, 43–64 (2003)
97. M.H.A. Davis, *Linear Estimation and Stochastic Control* (Chapman & Hall, London, 1977)
98. M.H.A. Davis, S.I. Marcus, An introduction to nonlinear filtering, in *Stochastic Systems: The Mathematics of Filtering and Identification and Applications*, vol. 2, ed. by M. Hazewinkel, J.C. Willems (Reidel, Dordrecht, 1981), pp. 53–75
99. M.H.A. Davis, P.P. Varaya, Information states in linear stochastic systems. Journal of Mathematical Analysis and Applications **37**, 384–402 (1972)
100. M.H.A. Davis, R.B. Vinter, *Stochastic Modelling and Control* (Chapman & Hall, London, 1985)
101. D.P. de Farias, J.C. Geromel, J.B.R. do Val, O.L.V. Costa, Output feedback control of Markov jump linear systems in continuous-time. IEEE Transactions on Automatic Control **45**, 944–949 (2000)
102. M.C. de Oliveira, D.P. de Farias, J.C. Geromel, LMISol—User's Guide, LAC/DT/FEEC/Unicamp (1997). http://www.dt.fee.unicamp.br/~carvalho
103. C.E. de Souza, M.D. Fragoso, On the existence of a maximal solution for generalized algebraic Riccati equation arising in the stochastic control. Systems & Control Letters **14**, 233–239 (1990)
104. C.E. de Souza, M.D. Fragoso, H_∞-control for linear systems with Markovian jumping parameters. Control, Theory and Advanced Technology **9**, 457–466 (1993)
105. C.E. de Souza, M.D. Fragoso, H_∞ filtering for Markovian jump linear systems. International Journal of Systems Science **33**, 909–915 (2002)
106. C.E. de Souza, M.D. Fragoso, Robust H_∞ filtering for uncertain Markovian jump linear systems. International Journal of Robust and Nonlinear Control **12**, 435–446 (2002)
107. C.E. de Souza, M.D. Fragoso, H_∞ filtering for discrete-time systems with Markovian jump parameters. International Journal of Robust and Nonlinear Control **14**, 1299–1316 (2003)
108. J.B.R. do Val, T. Başar, Receding horizon control of jump linear systems and a macroeconomic policy problem. Journal of Economic Dynamics & Control **23**, 1099–1131 (1999)
109. J.B.R. do Val, E.F. Costa, Numerical solution for linear-quadratic control problems of Markov jump linear systems and weak detectability concept. Journal of Optimization Theory and Applications **114**, 69–96 (2002)
110. J.B.R. do Val, J.C. Geromel, O.L.V. Costa, Uncoupled Riccati iterations for the linear quadratic control problem of discrete-time Markov jump linear systems. IEEE Transactions on Automatic Control **43**, 1727–1733 (1998)
111. J.B.R. do Val, J.C. Geromel, A.P.C. Gonçalves, The H_2-control for jump linear systems: Cluster observations of the Markov state. Automatica **38**, 343–349 (2002)
112. J.B.R. do Val, C. Nespoli, Y.R.Z. Cáceres, Stochastic stability for Markovian jump linear systems associated with a finite number of jump times. Journal of Mathematical Analysis and Applications **285**, 551–563 (2003)

113. J. Dong, G.H. Yang, Robust H_2 control of continuous-time Markov jump linear systems. Automatica **44**, 1431–1436 (2008)
114. U. Doraszelski, K.L. Judd, Avoiding the curse of dimensionality in dynamic stochastic games. Quantitative Economics **3**(1), 53–93 (2012)
115. A. Doucet, C. Andrieu, Iterative algorithms for state estimation of jump Markov linear systems. IEEE Transactions on Automatic Control **49**, 1216–1227 (2000)
116. A. Doucet, A. Logothetis, V. Krishnamurthy, Stochastic sampling algorithms for state estimation of jump Markov linear systems. IEEE Transactions on Automatic Control **45**, 188–202 (2000)
117. J.C. Doyle, K. Glover, P.P. Khargonekar, B.A. Francis, State-space solutions to standard H_2 and H_∞-control problems. IEEE Transactions on Automatic Control **34**, 831–847 (1989)
118. V. Dragan, T. Morozan, Stability and robust stabilization to linear stochastic systems described by differential equations with Markovian jumping and multiplicative white noise. Stochastic Analysis and Applications **20**, 33–92 (2002)
119. V. Dragan, T. Morozan, The linear quadratic optimization for a class of linear stochastic systems with multiplicative white noise and Markovian jumping. IEEE Transactions on Automatic Control **49**, 665–675 (2004)
120. V. Dragan, T. Morozan, Stochastic H_2 optimal control for a class of linear systems with periodic coefficients. European Journal of Control **11**, 619–631 (2005)
121. V. Dragan, T. Morozan, A.M. Stoica, *Mathematical Methods in Robust Control of Linear Stochastic Systems (Mathematical Concepts and Methods in Science and Engineering)* (Springer, New York, 2010)
122. V. Dragan, P. Shi, E.K. Boukas, Control of singularly perturbed systems with Markovian jump parameters: an H_∞ approach. Automatica **35**, 1369–1378 (1999)
123. F. Dufour, P. Bertrand, The filtering problem for continuous-time linear systems with Markovian switching coefficients. Systems & Control Letters **23**, 453–461 (1994)
124. F. Dufour, P. Bertrand, An image based filter for discrete-time Markovian jump linear systems. Automatica **32**, 241–247 (1996)
125. F. Dufour, R.J. Elliott, Adaptive control of linear systems with Markov perturbations. IEEE Transactions on Automatic Control **43**, 351–372 (1998)
126. F. Dufour, R.J. Elliott, Filtering with discrete state observations. Applied Mathematics & Optimization **40**, 259–272 (1999)
127. F. Dufour, D. Kannan, Discrete time nonlinear filtering with marked point process observations. Stochastic Analysis and Applications **17**, 99–115 (1999)
128. A. El Bouhtouri, K. El Hadri, Robust stabilization of jump linear systems subject to structured uncertainties in the state and input matrices. IMA Journal of Mathematical Control and Information **17**, 281–293 (2000)
129. E.O. Elliott, Estimates of error rates for codes on burst-error channels. Bell System Technical Journal **42**, 1977–1997 (1963)
130. R.J. Elliott, L. Aggoun, J.B. Moore, *Hidden Markov Models: Estimation and Control* (Springer, New York, 1995)
131. R.J. Elliott, F. Dufour, W.P. Malcolm, State and mode estimation for discrete-time jump Markov systems. SIAM Journal on Control and Optimization **44**, 1081–1104 (2005)
132. R.J. Elliott, F. Dufour, F. Sworder, Exact hybrid filters in discrete-time. IEEE Transactions on Automatic Control **41**, 1807–1810 (1996)
133. J.S. Evans, R.J. Evans, Image-enhanced multiple model tracking. Automatica **35**, 1769–1786 (1999)
134. M.H.C. Everdij, H.A.P. Blom, Embedding adaptive JLQG into LQ martingale control with a completely observable stochastic control matrix. IEEE Transactions on Automatic Control **41**, 424–430 (1996)
135. Y. Fang, A new general sufficient condition for almost sure stability of jump linear systems. IEEE Transactions on Automatic Control **42**, 378–382 (1997)
136. Y. Fang, K.A. Loparo, X. Feng, Almost sure and δ-moment stability of jump linear systems. International Journal of Control **59**, 1281–1307 (1994)

137. X. Feng, K.A. Loparo, Y. Ji, H.J. Chizeck, Stochastic stability properties of jump linear systems. IEEE Transactions on Automatic Control **37**, 38–53 (1992)
138. W.H. Fleming, Some Markovian optimization problems. Journal of Mathematics and Mechanics **12**, 131–140 (1963)
139. W.H. Fleming, Optimal control of partially observable diffusions. SIAM Journal on Control and Optimization **6**, 194–214 (1968)
140. W.H. Fleming, R.W. Rishel, *Deterministic and Stochastic Optimal Control* (Springer, Berlin, 1975)
141. A.K. Fletcher, S. Rangan, V.K. Goyal, Estimation from lossy sensor data: Jump linear modeling and Kalman filtering, in *Proc. Third Int. Symp. on Information Processing in Sensor Networks (IPSN)*, New York, NY, 2000, pp. 251–258
142. J.J. Florentin, Optimal control of continuous-time Markov stochastic systems. Journal of Electronics and Control **10**, 473–488 (1961)
143. M.D. Fragoso, On a partially observable LQG problem for systems with Markovian jumping parameters. Systems & Control Letters **10**, 349–356 (1988)
144. M.D. Fragoso, On a discrete-time jump LQG problem. International Journal of System Sciences **20**, 2539–2545 (1989)
145. M.D. Fragoso, A small random perturbation analysis of a partially observable LQG problem with jumping parameter. IMA Journal of Mathematical Control and Information **7**, 293–305 (1990)
146. M.D. Fragoso, J. Baczynski, Optimal control for continuous-time linear quadratic problems with infinite Markov jump parameters. SIAM Journal on Control and Optimization **40**, 270–297 (2001)
147. M.D. Fragoso, J. Baczynski, Lyapunov coupled equations for continuous-time infinite Markov jump linear systems. Journal of Mathematical Analysis and Applications **274**, 319–335 (2002)
148. M.D. Fragoso, J. Baczynski, Stochastic versus mean square stability in continuous-time linear infinite Markov jump parameters. Stochastic Analysis and Applications **20**, 347–356 (2002)
149. M.D. Fragoso, J. Baczynski, On an infinite dimensional perturbed Riccati differential equation arising in stochastic control. Linear Algebra and Its Applications **406**, 165–176 (2005)
150. M.D. Fragoso, O.L.V. Costa, A unified approach for mean square stability of continuous-time linear systems with Markovian jumping parameters and additive disturbances, in *Proceedings of the 39th Conference on Decision and Control*, Sydney, Australia, 2000, pp. 2361–2366
151. M.D. Fragoso, O.L.V. Costa, Mean square stabilizability of continuous-time linear systems with partial information on the Markovian jumping parameters. Stochastic Analysis and Applications **22**, 99–111 (2004)
152. M.D. Fragoso, O.L.V. Costa, A unified approach for stochastic and mean square stability of continuous-time linear systems with Markovian jumping parameters and additive disturbance. SIAM Journal on Control and Optimization **44**, 1165–1191 (2005)
153. M.D. Fragoso, O.L.V. Costa, A separation principle for the continuous-time LQ-problem with Markovian jump parameters. IEEE Transactions on Automatic Control **55**, 2692–2707 (2010)
154. M.D. Fragoso, O.L.V. Costa, J. Baczynski, The minimum linear mean square filter for a class of hybrid systems, in *Proceedings of the European Control Conference—ECC2001*, Porto, Portugal, 2001, pp. 319–322
155. M.D. Fragoso, O.L.V. Costa, J. Baczynski, N. Rocha, Optimal linear mean square filter for continuous-time jump linear systems. IEEE Transactions on Automatic Control **50**, 1364–1369 (2005)
156. M.D. Fragoso, O.L.V. Costa, C.E. de Souza, A new approach to linearly perturbed Riccati equations arising in stochastic control. Applied Mathematics & Optimization **37**, 99–126 (1998)
157. M.D. Fragoso, C.E. de Souza, H_∞ control for linear systems with Markovian jumping parameters, in *Proceedings of the 1991 European Control Conference*, Grenoble, France, 1991,

pp. 17–21

158. M.D. Fragoso, J.B.R. do Val, D.L. Pinto Jr., Jump linear H_∞-control: the discrete-time case. Control, Theory and Advanced Technology **10**, 1459–1474 (1995)
159. M.D. Fragoso, E.M. Hemerly, Optimal control for a class of noisy linear systems with Markovian jumping parameters and quadratic cost. International Journal of Systems Science **22**, 2553–3561 (1991)
160. M.D. Fragoso, N. Rocha, Stationary filter for continuous-time Markovian jump linear systems. SIAM Journal on Control and Optimization **44**, 801–815 (2005)
161. G. Freiling, A. Hochhaus, On a class of rational matrix differential equations arising in stochastic control. Linear Algebra and Its Applications **379**, 43–68 (2004)
162. K. Fujisawa, Y. Futakata, M. Kojima, S. Matsuyama, S. Nakamura, K. Nakata, M. Yamashita, SDPA-M (Semidefinite Programming Algorithm in Matlab) User's Manual—Version 6.2.0, Technical Report B-359, Dept. Math. & Comp. Sciences, Tokyo Institute of Technology, January 2000
163. P. Gahinet, P. Apkarian, A linear matrix inequality approach to H_∞ control. International Journal of Robust and Nonlinear Control **4**, 421–448 (1994)
164. Z. Gajic, R. Losada, Monotonicity of algebraic Lyapunov iterations for optimal control of jump parameter linear systems. Systems & Control Letters **41**, 175–181 (2000)
165. J. Gao, B. Huang, Z. Wang, LMI-based robust H_∞ control of uncertain linear jump systems with time-delay. Automatica **37**, 1141–1146 (2001)
166. J.C. Geromel, Optimal linear filtering under parameter uncertainty. IEEE Transactions on Signal Processing **47**, 168–175 (1999)
167. J.C. Geromel, A.P.C. Gonçalves, A.R. Fioravanti, Dynamic output feedback control of discrete-time Markov jump linear systems through linear matrix inequalities. SIAM Journal on Control and Optimization **48**, 573–593 (2009)
168. J.C. Geromel, P.L.D. Peres, S.R. Souza, H_2 guaranteed cost control for uncertain continuous-time linear-systems. Systems & Control Letters **19**, 23–27 (1992)
169. L. El Ghaoui, F. Oustry, M. Ait Rami, A cone complementarity linearization algorithm for static output-feedback and related problems. IEEE Transactions on Automatic Control **42**, 1171–1176 (1997)
170. L.E. Ghaoui, M.A. Rami, Robust state-feedback stabilization of jump linear systems via LMIs. International Journal of Robust and Nonlinear Control **6**, 1015–1022 (1996)
171. M.K. Ghosh, A. Arapostathis, S.I. Marcus, A note on an LQG regulator with Markovian switching and pathwise average cost. IEEE Transactions on Automatic Control **40**, 1919–1921 (1995)
172. E. Gilbert, Capacity of a burst-noise channel. Bell System Technical Journal **39**, 1253–1265 (1960)
173. A.P.C. Gonçalves, A.R. Fioravanti, J.C. Geromel, Markov jump linear systems and filtering through network transmitted measurements. Signal Processing **90**, 2842–2850 (2010)
174. W.S. Gray, O. González, Modelling electromagnetic disturbances in closed-loop computer controlled flight systems, in *Proceedings of the 1998 American Control Conference*, Philadelphia, PA, 1998, pp. 359–364
175. W.S. Gray, O.R. González, M. Doğan, Stability analysis of digital linear flight controllers subject to electromagnetic disturbances. IEEE Transactions on Aerospace and Electronic Systems **36**, 1204–1218 (2000)
176. B.E. Griffiths, K.A. Loparo, Optimal control of jump linear quadratic Gaussian systems. International Journal of Control **42**, 791–819 (1985)
177. C.-H. Guo, Newton's method for discrete algebraic Riccati equations when the closed-loop matrix has eigenvalues on the unit circle. SIAM Journal on Matrix Analysis and Applications **20**(2), 279–294 (1999)
178. C.-H. Guo, Iterative solution of a matrix Riccati equation arising in stochastic control, in *Linear Operators and Matrices*. Oper. Theory Adv. Appl., vol. 130 (Birkhäuser, Basel, 2002), pp. 209–221

179. R.Z. Hasminskii, *Stochastic Stability of Differential Equations* (Sijthoff & Noordhoff, Rockville, 1980)
180. D. Hinrichsen, A.J. Pritchard, Stability radii for linear systems. Systems & Control Letters **7**(1), 1–10 (1986)
181. D. Hinrichsen, A.J. Pritchard, Stability radius for structured perturbations and the algebraic Riccati equation. Systems & Control Letters **8**(2), 105–113 (1986)
182. D. Hinrichsen, A.J. Pritchard, *Mathematical Systems Theory I: Modelling, State Space Analysis, Stability and Robustness*. Texts in Applied Mathematics, vol. 48 (Springer, New York, 2005)
183. M.W. Hirsch, S. Smale, *Differential Equations, Dynamical Systems, and Linear Algebra* (Academic Press, San Diego, 1974)
184. T.-J. Ho, B.-S. Chen, Novel extended Viterbi-based multiple-model algorithms for state estimation of discrete-time systems with Markov jump parameters. IEEE Transactions on Signal Processing **54**, 393–404 (2006)
185. T. Hou, W. Zhang, H. Ma, Finite horizon H_2/H_∞ control for discrete-time stochastic systems with Markovian jumps and multiplicative noise. IEEE Transactions on Automatic Control **55**, 1185–1191 (2010)
186. S. Hu, W.-Y. Yan, Stability robustness of networked control systems with respect to packet loss. Automatica **43**, 1243–1248 (2007)
187. Y. Ji, H.J. Chizeck, Controllability, observability and discrete-time Markovian jump linear quadratic control. International Journal of Control **48**, 481–498 (1988)
188. Y. Ji, H.J. Chizeck, Optimal quadratic control of jump linear systems with separately controlled transition probabilities. International Journal of Control **49**, 481–491 (1989)
189. Y. Ji, H.J. Chizeck, Controllability, observability and continuous-time Markovian jump linear quadratic control. IEEE Transactions on Automatic Control **35**, 777–788 (1990)
190. Y. Ji, H.J. Chizeck, Jump linear quadratic Gaussian control: steady state solution and testable conditions. Control, Theory and Advanced Technology **6**, 289–319 (1990)
191. Y. Ji, H.J. Chizeck, X. Feng, K.A. Loparo, Stability and control of discrete-time jump linear systems. Control, Theory and Advanced Technology **7**, 247–270 (1991)
192. J. Jöfberg, YALMIP: A toolbox for modeling and optimization in MATLAB, in *Proceedings of the CACSD Conference*, Taipei, Taiwan, 2004. http://users.isy.liu.se/johanl/yalmip
193. P.D. Joseph, J.T. Tou, On linear control theory. AIEE Transactions on Applications and Industry **80**, 193–196 (1961)
194. G. Kallianpur, *Stochastic Filtering Theory* (Springer, New York, 1980)
195. N. Karmarkar, A new polynomial-time algorithm for linear programming. Combinatorica **4**, 373–395 (1984)
196. I.I. Kats, N.N. Krasovskii, On the stability of systems with random parameters. Applied Mathematics and Mechanics **24**, 1225–1246 (1960)
197. P.A. Kawka, A.G. Alleyne, Robust wireless servo control using a discrete-time uncertain Markovian jump linear model. IEEE Transactions on Control Systems Technology **17**, 733–742 (2009)
198. T. Kazangey, D.D. Sworder, Effective federal policies for regulating residential housing, in *Proceedings of the Summer Computer Simulation Conference*, 1971, pp. 1120–1128
199. F. Kozin, A survey of stability of stochastic systems. Automatica **5**, 95–112 (1969)
200. N.N. Krasovskii, E.A. Lidskii, Analytical design of controllers in systems with random attributes I. Automation and Remote Control **22**, 1021–1025 (1961)
201. N.N. Krasovskii, E.A. Lidskii, Analytical design of controllers in systems with random attributes II. Automation and Remote Control **22**, 1141–1146 (1961)
202. N.N. Krasovskii, E.A. Lidskii, Analytical design of controllers in systems with random attributes III. Automation and Remote Control **22**, 1289–1294 (1961)
203. H.J. Kushner, Optimal stochastic control. IRE Transactions on Automatic Control **7**, 120–122 (1962)
204. H.J. Kushner, Dynamical equations for nonlinear filtering. Journal of Differential Equations **3**, 179–190 (1967)

205. J. Lam, S. Xu, T. Chen, Robust H_∞-filtering for uncertain Markovian jump systems with mode-dependent time delays. IEEE Transactions on Automatic Control **48**, 900–907 (1999)
206. L. Li, V.A. Ugrinovskii, R. Orsi, Decentralized robust control of uncertain Markov jump parameter systems via output feedback. Automatica **43**, 1932–1944 (2007)
207. Z.G. Li, Y.C. Soh, C.Y. Wen, Sufficient conditions for almost sure stability of jump linear systems. IEEE Transactions on Automatic Control **45**, 1325–1329 (2000)
208. C. Lin, Z. Wang, F. Yang, Observer-based networked control for continuous-time systems with random sensor delays. Automatica **45**, 578–584 (2009)
209. M. Liu, D.W.C. Ho, Y. Niu, Stabilization of Markovian jump linear systems over networks with random communication delay. Automatica **45**, 416–421 (2009)
210. K.A. Loparo, Stochastic stability of coupled linear systems: A survey of methods and results. Stochastic Analysis and Applications **2**, 193–228 (1984)
211. K.A. Loparo, F. Abdel-Malek, A probabilistic mechanism to dynamic power system security. IEEE Transactions on Circuits and Systems **37**, 787–798 (1990)
212. K.A. Loparo, G.L. Blankenship, A probabilistic mechanism for small disturbance instabilities in electric power systems. IEEE Transactions on Circuits and Systems **32**, 177–184 (1985)
213. K.A. Loparo, M.R. Buchner, K. Vasudeva, Leak detection in an experimental heat exchanger process: a multiple model approach. IEEE Transactions on Automatic Control **36**, 167–177 (1991)
214. K.A. Loparo, Z. Roth, S.J. Eckert, Nonlinear filtering for systems with random structure. IEEE Transactions on Automatic Control **31**, 1064–1068 (1999)
215. D.G. Luenberger, *Introduction to Dynamic Systems: Theory, Models, and Applications* (Wiley, New York, 1979)
216. L. Ma, Z. Wang, Y. Bo, Z. Guo, Finite-horizon H_2/H_∞ control for a class of nonlinear Markovian jump systems with probabilistic sensor failures. International Journal of Control **84**, 1847–1857 (2011)
217. M.S. Mahmoud, P. Shi, Robust Kalman filtering for continuous time-lag systems with Markovian jump parameters. IEEE Transactions on Circuits and Systems **50**, 98–105 (2003)
218. M.S. Mahmoud, P. Shi, Robust stability, stabilization and H_∞ control of time-delay systems with Markovian jump parameters. International Journal of Robust and Nonlinear Control **13**, 755–784 (2003)
219. R. Malhamé, A jump-driven Markovian electric load model. Advances in Applied Probability **22**, 564–586 (1990)
220. R. Malhamé, C.Y. Chong, Electric load model synthesis by diffusion approximation of a high-order hybrid-state stochastic system. IEEE Transactions on Automatic Control **30**, 854–860 (1985)
221. M. Mariton, Almost sure and moments stability of jump linear systems. Systems & Control Letters **30**, 1145–1147 (1985)
222. M. Mariton, On controllability of linear systems with stochastic jump parameters. IEEE Transactions on Automatic Control **31**, 680–683 (1986)
223. M. Mariton, *Jump Linear Systems in Automatic Control* (Dekker, New York, 1990)
224. M. Mariton, P. Bertrand, Output feedback for a class of linear systems with jump parameters. IEEE Transactions on Automatic Control **30**, 898–900 (1985)
225. M. Mariton, P. Bertrand, Reliable flight control systems: Components placement and feedback synthesis, in *Proceedings of the 10th IFAC World Congress*, Munich, 1987, pp. 150–154
226. N. Meskin, K. Khorasani, Fault detection and isolation of discrete-time Markovian jump linear systems with application to a network of multi-agent systems having imperfect communication channels. Automatica **45**, 2032–2040 (2009)
227. S.P. Meyn, R.L. Tweedie, *Markov Chains and Stochastic Stability* (Springer, Berlin, 1993)
228. R. Morales-Menéndez, N. de Freitas, D. Poole, Estimation and control of industrial processes with particle filters, in *Proceedings of the 2003 American Control Conference*, Denver, 2003, pp. 579–584

229. T. Morozan, Optimal stationary control for dynamic systems with Markov perturbations. Stochastic Analysis and Applications **1**, 299–325 (1983)
230. T. Morozan, Stabilization of some stochastic discrete-time control systems. Stochastic Analysis and Applications **1**, 89–116 (1983)
231. T. Morozan, Stability and control for linear systems with jump Markov perturbations. Stochastic Analysis and Applications **13**, 91–110 (1995)
232. A.S. Morse, *Control Using Logic-based Switching* (Springer, London, 1997)
233. K.A. Narendra, A.M. Annaswamy, *Stable Adaptive Systems* (Prentice Hall, New York, 1989)
234. A.W. Naylor, G.R. Sell, *Linear Operator Theory in Engineering and Science*, 2nd edn. (Springer, Berlin, 1982)
235. Y. Nesterov, A. Nemirovskii, *Interior Point Polynomial Algorithms in Convex Programming* (SIAM, Philadelphia, 1994)
236. Z. Pan, T. Başar, Zero-sum differential games with random structures and applications in H_∞-control of jump linear systems, in *6th International Symposium on Dynamic Games and Applications*, Quebec, 1994
237. Z. Pan, T. Başar, H_∞-control of Markovian jump systems and solution to associated piecewise-deterministic differential game, in *Annals of Dynamic Games*, vol. 3, ed. by G.J. Olsder (Birkhäuser, Basel, 1995), pp. 61–94
238. Z. Pan, T. Başar, H_∞-control of large scale jump linear systems via averaging and aggregation. International Journal of Control **72**, 866–881 (1999)
239. E. Pardoux, *Markov Processes and Applications: Algorithms, Networks, Genome and Finance* (Wiley, Chichester, 2008)
240. B. Park, W.H. Kwon, Robust one-step receding horizon control of discrete-time Markovian jump uncertain systems. Automatica **38**, 1229–1235 (2002)
241. S. Patilkulkarni, H. Herencia-Zapana, W.S. Gray, O.R. González, On the stability of jump-linear systems driven by finite-state machines with Markovian inputs, in *Proceedings of the 2004 American Control Conference*, 2004, pp. 2534–2539
242. A. Pazy, *Semigroups of Linear Operators and Applications to Partial Differential Equations* (Springer, Berlin, 1983)
243. N.J. Ploplys, P.A. Kawka, A.G. Alleyne, Closed-loop control over wireless networks. IEEE Control Systems Magazine **24**, 58–71 (2004)
244. S. Primak, V. Kontorovich, V. Lyandres, *Stochastic Methods and Their Applications to Communications* (Wiley, Chichester, 2005)
245. M. Ait Rami, L. El Ghaoui, Robust stabilization of jump linear systems using linear matrix inequalities, in *IFAC Symposium on Robust Control Design*, Rio de Janeiro, 1994, pp. 148–151
246. M. Ait Rami, L. El Ghaoui, LMI optimization for nonstandard Riccati equations arising in stochastic control. IEEE Transactions on Automatic Control **41**, 1666–1671 (1996)
247. M. Ait Rami, L. El Ghaoui, H_∞ state-feedback control of jump linear systems, in *Proceedings of the 34th Conference on Decision and Control*, New Orleans, USA, December 1995, pp. 951–952
248. V.G. Robinson, D.D. Sworder, A computational algorithm for design of regulators for linear jump parameter systems. IEEE Transactions on Automatic Control **18**, 355–360 (1974)
249. S. Roy, A. Saberi, Static decentralized control of a single-integrator network with Markovian sensing topology. Automatica **41**, 1867–1877 (2005)
250. W. Rudin, *Real and Complex Analysis* (Tata McGraw-Hill, New York, 1974)
251. A. Saberi, P. Sannuti, B.M. Chen, *H_2-optimal Control* (Prentice Hall, New York, 1995)
252. P.A. Samuelson, Interactions between the multiplier analysis and the principle of acceleration. Review of Economics and Statistics **21**, 75–85 (1939)
253. T. Sathyan, T. Kirubarajan, Markov-jump-system-based secure chaotic communication. IEEE Transactions on Circuits and Systems. I, Fundamental Theory and Applications **53**, 1597–1609 (2006)
254. P. Seiler, R. Sengupta, An H_∞ approach to networked control. IEEE Transactions on Automatic Control **50**, 356–364 (2005)

255. M. Shen, G.H. Yang, H_2 state feedback controller design for continuous Markov jump linear systems with partly known information. International Journal of Systems Science **43**, 786–796 (2012)
256. P. Shi, E.K. Boukas, H_∞ control for Markovian jumping linear systems with parametric uncertainty. Journal of Optimization Theory and Applications **95**, 75–99 (1997)
257. P. Shi, E.K. Boukas, R.K. Agarwal, Control of Markovian jump discrete-time systems with norm bounded uncertainty and unknown delay. IEEE Transactions on Automatic Control **44**, 2139–2144 (1999)
258. P. Shi, E.K. Boukas, R.K. Agarwal, Robust control for Markovian jumping discrete-time systems. International Journal of Systems Science **30**, 787–797 (1999)
259. S.E. Shreve, *Stochastic Calculus for Finance II: Continuous-Time Models* (Springer, New York, 2004)
260. H.A. Simon, Dynamic programming under uncertainty with a quadratic criterion function. Econometrica **24**, 74–81 (1956)
261. A.A.G. Siqueira, M.H. Terra, Nonlinear and Markovian $\mathcal{H}_\infty$ controls of underactuated manipulators. IEEE Transactions on Control Systems Technology **12**, 635–648 (2004)
262. A.A.G. Siqueira, M.H. Terra, A fault-tolerant manipulator robot based on $\mathcal{H}_2$, $\mathcal{H}_\infty$, and mixed $\mathcal{H}_2/\mathcal{H}_\infty$ Markovian controls. IEEE/ASME Transactions on Mechatronics **14**, 257–263 (2009)
263. A.A.G. Siqueira, M.H. Terra, C. Buosi, Fault-tolerant robot manipulators based on output-feedback $\mathcal{H}_\infty$ controllers. Robotics and Autonomous Systems **55**, 785–794 (2007)
264. E.D. Sontag, *Mathematical Control Theory* (Springer, Berlin, 1990)
265. J. Speyer, J. Deyst, D. Jacobson, Optimization of stochastic linear systems with additive measurement and process noise using exponential performance criteria. IEEE Transactions on Automatic Control **19**, 358–366 (1974)
266. A. Stoica, I. Yaesh, Jump-Markovian based control of wing deployment for an uncrewed air vehicle. Journal of Guidance, Control, and Dynamics **25**, 407–411 (2002)
267. M. Sun, J. Lam, S. Xu, Z. Shu, Optimal time-weighted H_2 model reduction for Markovian jump systems. International Journal of Control **85**, 613–628 (2012)
268. L.E.O. Svensson, N. Williams, Optimal monetary policy under uncertainty: A Markov jump-linear-quadratic approach. Federal Reserve of St. Louis Review **90**, 275–293 (2008)
269. D.D. Sworder, On the stochastic maximum principle. Journal of Mathematical Analysis and Applications **24**, 627–640 (1968)
270. D.D. Sworder, Feedback control of a class of linear systems with jump parameters. IEEE Transactions on Automatic Control **14**, 9–14 (1969)
271. D.D. Sworder, J.E. Boyd, *Estimation Problems in Hybrid Systems* (Cambridge University Press, London, 1999)
272. D.D. Sworder, R.G. Hutchins, Image-enhanced tracking. IEEE Transactions on Aerospace and Electronic Systems **25**, 701–709 (1989)
273. D.D. Sworder, V.G. Robinson, Bayes' controllers with memory for a linear systems with jump parameters. IEEE Transactions on Automatic Control **17**, 118–121 (1972)
274. D.D. Sworder, V.G. Robinson, Feedback regulators for jump parameter systems with state and control dependent transition rates. IEEE Transactions on Automatic Control **19**, 47–49 (1973)
275. D.D. Sworder, R.O. Rogers, An LQG solution to a control problem with solar thermal receiver. IEEE Transactions on Automatic Control **28**, 971–978 (1983)
276. D.D. Sworder, P.F. Singer, R.G. Doria, R.G. Hutchins, Image-enhanced estimation methods. Proceedings of the IEEE **81**, 797–812 (1993)
277. H. Theil, A note on certainty equivalence in dynamic planning. Econometrica **25**, 346–349 (1957)
278. M.G. Todorov, M.D. Fragoso, Infinite Markov jump-bounded real lemma. Systems & Control Letters **57**, 64–70 (2008)
279. M.G. Todorov, M.D. Fragoso, Output feedback H_∞ control of continuous-time infinite Markovian jump linear systems via LMI methods. SIAM Journal on Control and Optimiza-

tion **47**, 950–974 (2008)
280. M.G. Todorov, M.D. Fragoso, Robust stability and stabilization of continuous-time infinite Markov jump linear systems, in *Proc. of the 2009 European Control Conference*, Budapest, Hungary, 2009, pp. 3227–3232
281. M.G. Todorov, M.D. Fragoso, On the robust control of continuous-time Markov jump linear systems subject to block-diagonal uncertainty, in *Proc. of the 2010 American Control Conference*, Baltimore, USA, 2010, pp. 4689–4694
282. M.G. Todorov, M.D. Fragoso, On the stability radii of continuous-time infinite Markov jump linear systems. Mathematics of Control. Signals and Systems **22**, 23–38 (2010)
283. M.G. Todorov, M.D. Fragoso, On the state-feedback robust control of continuous-time infinite Markov jump linear systems, in *Proc. of the 49th IEEE Conference on Decision & Control*, Atlanta, USA, 2010, pp. 6499–6504
284. M.G. Todorov, M.D. Fragoso, On the robust stability, stabilization, and stability radii of continuous-time infinite Markov jump linear systems. SIAM Journal on Control and Optimization **49**, 1171–1196 (2011)
285. K.C. Toh, M.J. Todd, R.H. Tütüncü, SDPT3—a Matlab software for semidefinite programming. Optimization Methods & Software **11**, 545–581 (1999)
286. P.N.V. Tu, *Dynamical Systems: An Introduction with Applications in Economics and Biology*, 2nd edn. (Springer, Berlin, 1994)
287. J.K. Tugnait, Adaptive estimation and identification for discrete systems with Markov jump parameters. IEEE Transactions on Automatic Control **27**, 1054–1064 (1982)
288. J.K. Tugnait, Detection and estimation for abruptly changing systems. Automatica **18**, 607–615 (1982)
289. R.H. Tütüncü, K.C. Toh, M.J. Todd, Solving semidefinite-quadratic-linear programs using SDPT3. Mathematical Programming Series A **95**, 189–217 (2003)
290. V.A. Ugrinovskii, Randomized algorithms for robust stability and guaranteed cost control of stochastic jump parameter systems with uncertain switching policies. Journal of Optimization Theory and Applications **124**, 227–245 (2005)
291. V.A. Ugrinovskii, H.R. Pota, Decentralized control of power systems via robust control of uncertain Markov jump parameter systems. International Journal of Control **78**, 662–677 (2005)
292. V.M. Ungureanu, Optimal control for linear discrete-time systems with Markov perturbations in Hilbert spaces. IMA Journal of Mathematical Control and Information **26**, 105–127 (2009)
293. J.G. van Antwerp, R.D. Braatz, A tutorial on linear and bilinear matrix inequalities. Journal of Process Control **10**, 363–385 (2000)
294. C. Van Loan, How near is a stable matrix to an unstable matrix? Contemporary Mathematics **47**, 465–478 (1985)
295. R. Vinter, *Optimal Control* (Birkhäuser, Basel, 2000)
296. H. Wang, T. Kirubarajan, Y. Bar-Shalom, Precision large scale air traffic surveillance using IMM/assignment estimators. IEEE Transactions on Aerospace and Electronic Systems **35**, 1597–1609 (1999)
297. Z. Wang, H. Qiao, K.J. Burnham, On stabilization of bilinear uncertain time-delay stochastic systems with Markovian jumping parameters. IEEE Transactions on Automatic Control **47**, 640–646 (2002)
298. J. Weidmann, *Linear Operators in Hilbert Spaces* (Springer, Berlin, 1980)
299. F.H. Westerhoff, Samuelson's multiplier–accelerator model revisited. Applied Economics Letters **13**, 89–92 (2006)
300. H.S. Witsenhausen, Separation of estimation and control for discrete time systems. Proceedings of the IEEE **59**, 1557–1566 (1971)
301. W.M. Wonham, Some applications of stochastic differential equations to optimal nonlinear filtering. SIAM Journal on Control and Optimization **3**, 347–369 (1965)
302. W.M. Wonham, On a matrix Riccati equation of stochastic control. SIAM Journal on Control **6**, 681–697 (1968)

303. W.M. Wonham, Random differential equations in control theory, in *Probabilistic Methods in Applied Mathematics*, vol. 2, ed. by A.T. Bharucha-Reid (Academic Press, San Diego, 1970), pp. 131–212
304. L. Xiao, A. Hassibi, J.P. How, Control with random communication delays via a discrete-time jump system approach, in *Proceedings of the 2000 American Control Conference*, Chicago, USA, 2000, pp. 2199–2204
305. J. Xiong, J. Lam, Robust H_2 control of Markovian jump systems with uncertain switching probabilities. International Journal of Systems Science **40**, 255–265 (2009)
306. J. Xiong, J. Lam, H. Gao, D.W.C. Ho, On robust stabilization of Markovian jump systems with uncertain switching probabilities. Automatica **41**, 897–903 (2005)
307. C. Yang, P. Bertrand, M. Mariton, Adaptive control in the presence of Markovian parameter jumps. International Journal of Control **52**, 473–484 (1990)
308. A.J. Yashin, Relative semi-invariants in the filtration of processes with step components. Avtomatika I Telemehanika **2**, 20–26 (1970)
309. G. Yin, Q. Zhang, *Continuous-Time Markov Chains and Applications: A Singular Perturbation Approach* (Springer, New York, 1998)
310. G. Yin, C. Zhu, *Hybrid Switching Diffusions* (Springer, New York, 2010)
311. J. Yong, X.Y. Zhou, *Stochastic Controls: Hamiltonian Systems and HJB Equations* (Springer, Berlin, 1999)
312. J. Zabczyk, *Mathematical Control Theory: An Introduction* (Birkhäuser, Boston, 1992)
313. G. Zames, Nonlinear operators for systems analysis, Sc.D. Dissertation, RLE Technical Rep. 370, MIT, Cambridge, 1960
314. G. Zames, Feedback and optimal sensitivity: Mode reference transformations, multiplicative seminorms and approximate inverses. IEEE Transactions on Automatic Control **26**, 301–320 (1981)
315. F. Zampolli, Optimal monetary policy in a regime-switching economy: the response to abrupt shifts in exchange rate dynamics. Journal of Economic Dynamics & Control **30**, 1527–1567 (2006)
316. L. Zhang, E.K. Boukas, Stability and stabilization of Markovian jump linear systems with partly unknown transition probabilities. Automatica **45**, 463–468 (2005)
317. L. Zhang, E.K. Boukas, H_∞ control for discrete-time Markovian jump linear systems with partly unknown transition probabilities. International Journal of Robust and Nonlinear Control **19**, 868–883 (2009)
318. L. Zhang, E.K. Boukas, J. Lam, Analysis and synthesis of Markov jump linear systems with time-varying delays and partially known transition probabilities. IEEE Transactions on Automatic Control **53**, 2458–2464 (2009)
319. L. Zhang, E.K. Boukas, P. Shi, H_∞ model reduction for discrete-time Markov jump linear systems with partially known transition probabilities. International Journal of Control **82**, 343–351 (2009)
320. Q. Zhang, Nonlinear filtering and control of a switching diffusion with small observation noise. SIAM Journal on Control and Optimization **36**, 1638–1668 (1998)
321. Q. Zhang, Optimal filtering of discrete-time hybrid systems. Journal of Optimization Theory and Applications **100**, 123–144 (1999)
322. Q. Zhang, Hybrid filtering for linear systems with non-Gaussian disturbances. IEEE Transactions on Automatic Control **45**, 50–61 (2000)
323. Q. Zhang, S.A. Kassam, Finite-state Markov model for Rayleigh fading channels. IEEE Transactions on Communications **47**, 1688–1692 (1999)
324. K. Zhou, J.C. Doyle, K. Glover, *Robust and Optimal Control* (Prentice Hall, New York, 1996)

Index

O.L.V. Costa et al., *Continuous-Time Markov Jump Linear Systems*, Probability and Its Applications, DOI 10.1007/978-3-642-34100-7,

MIX
Papier aus verantwortungsvollen Quellen
Paper from responsible sources
FSC® C105338

If you have any concerns about our products,
you can contact us on
ProductSafety@springernature.com

In case Publisher is established outside the EU,
the EU authorized representative is:
Springer Nature Customer Service Center GmbH
Europaplatz 3, 69115 Heidelberg, Germany

Printed by Libri Plureos GmbH
in Hamburg, Germany